Complements of Discriminants of Smooth Maps: Topology and Applications

Revised Edition

Recent Titles in This Series

133 **P. I. Naumkin and I. A. Shishmarev,** Nonlinear nonlocal equations in the theory of waves, 1994
132 **Hajime Urakawa,** Calculus of variations and harmonic maps, 1993
131 **V. V. Sharko,** Functions on manifolds: Algebraic and topological aspects, 1993
130 **V. V. Vershinin,** Cobordisms and spectral sequences, 1993
129 **Mitsuo Morimoto,** An introduction to Sato's hyperfunctions, 1993
128 **V. P. Orevkov,** Complexity of proofs and their transformations in axiomatic theories, 1993
127 **F. L. Zak,** Tangents and secants of algebraic varieties, 1993
126 **M. L. Agranovskiĭ,** Invariant function spaces on homogeneous manifolds of Lie groups and applications, 1993
125 **Masayoshi Nagata,** Theory of commutative fields, 1993
124 **Masahisa Adachi,** Embeddings and immersions, 1993
123 **M. A. Akivis and B. A. Rosenfeld,** Élie Cartan (1869–1951), 1993
122 **Zhang Guan-Hou,** Theory of entire and meromorphic functions: Deficient and asymptotic values and singular directions, 1993
121 **I. B. Fesenko and S. V. Vostokov,** Local fields and their extensions: A constructive approach, 1993
120 **Takeyuki Hida and Masuyuki Hitsuda,** Gaussian processes, 1993
119 **M. V. Karasev and V. P. Maslov,** Nonlinear Poisson brackets. Geometry and quantization, 1993
118 **Kenkichi Iwasawa,** Algebraic functions, 1993
117 **Boris Zilber,** Uncountably categorical theories, 1993
116 **G. M. Fel'dman,** Arithmetic of probability distributions, and characterization problems on abelian groups, 1993
115 **Nikolai V. Ivanov,** Subgroups of Teichmüller modular groups, 1992
114 **Seizô Itô,** Diffusion equations, 1992
113 **Michail Zhitomirskiĭ,** Typical singularities of differential 1-forms and Pfaffian equations, 1992
112 **S. A. Lomov,** Introduction to the general theory of singular perturbations, 1992
111 **Simon Gindikin,** Tube domains and the Cauchy problem, 1992
110 **B. V. Shabat,** Introduction to complex analysis Part II. Functions of several variables, 1992
109 **Isao Miyadera,** Nonlinear semigroups, 1992
108 **Takeo Yokonuma,** Tensor spaces and exterior algebra, 1992
107 **B. M. Makarov, M. G. Goluzina, A. A. Lodkin, and A. N. Podkorytov,** Selected problems in real analysis, 1992
106 **G.-C. Wen,** Conformal mappings and boundary value problems, 1992
105 **D. R. Yafaev,** Mathematical scattering theory: General theory, 1992
104 **R. L. Dobrushin, R. Kotecký, and S. Shlosman,** Wulff construction: A global shape from local interaction, 1992
103 **A. K. Tsikh,** Multidimensional residues and their applications, 1992
102 **A. M. Il'in,** Matching of asymptotic expansions of solutions of boundary value problems, 1992
101 **Zhang Zhi-fen, Ding Tong-ren, Huang Wen-zao, and Dong Zhen-xi,** Qualitative theory of differential equations, 1992
100 **V. L. Popov,** Groups, generators, syzygies, and orbits in invariant theory, 1992
99 **Norio Shimakura,** Partial differential operators of elliptic type, 1992

(*Continued in the back of this publication*)

Translations of
MATHEMATICAL MONOGRAPHS

Volume 98

Complements of Discriminants of Smooth Maps: Topology and Applications

Revised Edition

V. A. Vassiliev

American Mathematical Society
Providence, Rhode Island

В. А. ВАСИЛЬЕВ
ДОПОЛНЕНИЯ К ДИСКРИМИНАНТАМ ГЛАДКИХ ОТОБРАЖЕНИЙ: ТОПОЛОГИЯ И ПРИЛОЖЕНИЯ

Translated from the Russian by B. Goldfarb
Translation edited by Sergei Gelfand

1991 *Mathematics Subject Classification.* Primary 55P35, 57M25, 57R45.

ABSTRACT. The discriminant is the subset in a function space consisting of functions or maps having singularities of some distinguished types. Many key objects of mathematics can be described as the complements of suitably defined discriminants.
  We show how to investigate the topology of such complements. As applications of our method we get a new series of knot invariants (which includes all known polynomial knot invariants), the Smale-Hirsch homotopy principle for the spaces of functions without complicated singularities, and the stable cohomology rings of spaces of nondiscriminant holomorphic functions. These results imply and improve many known results in this area.
  The book also contains an introduction to configuration spaces and braid groups, as well as their applications to complexity theory.
  This book is intended for graduate students and professionals interested in algebraic topology and its application to various areas of mathematics.

Library of Congress Cataloging-in-Publication Data
Vasil′ev, V. A., 1956–
  [Dopolneniĭa k diskriminantam gladkikh otobrazheniĭ. English]
  Complements of discriminants of smooth maps: topology and applications/V. A. Vassiliev.— Rev. ed.
    p. cm. — (Translations of mathematical monographs, ISSN 0065-9282; v. 98)
  Includes bibliographical references.
  ISBN 0-8218-4618-3 (acid-free)
  1. Loop spaces. 2. Low-dimensional topology. I. Title. II. Series.
QA612.76.V3713 1994
514′.24—dc20   93-36963
                                                                                                  CIP

© Copyright 1994 by the American Mathematical Society. All rights reserved.
Translation authorized by the
All-Union Agency for Author's Rights, Moscow.
Reprinted and revised in 1993
The American Mathematical Society retains all rights
except those granted to the United States Government.
Printed in the United States of America.
Information on Copying and Reprinting can be found in the back of this volume.
♾ The paper used in this book is acid-free and falls within the guidelines
established to ensure permanence and durability.
♻ Printed on recycled paper.
This volume was typeset using $\mathcal{A}_{\mathcal{M}}\mathcal{S}$-TEX,
the American Mathematical Society's TEX macro system.

10 9 8 7 6 5 4 3 2   98 97 96 95 94

Table of Contents

Introduction	1
Chapter I. Cohomology of Braid Groups and Configuration Spaces	19
§1. Four definitions of Artin's braid group	19
§2. Cohomology of braid groups with constant coefficients	21
§3. Homology of symmetric groups and configuration spaces	28
§4. Cohomology of braid groups and configuration spaces with coefficients in the sheaf $\pm\mathbb{Z}$	37
§5. Cohomology of braid groups with coefficients in the Coxeter representation	41
Chapter II. Applications: Complexity of Algorithms, Superpositions of Algebraic Functions and Interpolation Theory	43
§1. The Schwarz genus	44
§2. Topological complexity of algorithms, and the genus of a fibration	46
§3. Estimates of the topological complexity of finding roots of polynomials in one variable	48
§4. Topological complexity of solving systems of equations in several variables	54
§5. Obstructions to representing algebraic functions by superpositions	69
§6. Dimension of the function spaces interpolating at any k points	73
Chapter III. Topology of Spaces of Real Functions without Complicated Singularities	77
§1. Statements of reduction theorems	77
§2. Spaces of functions on one-dimensional manifolds without zeros of multiplicity three	80
§3. Cohomology of spaces of polynomials without multiple roots	86
§4. Proof of the first main theorem	95
§5. Cohomology of spaces of maps from m-dimensional spaces to m-connected spaces	109

§6. On the cohomology and stable homotopy of complements of
   arrangements of planes in $\mathbb{R}^n$ 118

Chapter IV. Stable Cohomology of Complements of Discriminants
   and Caustics of Isolated Singularities of Holomorphic
   Functions 123
   §1. Singularities of holomorphic functions, their deformations and
      discriminants 127
   §2. Definition and elementary properties of the stable cohomology
      of complements of discriminants 132
   §3. Stable cohomology of complements of discriminants, and the
      loop spaces 136
   §4. Cohomological Milnor bundles 143
   §5. Proofs of technical results 145
   §6. Stable cohomology of complements of caustics, and other
      generalizations 149
   §7. Complements of resultants of polynomial systems in $\mathbb{C}^1$ 150

Chapter V. Cohomology of the Space of Knots 153
   §1. Definitions and notation 159
   §2. Basic spectral sequence 161
   §3. The term E_1 of the basic spectral sequence 166
   §4. Algorithms for computing the invariants and their values 176
   §5. The simplest invariants and their values for tabular knots 186
   §6. Conjectures, problems, and additional remarks 188
   §7. A guide to the recent results on the finite-order in variants 190
   §8. A Morse theory on the space of knots 192

Chapter VI. Invariants of Ornaments 195
   §1. Elementary theory 197
   §2. Elementary definition of finite-order invariants 202
   §3. Coding the finite-order invariants and calculating their values
      on ornaments 208
   §4. Discriminants and their resolutions 209
   §5. Complexes of connected hypergraphs 215
   §6. Structure of the space $\sigma_i - \sigma_{i-1}$ 218
   §7. Proof of Theorem 8 219
   §8. The first calculations in the stable spectral sequence 226
   §9. Generalized Fenn-Taylor and index-type invariants and
      Brunnean ornaments (A. B. Merkov) 229
   §10. Open problems and possible generalizations 238

Appendix 1. Classifying Spaces and Universal Bundles. Join 241

Appendix 2. Hopf Algebras and H-Spaces 245

Appendix 3. Loop Spaces 247

Appendix 4. Germs, Jets, and Transversality Theorems   249
Appendix 5. Homology of Local Systems   253
Bibliography   257
Added in Second Edition   264

Introduction

A discriminant is a subset of a function space consisting of functions (or maps) with singularities of some fixed type. Many objects in mathematics can be described as complements of (properly defined) discriminants. Among them are:

spaces of Morse and generalized Morse functions,

spaces of polynomials (or systems of polynomial equations) without multiple roots,

sets of nonsingular deformations of complex hypersurfaces,

iterated loop spaces of certain spaces,

spaces of knots (i.e. nonsingular imbeddings $S^1 \to S^3$ or $\mathbb{R}^1 \to \mathbb{R}^3$).

In this book we study the topology of such spaces and their various applications: in computational mathematics, algebraic geometry, differential topology, and other fields.

Here are some new results:

A) Asymptotically sharp estimates of the minimal topological complexity([1]) of algorithms for the approximate computation of roots of polynomials of degree d in one complex variable: this number lies in the interval

$$[d - \min D_p(d), d - 1],$$

where $D_p(d)$ is the sum of digits in the p-adic decomposition of the number d, and the minimum is taken over all primes p;

A′) Similar but somewhat weaker estimates for the complexity of solving systems of polynomial equations in several variables: in the most important cases the upper and lower estimates are again asymptotically (on the degree of equations) proportional to the dimensions of the spaces of the systems;

B) The Smale-Hirsch principle for spaces of smooth maps $M^m \to \mathbb{R}^n$ without complicated singularities: if the set of prohibited singularities forms a subset of codimension $\geq m + 2$ in the jet space $J^k(M^m, \mathbb{R}^n)$ (so that the corresponding discriminant has codimension ≥ 2 in the function space), then the set of maps without singularities of this type is homology equivalent to

---

([1])That is, the number of branching nodes in an algorithm: this number was introduced by Smale in [Smale$_4$], where its first estimates were obtained.

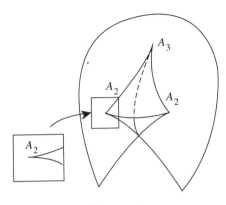

FIGURE 1

the space of admissible sections of the jet bundle; if the codimension of the prohibited set is $\geq m+3$, then these spaces are weakly homotopy equivalent;

C) Computation of stable cohomology rings of complements of discriminants and caustics for singularities of holomorphic functions on $\mathbb{C}^n$ ([2]): these rings are isomorphic to the cohomology ring of the space of $2n$-loops of the $2n+1$-dimensional sphere([3])

$$H^*(\Omega^{2n}S^{2n+1})$$

and to the ring

$$H^*(\Omega^{2n}\Sigma^{2n}\Lambda(n))$$

where $\Lambda(n)$ is the nth Lagrange Grassmannian, $\Lambda(n) = U(n)/O(n)$, respectively;

D) A new infinite sequence of invariants for knots and links (to which all known polynomial invariants reduce, as was shown recently by Joan Birman and Xiao-Song Lin; see [BL]);

E) An infinite series of invariants of *ornaments*; that is, of collections of plane curves, no three of which intersect at the same point.

The results B)–D) are obtained by using a spectral sequence whose various versions converge to the cohomology of the spaces mentioned in B), C) and also many other spaces including spaces of continuous maps of arbitrary m-dimensional polyhedra to m-connected polyhedra.

We also present many results of other authors published only in research papers: theorems of D. B. Fuchs, J. May, G. Segal, F. Cohen, and F. V. Vaĭnshteĭn on the cohomology of braid groups; the basics of genus theory

---

([2])The problem of computing these rings is problem #3 in the list of unsolved problems in singularity theory given in [Ar$_{13}$] and problem #17 in the list of [Ar$_{11+16}$]. An element of such a ring is a collection of cohomology classes defined for all deformations of all singularities of holomorphic functions on $\mathbb{C}^n$ and compatible with respect to imbeddings of discriminants (caustics) defined by adjacency of these singularities, see Figure 1.

([3])For $n = 1$ this result becomes the May-Segal theorem on the cohomology of the braid group with the infinite number of strings, see [Segal$_1$].

for fiber spaces, due to A. S. Schwarz; Smale's theorem on the connection between the topology of complements of discriminants and complexity of algorithms; Arnol'd's theorems on algebraic functions that cannot be represented as superpositions; N. H. V. Hung's computation of the $\mathbb{Z}_2$-cohomology of symmetric groups, and other results.

Let me mention some more results in the book:

An estimate for the dimensions of function spaces which interpolate any function on a manifold at any k points (e.g., for the real plane this dimension lies in the interval $[2k - D_2(k), 2k - 1])$;

An expression of the stable homotopy type of the complement of an affine plane arrangement in $\mathbb{R}^n$ in the terms of combinatorial (dimensional) characteristics of the arrangement;

A theorem which asserts that the space of monic polynomials of degree $d \cdot k$ in $\mathbb{C}^1$ without roots of multiplicity k is stable homotopy equivalent to the space of systems of k polynomials of degree d having no common roots (in the case $k = 2$ this result is due to F. Cohen, R. Cohen, B. Mann and J. Milgram, see [CCMM]);

Geometric realization of the $\mathbb{Z}_2$-homology and cohomology of configuration spaces $\mathbb{R}^m(d)$ (i.e. spaces of all subsets of cardinality d in $\mathbb{R}^m$) by smooth submanifolds;

Computation of the cohomology of braid groups with coefficients in some nontrivial representations (see §§4 and 5 of Chapter I).

Main results A, B, C, D, E are obtained in Chapters II, III, IV, V, and VI, respectively. The interdependence of the chapters is indicated by the following chart:

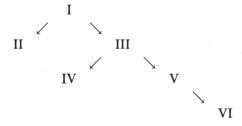

Now we give a detailed description of the content of each chapter.

Chapter I is introductory; we collect technical results on cohomology of configuration spaces used in later chapters.

A configuration space $\mathbb{R}^m(d)$ is the space of all cardinality d subsets in $\mathbb{R}^m$; for example, $\mathbb{R}^2(d)$ is the classifying space for the braid group on d strings $\text{Br}(d)$, and $\mathbb{R}^\infty(d)$ is the classifying space for the symmetric group $S(d)$.

In §2 we present results of D. B. Fuchs, G. Segal, F. Cohen, and F. V. Vaĭnshteĭn on the cohomology of braid groups with constant coefficients.

The $\mathbb{Z}_2$-cohomology of symmetric groups $S(d)$ (or, equivalently, of the spaces $\mathbb{R}^\infty(d)$) was found in [Hung$_1$]. In §3 we present these results and give a geometric realization of this cohomology (and also of $\mathbb{Z}_2$-homology and cohomology of all other configuration spaces $\mathbb{R}^m(d)$).

In §4 we study the cohomology of spaces $\mathbb{R}^m(d)$ with coefficients in the local system of groups $\pm\mathbb{Z}$; locally this system is isomorphic to $\mathbb{Z}$, but reverses orientation in traversing a path defining an odd permutation of d points in $\mathbb{R}^m$.

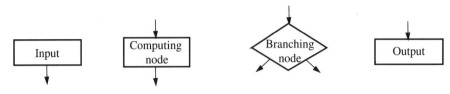

FIGURE 2

THEOREM 1. *For all integers $d, k, r > 0$ the map*
$$H^*(\mathbb{R}^{2k+r}(d), \pm\mathbb{Z}) \to H^*(\mathbb{R}^{2k}(d), \pm\mathbb{Z})$$
defined by the obvious imbedding $\mathbb{R}^{2k} \to \mathbb{R}^{2k+r}$ is an epimorphism. In particular, the map $H^*(S(d), T) \to H^*(\mathrm{Br}(d), T)$ is an epimorphism, where T is the alternating representation of groups $S(d)$, $\mathrm{Br}(d)$ in $\mathbb{Z}$.

THEOREM 2. *The group $H^{d-1}(\mathbb{R}^2(d), \pm\mathbb{Z})$ is trivial if d is not a power of a prime number, and isomorphic to $\mathbb{Z}_p$ if $d = p^t$, p prime.*

(For $i > d-1$ the group $H^i(\mathbb{R}^2(d))$ with any coefficients is trivial because of dimensional considerations.)

In Chapter II these results are applied to the theory of computational complexity, to the theory of algebraic functions, and to interpolation theory.

The space $\mathbb{R}^2(d) \equiv K(\mathrm{Br}(d), 1)$ can be considered as the space of all complex polynomials of the form
$$z^d + \lambda_1 z^{d-1} + \cdots + \lambda_{d-1} z + \lambda_d \tag{1}$$
without multiple roots: to each of these polynomials corresponds the set of its roots in $\mathbb{C}^1 \simeq \mathbb{R}^2$. The set of polynomials (1) with multiple roots is called the discriminant and denoted by Σ; this is a singular hypersurface in $\mathbb{C}^d$. Over the space $\mathbb{C}^d - \Sigma$ there are two coverings: the $d!$-fold covering $f_d: M^d \to \mathbb{C}^d - \Sigma$ whose points are ordered collections of d distinct points in $\mathbb{C}^1$ and the d-fold covering $\varphi_d: N^d \to \mathbb{C}^d - \Sigma$ whose points are pairs of the form (a polynomial $p \in \mathbb{C}^d - \Sigma$; one of the roots of this polynomial).

S. Smale in [Smale$_4$] used topological properties of the covering f_d to estimate the algorithmic complexity of the problem of calculating the roots of polynomials.

According to Smale, an algorithm is a finite oriented tree with nodes of the following four types (see Figure 2): a single input node that accepts the initial data (in our case the real and imaginary parts of the numbers λ_i); computing nodes where arithmetic computations with this data and other numbers obtained before in the algorithm are performed; branching nodes in which some previously calculated numbers are compared to zero and control is transferred depending on the result; finally, output nodes in which some of the numbers calculated earlier are declared to be the required values, and the algorithm stops.

DEFINITION (see [Smale$_4$]). The *topological complexity* of an algorithm is the number of its branching nodes. The topological complexity of a computational problem is the minimal topological complexity of algorithms that solve it.

Let $\mathfrak{A}$ be the unit polydisk in the space of polynomials (1), and ε be a positive number. The problem $P(d, \varepsilon)$ is that of approximate computation (with accuracy ε) of all the roots of any polynomial from the domain $\mathfrak{A}$.

Smale has proved that for sufficiently small ε the topological complexity $\tau(d, \varepsilon)$ of this problem satisfies the inequality

$$\tau(d, \varepsilon) \geq (\log_2 d)^{2/3}. \tag{2}$$

We prove the following two-sided estimate for $\tau(d) \equiv \lim_{\varepsilon \to +0} \tau(d, \varepsilon)$.

THEOREM 3. *The number $\tau(d)$ satisfies the inequality*

$$d - \min D_p(d) \leq \tau(d) \leq d - 1, \tag{3}$$

where the minimum is taken over all prime p, and $D_p(d)$ is the sum of all digits in the expansion of d in base p.

In particular, we always have $\tau(d) \geq d - \log_2 d$, and if d is a power of a prime number then $\tau(d) = d - 1$. In the latter case the following much stronger result can be proved.

THEOREM 4. *If d is a power of a prime number, then for sufficiently small ε the topological complexity $\tau_1(d, \varepsilon)$ of the problem of finding (with accuracy ε) only one of the roots of any polynomial from $\mathfrak{A}$ is equal to $d - 1$.*

THEOREM 4'. *For an arbitrary d and small ε the number $\tau_1(d, \varepsilon) + 1$ is not less than the greatest divisor of d that is a power of a prime.*

Remark that even this estimate is better than (2): it follows from the distribution law of prime numbers that for $d \to \infty$ the last value is asymptotically greater than $\ln d$.

The lower estimate of (3), like the estimate of Smale, is proved using the Schwarz genus of the $d!$-fold covering $f_d: M^d \to \mathbb{C}^d - \Sigma$. We recall this notion.

DEFINITION (see [Schwarz]). *The genus of a fibration $\pi: X \to Y$ is the smallest possible cardinality of the cover of its base Y by open domains such that there is a continuous section of the fibration over each of these domains. The genus is denoted by $g(\pi)$.*

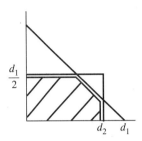

FIGURE 3

THEOREM (Smale Principle, see [Smale₄]). *The topological complexity of the problem $P(d, \varepsilon)$ for sufficiently small $\varepsilon > 0$ is at least $g(f_d) - 1$.*

(A similar inequality relates the topological complexity of the problem of finding one root (see Theorem 4) and the genus of the covering φ_d). The estimate (2) follows from this theorem and a homological estimate for the genus based on the work of D. B. Fuchs [Fuchs₂] concerning the rings $H^*(\mathrm{Br}(d), \mathbb{Z}_2)$. The lower estimate in our formula (3) is based on more effective homological estimates of the genus (see [Schwarz]) and on new results about cohomology of braid groups, in particular Theorems 1 and 2 above.

In order to prove Theorem 4 (and 4') we explicitly compute the (unique) topological obstruction to the existence of a cover of $\mathbb{C}^d - \Sigma$ by $d - 1$ domains with sections of the covering φ_d over each of them; for $m = p^k$ this obstruction is surely nontrivial.

The fourth section of Chapter II is devoted to multidimensional analogues of the previous problem: we estimate the topological complexity of solving polynomial systems in several complex variables. Here are some basic results.

Let $\Phi = \{\varphi_1, \ldots, \varphi_n\}$ be a system of homogeneous polynomials on $\mathbb{C}^n$ of degrees $d_1, \ldots, d_n$ such that 0 is the only solution of the system $\varphi_1 = \cdots = \varphi_n = 0$. Consider the space of polynomial systems of the form $F = \{f_1, \ldots, f_n\}$, where f_i is a polynomial of degree d_i with the principal homogeneous part φ_i. Let $\mathfrak{A}$ be a closed ball in this space with center at Φ.

The problem $P(\Phi, \mathfrak{A}, \varepsilon)$ consists in the approximate computation (with accuracy ε) of all $d_1 \cdots d_n$ solutions of any system F from the set $\mathfrak{A}$; the topological complexity of this problem is denoted by $\tau(\Phi, \mathfrak{A}, \varepsilon)$.

THEOREM 5. *For any $\varepsilon > 0$ the number $\tau(\Phi, \mathfrak{A}, \varepsilon)$ does not exceed $\dim_{\mathbb{R}} \mathfrak{A} = 2\sum_{i=1}^{n} \binom{n+d_i-1}{n}$.*

In order to formulate an inequality in the other direction, we assume without loss of generality that $d_1 = \max(d_i)$ and for each n-tuple of integers $D = (d_1, \ldots, d_n)$ define $W(D)$ as the number of points $(a_1, \ldots, a_n) \in \mathbb{Z}$ satisfying the inequalities $0 \leq a_1 < [d_1/2]$, $0 \leq a_2 < d_2, \ldots, 0 \leq a_n < d_n$, $a_1 + \cdots + a_n < d_1$; see Figure 3.

For example, for $d_1 = \cdots = d_n = d$ we have
$$W(d,\ldots,d) = \binom{n+d-1}{n} - \binom{n+d-[d/2]-1}{n};$$
for $d \to \infty$ and n fixed this number is $(1/n!)(1 - 2^{-n})d^n(1 + O(1/d))$.

THEOREM 6. *For almost every system of homogeneous polynomials* $\Phi = (\varphi_1, \ldots, \varphi_n)$ *of fixed degrees* $d_1, \ldots, d_n$ *we have the estimate* $\tau(\Phi, \mathfrak{A}, \varepsilon) \geq W(D)$. *In particular, this estimate is valid for the system* $\{\varphi_1 = x_1^{d_1}, \ldots, \varphi_n = x_n^{d_n}\}$.

REMARKS. 1. For $d_1 = \cdots = d_n = d \to \infty$ and n fixed the leading terms in the asymptotics of the upper and lower estimates from Theorem 5 and 6 differ by the factor $2n/(1 - 2^{-n})$.

2. For $n = 1$ the upper and lower estimates from Theorems 5 and 6 are respectively twice as large and twice as small, asymptotically, as the true asymptotics $\tau(d) \sim d$ from formula (3).

The proof of Theorem 6 is also (similarly to the lower estimate from (3)) based on studying the topology of complements of discriminants: we remove from the space $\mathfrak{A}$ the set of systems with multiple roots and consider the natural $(d_1 \cdot \ldots \cdot d_n)!$-fold covering over its complement; the homological estimate of the genus of this covering gives Theorem 6.

Theorem 5 is proved (similarly to the upper bound from (3)) by presenting an algorithm of the required complexity.

Now let us consider the multidimensional analogues of Theorems 4 and 4' about the complexity of the problem of the computation of one root. Let $D = (d_1, \ldots, d_k)$ and $\mathfrak{A}(D, n)$ be the space of all polynomial systems $F = (f_1, \ldots, f_k)$, $\deg f_i = d_i$, in $\mathbb{C}^n$. Let U be a nonempty domain in $\mathbb{C}^n$. Denote by $\tau_1(D, n, U, \varepsilon)$ the topological complexity of the problem of finding, with accuracy ε, a root of any system $F \in \mathfrak{A}(D, n)$ in the domain U. (An algorithm that solves such a problem should either exhibit a root in the ε-neighborhood of U or give the correct statement that no such root exists in U.) Let $\tau_1(D, n, U) = \lim_{\varepsilon \to +0} \tau_1(D, n, U, \varepsilon)$.

THEOREM 7. A) *For any fixed* k, n ($k \leq n$) *and* $U \neq \varnothing$ *there is an asymptotic* (*with respect to* $d \to \infty$) *estimate*
$$\tau_1(\{\underbrace{d, \ldots, d}_{k}\}, n, U) \geq (d/k)^k (1 + O(1/d)); \tag{4}$$

B) *If the domain* $U \subset \mathbb{C}^n$ *is semialgebraic and bounded, then for any* $D \in N^k$ *all numbers* $\tau_1(D, n, U, \varepsilon)$ *are bounded from above by a constant* $c(D, n, U)$ *independent of* ε.

Together with the problem of finding a root, we consider a more natural problem of finding an ε-root, that is, a point $z \in \mathbb{C}^n$ at which $|f_1(z)| \leq \varepsilon, \ldots, |f_k(z)| \leq \varepsilon$. For each subset $B \subset \mathfrak{A}(D, n)$ denote by $\tau_\approx(B, U, \varepsilon)$

the topological complexity of the problem of finding an ε-root in U for any system from the set B. Set $\tau_{\approx}(B, U) = \lim_{\varepsilon \to +0} \tau(B, U, \varepsilon)$; in particular, $\tau_{\approx}(D, n, U) \equiv \tau_{\approx}(\mathfrak{A}(D, n), U)$.

THEOREM 8. A) *Under the hypotheses of Theorem 7 we can change* $\tau_1(\{d, \ldots, d\}, n, U)$ *to* $\tau_{\approx}(\{d, \ldots, d\}, n, U)$ *in formula* (4) *and, moreover, to* $\tau_{\approx}(B_d, U)$, *where* B_d *are arbitrary nonempty domains in* $\mathfrak{A}(\{d, \ldots, d\}, n)$ *containing identically zero systems.*

B) *If the domain U is bounded in $\mathbb{C}^n$, and the set B is bounded in $\mathfrak{A}(D, n)$ then* $\tau_{\approx}(B, U, \varepsilon) \leq \dim_{\mathbb{R}} \mathfrak{A}(D, n) + 1$ *for all* $\varepsilon > 0$.

In particular, if $k = n$ and $d_1 = \cdots = d_n = d \to \infty$ then the upper and lower bounds from parts B) and A) of this theorem differ by the factor $((n-1)!/2n^n)(1 + O(1/d))$.

The study of the cohomology of braid groups was initiated in [Ar$_{4-6}$] and [Fuchs$_2$] in connection with the problem of representability of algebraic functions by superpositions of functions in lower number of variables. In §5 of Chapter II we give a new construction of obstructions to such representations, based on the notion of the genus of a fibration.

In §6, we apply the topology of configuration spaces to interpolation theory.

Let M be a topological space and L a finite dimensional vector subspace in the space of real continuous functions on M. The space L is called k-interpolating if any real function on M can be interpolated at any k points of M by an appropriate function from the space L. The lowest possible dimension of the k-interpolating functions on M is denoted by $I(M, k)$. For instance, $I(\mathbb{R}^1, k) = k$ for any k by Newton's interpolation theorem. Also, $I(S^1, k) = k$ for odd k: for L we can take the space of trigonometric polynomials of degree $\leq [k/2]$.

THEOREM 9. *If k is even, then $I(S^1, k) = k + 1$. For any k,*
$$2k - d(k) \leq I(\mathbb{R}^2, k) \leq 2k - 1,$$
where $d(k) \equiv D_2(k)$ is the number of units in the binary reprsentation of k. In particular, $I(\mathbb{R}^2, k) = 2k - 1$ if k is a power of 2.

Similar estimates are formulated for other M: they are expressed in the terms of certain characteristic classes of the configuration spaces $M(k)$.

In Chapter III we study spaces of smooth maps of manifolds into $\mathbb{R}^n$ without complicated singularities. The topological structure of such spaces is an important characteristic of the manifold; their study was initiated by the work of Smale and Cerf in differential topology, see [Smale$_{1,3}$], [Cerf$_1$]; concerning these results and their applications see [CL], [Serg], [Wagoner], [Volodin], [Hatcher], [HW], [Sharko], [Igusa$_{1-3}$], [Cerf$_2$], [Gromov], [GZ], and [Ar$_{18}$].

In order to formulate the main result of Chapter III we introduce some notation. For a smooth real manifold M denote by $J^k(M, \mathbb{R}^n)$ the space of

k-jets of smooth maps $M \to \mathbb{R}^n$. Let $\mathfrak{A}$ be an arbitrary subset in $J^k(\mathbb{R}^m, \mathbb{R}^n)$ invariant with respect to the obvious action of the group $\text{Diff}(\mathbb{R}^m)$. Then for any m-dimensional manifold M there is an invariantly defined subset $\mathfrak{A}(M) \subset J^k(M, \mathbb{R}^n)$ consisting of k-jets that belong to the set $\mathfrak{A}$ for any choice of local coordinates in M. Let $\varphi: M \to \mathbb{R}^n$ be an arbitrary map without singular points of type $\mathfrak{A}$ near ∂M.

Denote by $A(M, \mathfrak{A}, \varphi)$ the space of all smooth maps $M \to \mathbb{R}^n$ coinciding with φ in some neighborhood of ∂M and such that their k-jets do not belong to $\mathfrak{A}(M)$ at any point of M; let $B(M, \mathfrak{A}, \varphi)$ be the space of smooth sections of the obvious fiber bundle $J^k(M, \mathbb{R}^n) \to M$ disjoint from the set $\mathfrak{A}(M)$ and coinciding with the k-jet extension of the map φ in some neighborhood of ∂M. The operation of k-jet extension defines an imbedding

$$j^k: A(M, \mathfrak{A}, \varphi) \to B(M, \mathfrak{A}, \varphi). \tag{5}$$

THEOREM 10. *Suppose M^m is a compact manifold, and $\mathfrak{A}$ is a semialgebraic $\text{Diff}(\mathbb{R}^m)$-invariant closed subset in $J^k(\mathbb{R}^m, \mathbb{R}^n)$ whose codimension is at least $m+2$. Then the imbedding (5) defines an isomorphism of cohomology rings. If moreover, $\text{codim}\,\mathfrak{A} > m+2$, then the imbedding (5) is a weak homotopy equivalence.*

REMARKS. 1. The last assertion of this theorem is a direct corollary of the first assertion and the Whitehead's theorem: in this case both spaces (5) are simple-connected.

2. K. Igusa considered the case when $n = 1$ and the forbidden class consists of singularities of type A_3 or more complicated ones. He proved that in this case the imbedding (5) is m-connected (see [I$_1$], [Cerf$_2$]). Since the codimension of this class equals $m+2$, this result does not follow from Theorem 10.

EXAMPLE. The space $F - \Sigma_k$ of smooth functions $\mathbb{R}^1 \to \mathbb{R}^1$ that are without roots of multiplicity k ($k \geq 3$) and are equal to 1 outside some compact set is weakly homology equivalent to the loop space ΩS^{k-1}. The space of functions $S^1 \to \mathbb{R}^1$ without zeros of multiplicity k is homology equivalent to the space of continuous maps $S^1 \to S^{k-1}$. Moreover, if $k > 3$, then these homology equivalences are induced by homotopy equivalences.

The study of such spaces was initiated in [Ar$_{18}$], where their homology groups and (for $k = 3$) fundamental groups were calculated.

Finite-dimensional analogues of $F - \Sigma_k$ are the spaces $P_d - \Sigma_k$ of real polynomials of the form (1) without roots of multiplicity k.

THEOREM 11. *For any $d \geq k \geq 3$ the ring $H^*(P_d - \Sigma_k)$ is naturally isomorphic to the ring $H^*(\Omega S^{k-1})$ factored by the terms of degree greater than $d(k-2)/k$. If $[d/k] = [c/k]$ then the spaces $P_d - \Sigma_k$ and $P_c - \Sigma_k$ are homotopy equivalent.*

Theorem 9 is a side result of an efficient method of computing cohomology of both spaces in (5), namely the construction of a spectral sequence two different versions of which converge to the cohomology of these spaces and are isomorphic beginning from the term E_1, see §4 of Chapter III.

Numerous other versions of this spectral sequence are used in computing cohomology of complements of any reasonably defined discriminants in function spaces: the only essential restriction is that the codimensions of these discriminants should be at least 2.

EXAMPLE. The simplest version of our spectral sequence converges to the cohomology of the iterated loop space of the sphere $\Omega^m S^n$, $m < n$. Indeed, this space is homotopy equivalent to the space $\Omega^m(\mathbb{R}^{n+1} - 0)$ which can be described as the complement of a discriminant: consider the space of all smooth maps $(S^m, \text{pt}) \to (\mathbb{R}^{n+1}, (1, 0, \ldots, 0))$ and let the set of maps whose image contains the point 0 be the discriminant. In this case the term E_1 of our spectral sequence $E_r^{p,q} \to H^{p+q}(\Omega^m(\mathbb{R}^{n+1} - 0))$ has the following form: $E_1^{p,q} = 0$ for $p > 0$ and

$$E_1^{p,q} = H^{q+p(n+1-m)}(\mathbb{R}^m(-p), (\pm\mathbb{Z})^{\otimes(n-m)}) \tag{6}$$

for $p \leq 0$, where $\mathbb{R}^m(-p)$, $\pm\mathbb{Z}$ are the same as in Theorem 1. This spectral sequence degenerates at the term E_1 for odd n (or for any n if we calculate the $\mathbb{Z}_2$-cohomology of $\Omega^m S^n$ and take $\mathbb{Z}_2$ instead of $\pm\mathbb{Z}$ on the right-hand side of formula (6)). In particular, we get a new proof of the followingsplitting formula:

$$H^i(\Omega^\infty S^{\infty+j}) \cong \bigoplus_{t=0}^{\infty} H^{i-tj}(S(t), (\pm\mathbb{Z})^{\otimes j}),$$

cf. [CMT$_1$], [May$_2$], [Snaith].

A simple generalization of spectral sequence (6) gives spectral sequences that converge to the cohomology of spaces of maps of m-dimensional simplicial polyhedra into m-connected ones, see §5 of Chapter III.

Another application of this spectral sequence is a short proof of the Goresky-MacPherson formula [GM] for the cohomology of the complement of an arrangement of affine planes in $\mathbb{R}^n$: this formula expresses the cohomology in terms of dimensions of the planes and their various intersections. Moreover, these data determine uniquely the stable homotopy type of such complements. About all this see §6 of Chapter III.

In Chapter IV we consider three more classes of discriminants: bifurcation diagrams of zeros, caustics of holomorphic functions in $\mathbb{C}^n$, and resultants of collections of polynomials in $\mathbb{C}^1$; following tradition, we call bifurcation diagrams of zeros simply discriminants, see [AGV$_1$], [AVGL$_1$].

Let $f: (\mathbb{C}^n, 0) \to (\mathbb{C}, 0)$ be a holomorphic function with a singularity at 0, and $F: (\mathbb{C}^n \times \mathbb{C}^l, 0) \to (\mathbb{C}, 0)$ be a deformation of f, $f \equiv F(\cdot, 0)$. The discriminant of the deformation F is the set of values of the parameter $\lambda \in \mathbb{C}^l$ for which the function $f_\lambda \equiv F(\cdot, \lambda)$ has a critical point with critical value 0. (In the special case when $n = 1$, $f = z^d$, and the deformation $F(z, \lambda)$ has the form (1) this definition coincides with the one used before the statement of Theorem 3.)

The caustic of the deformation F is the set of λ such that the function f_λ has a non-Morse critical point near 0.

Example: the complement of the discriminant (of the caustic) for the deformation (1) is the space $K(\pi, 1)$ for the braid group with d (with $d - 1$) strings.

The main result of Chapter IV is the computation of stable cohomology rings of complements of discriminants and caustics of singularities of holomorphic functions. The definition of the first of these rings consists in the following.

Every deformation of any finite-dimensional singularity of a holomorphic function in $\mathbb{C}^n$ determines a ring, namely the cohomology ring of the complement of its discriminant. Adjacencies of singularities determine ring homomorphisms: the complement of the discriminant for a deformation of a simpler singularity is mapped to the complement of the discriminant for any sufficiently large deformation of the more complicated singularity (for example, any of its versal deformations, see [AGV$_1$], [AVGL$_1$]). Figure 1 shows a typical imbedding of complements of discriminants for versal deformations corresponding to the adjacency of real singularities $A_3 \to A_2$.

This system of rings and homomorphisms between them allows us to define the limit object, the stable cohomology ring. By definition, an element of this ring is a rule which assigns to each deformation of any singularity the cohomology class of the complement of its discriminant, in such a way that, if two singularities are adjacent, the class corresponding to the simpler singularity is induced from the class corresponding to a more complicated one by any admissible imbedding of the complements of their discriminants.

A similar procedure defines the stable cohomology ring for complements of caustics.

For example, for $n = 1$ both stable rings are isomorphic to the cohomology ring of the braid group with infinitely many strings.

The problem of computing (and giving a precise definition of) stable cohomology rings for complements of discriminants for all n was formulated in [Ar$_{11+16}$] and [Ar$_{13}$]. Our solution of this problem consists in the following.

THEOREM 13. *The stable cohomology ring ${}_n\mathscr{H}^*$ of complements of discriminants of holomorphic functions in n complex variables is isomorphic to the cohomology ring of the space of $2n$-fold loops of the $2n + 1$-dimensional sphere*:

$${}_n\mathscr{H}^* \cong H^*(\Omega^{2n} S^{2n+1}). \tag{7}$$

COROLLARIES. 1. *For $n = 1$ Theorem 13 gives the May-Segal theorem on* *the cohomology of the stable braid group*:

$$H^*(\mathrm{Br}(\infty)) \cong H^*(\Omega^2 S^3).$$

2. *All groups ${}_n\mathscr{H}^*$ except ${}_n\mathscr{H}^0 \cong {}_n\mathscr{H}^1 \cong \mathbb{Z}$ are finite.*

REMARK. The homology isomorphism (7) is not a corollary of the homotopy isomorphism: already the stable fundamental groups of complements of discriminants are not equal to $\pi_1(\Omega^{2n} S^{2n+1}) \cong \mathbb{Z}$; for example, for $n = 1$ this group is $\mathrm{Br}(\infty)$.

The isomorphism (7) is given by natural imbeddings of complements of discriminants into the space $\Omega^{2n} S^{2n+1}$. As an example, consider again the case $n = 1$ when the deformation F is of the form (1). To each value of the parameter $\lambda \in \mathbb{C}^d$ corresponds a map $\chi_\lambda \colon \mathbb{C}^1 \to \mathbb{C}^2$ given by the formula $\chi_\lambda(\tilde{z}) = \{F(\tilde{z}, \lambda); \frac{\partial}{\partial \bar{z}}|_{\tilde{z}} F(\cdot, \lambda)\}$. The image of this map contains the point 0 if and only if λ belongs to the discriminant Σ. Hence for any $\lambda \in \mathbb{C}^d - \Sigma$ we have a map $\overline{\chi}_\lambda \colon \mathbb{C}^1 \to S^3$ given by the formula $\overline{\chi}_\lambda(z) = \chi_\lambda(z)/|\chi_\lambda(z)|$. Using stereographic projection, let us identify $\mathbb{C}^1$ with the open lower hemisphere in S^2. Then the induced map $\tilde{\chi}_\lambda$ of this hemisphere into S^3 extends continuously to the equator, and on the equator this extension does not depend on λ. Assign to each value $\lambda \in \mathbb{C}^d - \Sigma$ a map $S^2 \to S^3$ whose restriction to the lower hemisphere coincides with $\tilde{\chi}_\lambda$ and the restriction to the upper hemisphere is an arbitrary map independent of λ which agrees with all $\tilde{\chi}_\lambda$ on the equator and maps the north pole to the base point in S^3.

Hence, we have constructed an imbedding of the set $\mathbb{C}^d - \Sigma$ into the space of continuous maps $(S^2, *) \to (S^3, *)$, that is, into the space $\Omega^2 S^3$.

PROPOSITION. *The map $H^i(\Omega^2 S^3) \to H^i(\mathbb{C}^d - \Sigma)$ induced by the above imbedding is an isomorphism for all $i \leq [d/2] + 1$.*

Similar imbeddings can be constructed (in a more complicated way) for all n and F. These imbeddings homotopy commute with the maps of complements of discriminants given by adjacencies of singularities (see Figure 1); passing to the projective limit with respect to F we get a homomorphism $H^*(\Omega^{2n} S^{2n+1}) \to {}_n\mathscr{H}^*$. The fact that this is an isomorphism follows from one more comparison theorem for spectral sequences converging to these groups.

In a similar way we compute the stable cohomology of complements of caustics.

THEOREM 14. *The stable cohomology ring of complements of caustics of holomorphic functions in n complex variables is isomorphic to the ring $H^*(\Omega^{2n} \Sigma^{2n} \Lambda(n))$, where Σ^{2n} denotes the 2n-fold suspension, and $\Lambda(n)$ is the nth Lagrange Grassmannian, $\Lambda(n) = U(n)/O(n)$.*

This isomorphism is realized in the same way as the isomorphism from Theorem 13: note that the space $\Sigma^{2n} \Lambda(n)$ is homotopy equivalent to the space of all germs of functions $(\mathbb{C}^n, 0) \to \mathbb{C}^1$ having either a regular point or a Morse singularity at 0.

In the same chapter we prove the conjecture from [Ar$_{11,16}$] about the stable irreducibility of strata of singularities.

Let f, g be two isolated singularities of holomorphic functions in $\mathbb{C}^n$, and let G be a versal deformation of g. By definition g is adjacent to f (or "is more complicated than f") if in any neighborhood of the origin the subset $\{f\}$ of the parameter space of the deformation G corresponding to functions g_λ with a singular point equivalent to f, is nonempty. In general, this set is reducible: for example, in the versal deformation of type D_4 the stratum $\{A_3\}$ is represented by three components. But it turns out that if the singularity g is "sufficiently complicated" compared to f, such a component is unique.

THEOREM 15. *For any isolated singularity $f: (\mathbb{C}^n, 0) \to (\mathbb{C}, 0)$ there is an isolated singularity g such that the set $\{f\}$ is irreducible in the base of the versal deformation of g and of any singularity adjacent to g.*

This statement was conjectured in [Ar$_{11+16}$].

Another application of the basic spectral sequence is the comparison theorem for the spaces of polynomial systems without common roots in $\mathbb{C}^1$ and polynomial systems without roots of high multiplicity.

Denote by $\mathbb{C}^{d_1+\cdots+d_k}$ the space of systems of k complex polynomials

$$x^{d_1} + a_{1,1}x^{d_1-1} + \cdots + a_{1,d_1},$$
$$\cdots\cdots\cdots\cdots\cdots\cdots\cdots\cdots\cdots\cdots\cdots\cdots\cdots\cdots\cdots \quad (8)$$
$$x^{d_k} + a_{k,1}x^{d_k-1} + \cdots + a_{k,d_k}.$$

The resultant $\Sigma_{d_1,\ldots,d_k}$ is the set of systems (8) having a common root. It is easy to check that for $d_1 \geq d_2 \geq \cdots \geq d_k$ the space $\mathbb{C}^{d_1+\cdots+d_k} - \Sigma_{d_1,\ldots,d_k}$ of nonresultant systems is homotopyy equivalent to $\mathbb{C}^{d_k+\cdots+d_k} - \Sigma_{d_k,\ldots,d_k}$.

THEOREM 16. *The space $\mathbb{C}^{d_1+\cdots+d_k} - \Sigma_{d_1,\ldots,d_k}$ (where $d_1 \geq \cdots \geq d_k$) is stable homotopy equivalent to the space of complex polynomials of the form* (1) *of degree $k \cdot d_k$ having no roots of multiplicity k.*

In the case $k = 2$ this reuslt was previously obtained in [CCMM].

In Chapter V we construct a new series of numerical invariants of knots. This construction is based on the study of the discriminant (which in this case is defined as the set of maps $S^1 \to S^3$ with singularities or self-intersections). The discriminant is a hypersurface in the space of all maps, and its nonsingular points correspond to maps with a single point of transversal self-intersection, while its singularities are maps with zeros of the derivative and nontransversal or multiple self-intersections.

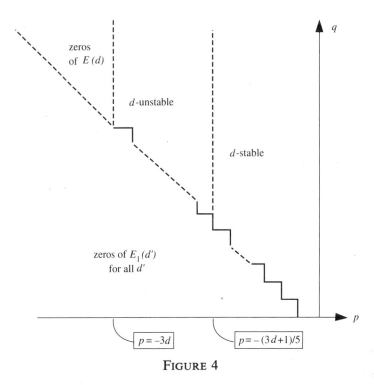

FIGURE 4

The numerical invariant is tautologically a class in the 0-dimensional cohomology group of the complement of the discriminant in the space of all maps $S^1 \to S^3$. Each of these invariants can be described in terms of the discriminant: to each nonsingular piece of the discriminant (i.e. a connected component of the set of its nonsingular points) we have to assign an index which is the difference of the values of the invariant for neighboring knots separated by this piece.

On the other hand, suppose that to each nonsingular component of the discriminant we have assigned a numerical index. In order for this collection to define an invariant of the isotopy type of knots it should satisfy the following homological condition: a linear combination of these components with the appropriate coefficients should not have a boundary in the space of all maps $S^1 \to S^3$. Enumerating such admissible collections is a problem in homology theory and is achieved by standard methods of this theory.

We construct a spectral sequence that produces such collections of indices. This spectral equation $E_r^{p,q}$ is generated by a natural stratification of the discriminant by the types of degeneration of the corresponding maps and generalizes our spectral sequences from Chapters III and IV. For $r \geq 1$ it lies in the region $\{(p, q) | p < 0, p+q \geq 0\}$, see Figure 4. The knot invariants correspond to the elements of the groups $E_\infty^{-i,i}$, $i \geq 1$; in general, to each element of the group $E_\infty^{p,q}$ corresponds a $(p+q)$-dimensional cohomology class. We define explicitly the term E_1 of this spectral sequence and produce algorithms (ready to be turned into a computer program) that compute all terms $E_\infty^{-i,i}$, the corresponding invariants, and the values of these invariants on arbitrary tame knots. (Such a computer program was recently written by T. Stanford of Columbia University.)

The natural question about completeness of our system of invariants is the question about convergence of the spectral sequence (at least on the diagonal $\{p+q=0\}$). I hope that the answer to this question is positive.

NOTE IN PROOF. Many classical invariants of knots and links can be reduced to the ours. So Dror Bar-Natan [BN_1, BN_2] showed that all coefficients of the Conway polynomials for knots occur among our invariants. Joan Birman and Xiao-Song Lin proved that all other known polynomial invariants for knots (in particular, those of Jones, Kauffman, and HOMFLY) can be reduced to our invariants; see [BL].

Our invariants can be easily generalized to the invariants of links, see section V.6. Lin [Lin^2] and Bar-Natan [BN_3] proved that all the Milnor's higher linking indices can be obtained in this way.

Our first nontrivial invariant, coming from the cell $E^{-2,2}$, coincides with the coefficient in the Conway polynomial at the monomial x^2, but already the second one (from $E^{-3,3}$) cannot be reduced to the Conway polynomial: it distinguishes between two mirror images of the trefoil knot. The term $E^{-4,4}$ gives three new invariants. The values of these five invariants for tabular simple knots with ≤ 7 overlaps and for two of the simplest nonsimple knots are given in Figure 5 (next page) in square brackets; for comparison, we also give their Conway polynomials. For the missing knots that are mirror images of the knots in Figure 5, the values of the invariants are obtained by the following rule: all our invariants encoded in the cells $E_\infty^{-i,i}$ with even (odd) i take the same (opposite) values on the corresponding mirror images of knots. These properties suggest another application of our invariants: odd invariants can help in testing whether a knot is a mirror image of itself. This poses one more problem: is it true that this test is complete, i.e. is a knot equivalent to its mirror image if and only if all odd invariants take zero value on it?

Our spectral sequence has numerous generalizations: for example, its versions converge to the cohomology of the space of imbeddings $M^m \to \mathbb{R}^n$, $n \geq 2m+2$, or give invariants for the imbeddings $M^m \to \mathbb{R}^{2m+1}$; for the problems of this type see §6.5 of Chapter V.

The classification of knots and links is only one of a great family of problems in which the methods developed in Chapter V work. This family of problems was formulated in [FNRS] as a byproduct of an attempt to define natural multidimensional analogues of the Chern-Simons theory. Namely, given m manifolds $M_1, \ldots, M_m$ and a natural N, the problem is to investigate the space of maps (or imbeddings) of these manifolds into $\mathbb{R}^N$ having no common points of images (or no points where some k of these images intersect for a certain k, taking or not the self-intersection points into account, etc.). For instance, if $\sum_{i=1}^{m}(N - \dim M_i) = N + 1$, then the problem of the classification of maps having no common points is well-posed; the first and obvious invariant of such maps (in the case when all M_i are orientable) is the linking number (see [FNRS]), which is an immediate generalization of that from the classical link theory.

In Chapter VI, we consider a model problem of this series: the classification of the ornaments, i.e., of the collections of closed plane curves, no three of which have common points. We construct a series of invariants of ornaments; many of them can be described in very classical terms (see Subsections 1.4 and 9.10).

All these invariants arise from a spectral sequence, which is one more generalization of the spectral sequence from Chapters III-V. The explicit calculations in it lead naturally to many problems in modern homological combinatorics (say, the homology groups of complexes of connected graphs and multipgrahs, investigated in [BLY], [Blov], [BW] in a connection with problems of complexity theorey, arise as a sufficient part of this spectral sequence).

All these constructions can be generalized without difficulty to the case of collections of multidimensional manifolds and spaces. There are also several other areas of modern mathematics closely related to these constructions; in particular, the equations of higher dimensional simplices (see [MSh], [Zam], [CS]) and theory of operads: this relation is of the same nature, as the relation of the Yang-Baxter equation and the Lie algebras to the standard theory of knots and links.

One more notation: the square □ means the end or the absence of proof.

I am grateful to:

V. I. Arnol'd who taught me singularity theory; all the main results of Chapters III and IV were obtained in the process of solving problems posed by him and thinking over their solutions;

D. B. Fuchs who taught me topology and whose topological insight helped me in difficult situations;

I. M. Gelfand, who in 1989 invited me to his laboratory and thus provided a possibility to work intensively in mathematics;

A. B. Merkov for his work described in §9 of Chapter VI;

S. Smale for the formulation of the problem of the topological complexity of solving systems of polynomial equations;

A. A. Beilinson, J. Birman, A. Björner, I. M. Gelfand, A. B. Giventhal, M. M. Kapranov, A. G. Khovanskii, X.-S. Lin, A. B. Sossinsky, A. N. Varchenko, and O. Ya. Viro for discussions and their interest in this work, and also

F. Cohen, H. Levin, J. Milgram, and G. Ziegler, who sent me preprints of their papers.

During my work on Chapter VI (added in the second edition) I was supported in part by the American Mathematical Society's fSU Fund.

INTRODUCTION

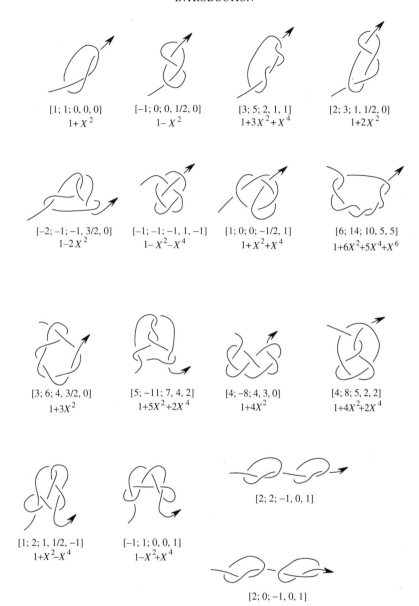

FIGURE 5

CHAPTER I

Cohomology of Braid Groups and Configuration Spaces

A configuration space $\mathbb{R}^m(t)$ is the space of all subsets of cardinality t in $\mathbb{R}^m$. The space $\mathbb{R}^2(t)$ is a space of type $K(\pi, 1)$ for the braid group on t strings, the space $\mathbb{R}^\infty(t)$ is $K(\pi, 1)$ for the symmetric group $S(t)$.

In §§1 and 2 of this chapter we list basic properties of the braid group $\text{Br}(t) = \pi_1(\mathbb{R}^2(t))$ and describe its cohomology groups with constant coefficients (calculated in [Fuchs$_2$], [C$_1$], [Vaĭn]).

The $\mathbb{Z}_2$-homology and cohomology groups of symmetric groups (or, equivalently, of spaces $\mathbb{R}^\infty(t)$) were computed in [Hung$_1$]; in §3 we present these results and realize this homology by imbedded submanifolds in $\mathbb{R}^\infty(t)$. Moreover, we give a similar realization of $\mathbb{Z}_2$-homology and cohomology groups of all spaces $\mathbb{R}^m(t)$.

In §4 we study cohomology of spaces $\mathbb{R}^m(t)$ with coefficients in the local system of groups $\pm\mathbb{Z}$, which are locally isomorphic to $\mathbb{Z}$ but reverse orientation along the paths which define odd permutations of t points in $\mathbb{R}^m$; see, in particular, Theorems 1 and 2 from the introduction.

Results of §§3 and 4 are based on the natural cellular decomposition of the one-point compactifications of spaces $\mathbb{R}^m(t)$ generalizing the decomposition of spaces $\mathbb{R}^2(t)$ constructed in [Fuchs$_2$].

In §5 we compute the weak cohomology groups of the braid group $\text{Br}(t)$ with coefficients in the representation $\text{Br}(t) \to \text{Aut}(\mathbb{Z}^t)$ that maps each braid to the permutation of basis vectors in $\mathbb{Z}^t$ coinciding with the permutation of the ends of strings for this braid.

§1. Four definitions of Artin's braid group

1.1. Geometric definition. Imbed a rectangle ("frame") in $\mathbb{R}^3$ and mark m points on each of two opposite sides (say, left and right). A *braid on* m *strings* is a collection of m nonintersecting piecewise smooth paths in $\mathbb{R}^3$ connecting points on the opposite sides so that the orthogonal projection of each string to any of the other two sides of the frame is a homeomorphism: see an example of braid in Figure 6 (next page).

Identify all braids homotopic to each other in the space of all braids. The

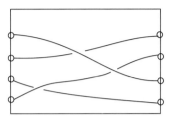

FIGURE 6

set of classes of homotopic braids on m strings forms a group. In order to multiply braids a and b one has to identify the right side of the frame with braid a with the left side of the frame b and wipe off the segment resulting from this identification. The braid inverse to the given one is its reflection, and the identity element is the braid consisting of m horizontal strings.

Geometric definition of the braid group: the set of all possible braids on m strings, considered up to homotopy, with the described group structure is called *Artin's braid group on m strings* and is denoted by $\operatorname{Br}(m)$.

1.2. Topological definition. Let X be an arbitrary topological space.

DEFINITION. The rth *configuration space* for X is the set of all unordered collections of r distinct points of the set X; it is denoted by $X(r)$. The rth *ordered configuration space* $X[r]$ is the analogous set of ordered collections.

An equivalent but often more useful definition is the following. Consider the rth (Cartesian) power X^r of the space X. On this set there is an obvious action of the group of permutations of factors (isomorphic to the symmetric group $S(r)$). The space $X(r)$ is defined as the set of regular orbits (those of cardinality $r!$) of this action, and the space $X[r]$ as the union of these orbits; in particular, $X(r) = X[r]/S(r)$.

DEFINITION. The *braid group on m strings* is the fundamental group of the mth configuration space of the real plane: $\operatorname{Br}(m) = \pi_1(\mathbb{R}^2(m))$.

THEOREM ([FoN]). *The spaces $\mathbb{R}^2(m)$ and $\mathbb{R}^2[m]$ are of type $K(\pi, 1)$ for each m, i.e. $\pi_i(\mathbb{R}^2(m)) = \pi_i(\mathbb{R}^2[m]) = 0$ for $i > 1$.*

PROOF. The exact sequence of the covering $\mathbb{R}^2[m] \to \mathbb{R}^2(m)$ shows that it suffices to prove the claim for the space $\mathbb{R}^2[m]$. But this space is a fiber bundle with base $\mathbb{R}^2[m-1]$ and fiber $\mathbb{R}^2$ less $m-1$ points. By the induction (on m) hypothesis both of these spaces have the type $K(\pi, 1)$, and it remains to apply the exact homotopy sequence of the fibration. □

DEFINITION. The fundamental group of the space $\mathbb{R}^2[m]$ is called the *group of colored braids on m strings* and is denoted by $I(m)$. This group is a subgroup of index $m!$ of the group $\operatorname{Br}(m)$ and has the following description.

Every element of the group $\operatorname{Br}(m)$ defines a permutation of m points (especially clear in Figure 6). So we obtain a homomorphism $\operatorname{Br}(m) \to S(m)$. The group $I(m)$ is just the kernel of this homomorphism.

1.3. Algebraic-geometric definition. Consider the space of all polynomials of degree m in one complex variable z of the form

$$z^m + a_1 z^{m-1} + \cdots + a_{m-1} z + a_m, \qquad a_i \in \mathbb{C}. \tag{1}$$

This space is isomorphic to $\mathbb{C}^m$ where coordinates are coefficients a_i. The subspace of $\mathbb{C}^m$ consisting of polynomials with multiple roots is called its *discriminant* and denoted by Σ, or $\Sigma(m)$.

THEOREM ([Ar$_4$]). *The complement of the discriminant $\Sigma(m)$ in $\mathbb{C}^m$ is diffeomorphic to the space $\mathbb{R}^2(m)$.*

Indeed, this diffeomorphism maps each polynomial to the collection of its roots in $\mathbb{C}^1 \sim \mathbb{R}^2$.

In particular, the group $\mathrm{Br}(m)$ can be defined as the fundamental group of the space $\mathbb{C}^m - \Sigma(m)$.

1.4. Algebraic definition. The braid group with m strings is the group with $m-1$ generators $\alpha_1, \ldots, \alpha_{m-1}$ and the following generating relations: $\alpha_i \alpha_j = \alpha_j \alpha_i$ for $|i-j| \geq 2$, $\alpha_i \alpha_{i+1} \alpha_i = \alpha_{i+1} \alpha_i \alpha_{i+1}$. In the geometric description of 1.1, to the element α_i corresponds the braid with all strings except the ith and the $(i+1)$th being horizontal, and these two strings cross each other so that the ith string passes over the $(i+1)$th one. For example, Figure 6 shows the braid $\alpha_3^{-1} \alpha_1 \alpha_2$.

1.5. The stable braid group. The group $\mathrm{Br}(m)$ is naturally imbedded into all groups $\mathrm{Br}(m+k)$; this imbedding is given by adding horizontal strings at the bottom of the braid.

DEFINITION. The *stable braid group* Br is the limit group $\varinjlim \mathrm{Br}(m)$ with respect to the above imbeddings.

§2. Cohomology of braid groups with constant coefficients

The cohomology ring of a discrete group π can be defined as the cohomology ring of any of its classifying spaces $K(\pi, 1)$; in the case of a braid group, it is convenient to choose the space $\mathbb{C}^1(m) \sim \mathbb{R}^2(m)$ considered in 1.3 as a classifying space. D. B. Fuchs [Fuchs$_2$] has computed the cohomology of this space with coefficients in $\mathbb{Z}_2$ using the following natural cellular decomposition of its one-point compactification.

2.1. The decomposition of the space $\mathbb{C}^1(m)$. Fix a coordinate z in $\mathbb{C}^1$ so that $\mathrm{Re}\, z$ and $\mathrm{Im}\, z$ are the real coordinates in $\mathbb{C}^1$.

Suppose we are given an arbitrary point $\xi \in \mathbb{C}^1(m)$, that is, an unordered collection $\{z_1, \ldots, z_m\} \subset \mathbb{C}^1$. To this collection corresponds a collection of positive integers $m_1, \ldots, m_t$ with $\sum m_i = m$, called the index of ξ. Namely, let t be the number of distinct values of the function $\mathrm{Re}\, z$ on the collection ξ, m_1 be the number of points from this collection with the minimum value of this function, m_2 be the number of points with the next

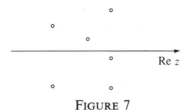

FIGURE 7

higher value, etc.; for example, the index of the collection in Figure 7 is $(2, 1, 3)$. For each finite collection $(m_1, \ldots, m_t)$ with the sum equal to m, consider the set $e(m_1, \ldots, m_t) \subset \mathbb{C}^1(m)$ consisting of all the points with index $(m_1, \ldots, m_t)$. The dimension of this set is $m + t$.

Denote by $\widetilde{\mathbb{C}}^1(m)$ the one-point compactification of the space $\mathbb{C}^1(m)$. By Poincaré-Lefschetz duality,

$$H^*(\mathbb{C}^1(m), \mathbb{Z}_2) \cong H_{2m-*}(\widetilde{\mathbb{C}}^1(m), \mathbb{Z}_2).$$

We will study the latter group.

PROPOSITION (see [Ar$_4$]). *If $i \geq m$ then the ith cohomology group of the space $\mathbb{C}^1(m)$ with any coefficients is trivial.*

PROOF. The space $\mathbb{C}^1(m)$ is homotopy equivalent to its subset consisting of all collections with center of mass at zero. This subset is diffeomorphic to the set of polynomials (1) without multiple roots and with $a_1 = 0$. The latter set is an $(m-1)$-dimensional Stein manifold and, therefore, is homotopy equivalent to an $(m-1)$-dimensional complex. □

LEMMA ([Fuchs$_2$]). *The space $\widetilde{\mathbb{C}}^1(m)$ can be represented as a finite CW-complex whose cells are various sets $e(m_1, \ldots, m_t)$ and the added point.* □

LEMMA. *The boundary of the chain $e(m_1, \ldots, m_t) \in C_{m+1}(\widetilde{\mathbb{C}}^1(m), \mathbb{Z}_2)$ is equal to*

$$\sum_{i=1}^{t-1} \binom{m_i + m_{i+1}}{m_i} e(m_1, \ldots, m_{i-1}, m_i + m_{i+1}, m_{i+2}, \ldots, m_t), \qquad (2)$$

where all binomial coefficients $\binom{a}{b}$ are reduced $\bmod 2$.

Indeed, two cells whose dimensions differ by 1 are geometrically incident only if the smaller one is obtained as a degeneration of the larger one. Figure 8 shows two (of the $\binom{4}{2} = 6$ possible) cases of degeneration of the cell $e(1, 2, 2)$ to $e(1, 4)$.

2.2. The rings $H^*(\mathrm{Br}(m), \mathbb{Z}_2)$.

THEOREM ([Fuchs$_2$]). *The ring $H^*(\mathrm{Br}, \mathbb{Z}_2)$ is multiplicatively generated by generators $a_{r,k}$ ($r \geq 1$, $k \geq 0$), $\dim a_{r,k} = 2^k(2^k - 1)$ with relations*

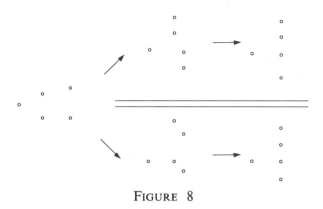

FIGURE 8

generated by the usual (skew)commutativity relations and the relations $a_{r,k}^2 = 0$. *The ring* $H^*(\mathrm{Br}(m), \mathbb{Z}_2)$ *is obtained from* $H^*(\mathrm{Br}, \mathbb{Z}_2)$ *by adding the relations*

$$a_{r_1,k_1} \cdot \ldots \cdot a_{r_q,k_q} \quad \textit{for} \quad 2^{r_1+k_1+\cdots+r_q+k_q} > m. \quad \square$$

First, we describe the additive generators of these rings in terms of the cellular decomposition from 2.1.

DEFINITION (see [Fuchs$_2$]). A chain of the CW-complex $C_*(\widetilde{\mathbb{C}}^1(m), \mathbb{Z}_2)$ is called *symmetric* if the cells $e(m_1, \ldots, m_t)$, $e(m_{\sigma(1)}, \ldots, m_{\sigma(t)})$, where σ is a permutation of t elements, occur in this chain with equal coefficients for all $m_1, \ldots, m_t, \sigma$. A cell $e(m_1, \ldots, m_t)$ is called a 2-*cell* if all m_i are powers of 2. A chain is called a 2-*chain* if it is a sum of 2-cells.

PROPOSITION ([Fuchs$_2$]). *Each symmetric 2-chain is a cycle. Each cycle of the complex* $C_*(\widetilde{\mathbb{C}}^1(m), \mathbb{Z}_2)$ *is homologous to a unique symmetric 2-chain. Hence, the graded group* $H_*(\widetilde{\mathbb{C}}^1(m), \mathbb{Z}_2)$ *is isomorphic to the subgroup of the graded group* $C_*(\widetilde{\mathbb{C}}^1(m), \mathbb{Z}_2)$ *consisting of symmetric 2-chains.*

This follows from formula (2). $\square$

COROLLARY. *The dimension of the group* $H^k(\mathrm{Br}(m), \mathbb{Z}_2)$ *over* $\mathbb{Z}_2$ *is equal to the number of representations of* m *as a sum of* $m-k$ *powers of 2 (representations that differ only by the order of summation are considered to be equivalent).*

COROLLARY ([Fuchs$_2$]). *The group* $H^{m-1}(\mathrm{Br}(m), \mathbb{Z}_2)$ *is trivial if* m *is not a power of 2, and is generated by one element (corresponding to the cell* $e(m)$*) otherwise.*

Let $s_1, \ldots, s_q$ be a nonincreasing sequence of natural numbers with $2^{s_1} + \cdots + 2^{s_q} \leq m$. Denote by $\langle s_1, \ldots, s_q \rangle$ the symmetric 2-chain in $\widetilde{\mathbb{C}}^1(m)$ that contains exactly one copy of each cell of the form $e(m_1, \ldots, m_t)$, where

$\{m_1, \ldots, m_t\}$ is a permutation of the numbers $2^{s_1}, \ldots, 2^{s_q}, 1, \ldots, 1$. For example, if $m = 5$, then $\langle 1, 1 \rangle = e(2, 2, 1) + e(2, 1, 2) + e(1, 2, 2)$.

It follows directly from the proposition that the chains of the form $\langle s_1, \ldots, s_q \rangle$ form a basis of the group $H_*(\widetilde{\mathbb{C}}^1(m), \mathbb{Z}_2)$, and the dual cocycles form a basis of the group $H^*(\mathbb{C}^1(m), \mathbb{Z}_2)$.

Let us see what happens to these generators under stabilization of the groups $\text{Br}(m)$ with respect to m.

The canonical imbedding $\text{Br}(m) \to \text{Br}(m + 1)$ (see 1.5) determines a unique (up to homotopy) map of classifying spaces, hence a map

$$H^*(\text{Br}(m + 1), \mathbb{Z}_2) \to H^*(\text{Br}(m), \mathbb{Z}_2). \tag{3}$$

LEMMA (see [Fuchs$_2$]). *The map (3) is an epimorphism. Namely, it sends a generator $\langle s_1, \ldots, s_q \rangle \in H^*(\text{Br}(m + 1), \mathbb{Z}_2)$ to the generator $\langle s_1, \ldots, s_q \rangle$ of the group $H^*(\text{Br}(m), \mathbb{Z}_2)$ if $2^{s_1} + \cdots + 2^{s_q} \leq m$, and to 0 if $2^{s_1} + \cdots + 2^{s_q} = m + 1$.*

To prove this, let us note that the map of the corresponding classifying spaces can be chosen as follows. Let $\xi \in \mathbb{C}^1(m)$ be a collection of m points $z_1, \ldots, z_m$. Define the corresponding point $z(\xi)$ as the point with coordinates $\text{Im } z(\xi) = 0$, $\text{Re } z(\xi) = 1 + \max(\text{Re } z_i)$. By adding to each collection ξ the point $z(\xi)$ we get an imbedding $\varphi: \mathbb{C}^1(m) \to \mathbb{C}^1(m + 1)$. This imbedding can be extended to a map of one-point compactifications, and the cell $e(m_1, \ldots, m_t)$ is mapped regularly to the cell $e(m_1, \ldots, m_t, 1)$. In particular, if c is a compact cycle in $\mathbb{C}^1(m)$ then the intersection index of the cycle $\varphi(c)$ with the chain $\langle s_1, \ldots, s_q \rangle \subset \mathbb{C}^1(m + 1)$ is equal to 0 if $2^{s_1} + \cdots + 2^{s_q} = m + 1$, and to the intersection index of the cycle c with the chain $\langle s_1, \ldots, s_q \rangle \subset \mathbb{C}^1(m)$ otherwise. This, together with the definition of the Poincaré isomorphism, implies the lemma. □

COROLLARY. *A basis in the stable group $H^*(\text{Br}, \mathbb{Z}_2)$ is given by (co)chains $\langle s_1, \ldots, s_q \rangle$ with arbitrary $s_1, \ldots, s_q \geq 1$.*

Now we are ready to give a geometric description of the generators $a_{r,k}$ from the main theorem of 2.2. Namely, the element $a_{r,k}$ is given by the (co)chain $\langle r, \ldots, r \rangle$ (2^k elements).

2.3. The homomorphism of the braid group into the orthogonal group.

The symmetric group $S(m)$ acts on the space $\mathbb{R}^m$ by permutations of the basis vectors; this action defines a homomorphism $S(m) \to O(m)$, and hence, the composition

$$\text{Br}(m) \to S(m) \to O(m). \tag{4}$$

The corresponding maps of classifying spaces

$$K(\text{Br}(m), 1) \to K(S(m), 1) \to BO(m) \tag{5}$$

§2. COHOMOLOGY OF BRAID GROUPS WITH CONSTANT COEFFICIENTS   25

induce the homomorphisms
$$H^*(BO(m), A) \to H^*(S(m), A) \to H^*(\mathrm{Br}(m), A) \qquad (6)$$
on cohomology with any coefficient ring A. For $A = \mathbb{Z}_2$, the ring $H^*(BO(m), A)$ is generated by the universal Stiefel-Whitney classes $w_1, \ldots, w_m$; see Appendix 1.

THEOREM ([Fuchs$_2$]). *For any $k = 1, \ldots, m$ the composition (6) maps the kth universal Stiefel-Whitney class w_k to the class dual to the sum of all cells $e(m_1, \ldots, m_t)$ of dimension $2m - k$ (that is, all the cells with $t = m - k$).*

In particular, the class w_m is mapped to 0.

PROOF (see [Fuchs$_2$]). First we give an equivalent description of map (6). Take $\mathbb{C}^1(m)$ as $K(\mathrm{Br}(m), 1)$ and define an m-dimensional vector bundle $T(m)$ over it whose fiber over a collection $\xi \in \mathbb{C}^1(m)$ is the space of functions on the points of this collection. This vector bundle coincides with the bundle induced by the composition (5) from the tautological bundle over the Grassmannian $G_m \sim BO(m)$; in particular, all Stiefel-Whitney classes of the bundle $T(m)$ coincide with the classes induced by the homomorphism (6) from the universal classes of $BO(m)$. The Stiefel-Whitney classes of a smooth m-dimensional vector bundle can be described as follows: construct $m - k + 1$ smooth sections of this bundle in general position and take the set of those points in the base over which these sections are linearly dependent. This set is a cycle which is mapped to the class w_k under the Poincaré isomorphism.

A section of the bundle $T(m)$ is a function that sends an unordered collection $\{z_1, \ldots, z_m\}$ of distinct complex numbers to the unordered collection $\{(z_1, x_1), \ldots, (z_m, x_m)\}$ with the same z_i and real x_i. Define $m - k + 1$ sections of the bundle $T(m)$ as follows:
$$\gamma_i(z_1, \ldots, z_m) = \{(z_1, (\mathrm{Re}\, z_1)^i), \ldots, (z_m, (\mathrm{Re}\, z_m)^i)\},$$
$i = 0, 1, \ldots, m - k$. It is easy to check that they are in general position. These sections are linearly dependent at $\{z_1, \ldots, z_m\} \in \mathbb{C}^1(m)$ if and only if there exist real numbers $a_0, a_1, \ldots, a_{m-k}$ such that $a_0(\mathrm{Re}\, z_j)^{m-k} + \cdots + a_{m-k-1}\mathrm{Re}\, z_j + a_{m-k} = 0$ for $j = 1, \ldots, m$, i.e. if among the numbers $\mathrm{Re}\, z_1, \ldots, \mathrm{Re}\, z_m$ there are no more than $m - k$ distinct ones: the points $z_1, \ldots, z_m$ must lie on $m - k$ vertical lines. But this condition is satisfied exactly when the collection $\{z_1, \ldots, z_m\}$ belongs to the union of cells of codimension k. □

The bundle $T(m)$ splits into the sum of a one-dimensional bundle (spanned by the function with value 1 at all the points of collections $\xi \in \mathbb{C}^1(m)$) and the $(m-1)$-dimensional bundle $T'(m)$ orthogonal to it; this implies that the map (5) reduces to the map
$$K(\mathrm{Br}(m), 1) \to BO(m - 1). \qquad (7)$$

It is clear from the construction of the imbedding $\mathbb{C}^1(m) \to \mathbb{C}^1(m+1)$ corresponding to the imbedding $\mathrm{Br}(m) \to \mathrm{Br}(m+1)$ that the bundle over $\mathbb{C}^1(m)$ induced from the bundle $T(m+1)$ is isomorphic to the direct sum of the bundle $T(m)$ and the trivial one-dimensional bundle. Therefore, homomorphism (6) maps all Stiefel-Whitney classes of the bundle $T(m+1)$ to the corresponding classes of $T(m)$. This allows us to define stable Stiefel-Whitney classes of the sequence of bundles $T(m)$ in the ring $H^*(\mathrm{Br}, \mathbb{Z}_2)$.

When we stabilize the groups $\mathrm{Br}(m)$ with respect to m, homomorphisms (4) define the map
$$\mathrm{Br} \to S(\infty) \to O(\infty), \tag{8}$$
and the above-mentioned stable classes coincide with the classes induced from the universal Stiefel-Whitney classes $w_i \in H^i(BO(\infty), \mathbb{Z}_2)$.

THEOREM ([Fuchs$_2$]). *For each m the homomorphism $H^*(BO(m), \mathbb{Z}_2) \to H^*(\mathrm{Br}(m), \mathbb{Z}_2)$ is an epimorphism. The same is true for the stable homomorphism $H^*(BO(\infty), \mathbb{Z}_2) \to H^*(\mathrm{Br}, \mathbb{Z}_2)$.* □

In other words, for each m the ring $H^*(\mathrm{Br}(m), \mathbb{Z}_2)$ is multiplicatively generated by the Stiefel-Whitney classes of the bundle $T(m)$, and the ring $H^*(\mathrm{Br}, \mathbb{Z}_2)$ is generated by the Stiefel-Whitney classes of the stabilized bundle $T(\infty)$.

This theorem easily enables us to define the action of the Steenrod algebra on the ring $H^*(\mathrm{Br}(m), \mathbb{Z}_2)$ and, in particular, the following result.

THEOREM (see [Fuchs$_2$], [Lin$_3$]). *The Bockstein homomorphism*
$$\beta_2 = \mathrm{Sq}^1 : H^i(\mathrm{Br}(m), \mathbb{Z}_2) \to H^{i+1}(\mathrm{Br}(m), \mathbb{Z}_2)$$
is given by the formula $\beta_2 a_{r,k} = a_{r+1,0} \cdot a_{r,1} \cdot \ldots \cdot a_{r,k-1}$. □

2.4. Hopf algebra. The obvious imbedding
$$\mathrm{Br}(m) \times \mathrm{Br}(k) \to \mathrm{Br}(m+k) \tag{9}$$
induces the homomorphism
$$H^*(\mathrm{Br}(m+k), \mathbb{Z}_2) \to H^*(\mathrm{Br}(m), \mathbb{Z}_2) \otimes H^*(\mathrm{Br}(k), \mathbb{Z}_2).$$
Under the stabilization $m, k \to \infty$ these homomorphisms define a map
$$\Delta : H^*(\mathrm{Br}, \mathbb{Z}_2) \to H^*(\mathrm{Br}, \mathbb{Z}_2) \otimes H^*(\mathrm{Br}, \mathbb{Z}_2),$$
which, together with the usual cohomology product, defines the Hopf algebra structure on $H^*(\mathrm{Br}, \mathbb{Z}_2)$.

THEOREM ([Fuchs$_2$]). *The map Δ sends the generator $\langle s_1, \ldots, s_q \rangle$ of the ring $H^*(\mathrm{Br}, \mathbb{Z}_2)$ to the element*
$$1 \otimes \langle s_1, \ldots, s_q \rangle + \langle s_1, \ldots, s_q \rangle \otimes 1 + \sum \langle s_{i_1}, \ldots, s_{i_r} \rangle \otimes \langle s_{j_1}, \ldots, s_{j_{q-r}} \rangle,$$

where the sum is taken over all subdivisions of the collection $s_1, \ldots, s_q$ into two nonempty collections. □

(For example, $\Delta\langle 1, 1\rangle = 1 \otimes \langle 1, 1\rangle + \langle 1, 1\rangle \otimes 1 + \langle 1\rangle \otimes \langle 1\rangle$.) This formula becomes evident if we use the following realization of the map

$$K(\mathrm{Br}(m), 1) \times K(\mathrm{Br}(k), 1) \to K(\mathrm{Br}(m+k), 1)$$

corresponding to the imbedding (9). Fix two orientation preserving diffeomorphisms $i_+ : \mathbb{R} \to \mathbb{R}_+$, $i_- : \mathbb{R} \to \mathbb{R}_-$, where $\mathbb{R}_\pm$ are the sets of positive and negative numbers. They define the obvious diffeomorphisms $I_\pm : \mathbb{C}^1 \to \mathbb{R}_\pm \times \mathbb{R}$ preserving the imaginary part and acting on the real part according to $i_\pm$. Finally, to each point $\zeta \times \eta \in \mathbb{C}^1(m) \times \mathbb{C}^1(k)$ we assign the union of collections $I_-(\zeta)$ and $I_+(\eta)$.

Note that we have actually constructed a map of configuration spaces $X(m) \times X(k) \to X(m+k)$ for each space X that can be represented in the form $\widetilde{X} \times \mathbb{R}$.

There is a dual Hopf algebra structure on the group $H_*(\mathrm{Br}, \mathbb{Z}_2)$: for example, multiplication is given by the stabilization of imbeddings (9).

THEOREM (see [Fuchs$_2$]). 1. *The Hopf algebra $H_*(\mathrm{Br}, \mathbb{Z}_2)$ is a polynomial algebra over $\mathbb{Z}_2$ with generators $\varkappa_i$, $i = 1, 2, \ldots$, of degree $2^i - 1$ and comultiplication given by the formulas $\mu(\varkappa_i) = 1 \otimes \varkappa_i + \varkappa_i \otimes 1$.*

2. For each m the group $H_*(\mathrm{Br}(m), \mathbb{Z}_2)$ is a subcoalgebra in $H_*(\mathrm{Br}, \mathbb{Z}_2)$ with basis consisting of monomials $\varkappa_1^{k_1} \cdot \ldots \cdot \varkappa_r^{k_r}$ such that $\sum k_i 2^i \leq m$.

3. The homomorphism of Hopf algebras $H_*(\mathrm{Br}, \mathbb{Z}_2) \to H_*(BO(\infty), \mathbb{Z}_2)$ given by the limit of the maps (4) is a monomorphism. □

2.5. Cohomology of stable braid groups.

THEOREM (May, Segal; see, for example, [Segal$_1$]). *For any coefficient ring A there is an algebra isomorphism $H_*(\mathrm{Br}, A) \cong H_*(\Omega^2 S^3, A)$, where $\Omega^2 S^3$ is the double loop space of the sphere S^3.* □

The algebras $H_*(\Omega^2 S^3, \mathbb{Z}_p)$ were calculated in [AK$_1$] for $p = 2$ and in [DL] for $p > 2$. Combined with the finiteness theorem for allgroups $H_i(\mathrm{Br}, \mathbb{Z})$ with $i \geq 2$ this gives us a complete description of algebras $H_*(\mathrm{Br}, A)$ with finitely generated A.

The corresponding results for the dual algebras $H^*(\mathrm{Br}, \mathbb{Z}_p)$ and unstable rings $H^*(\mathrm{Br}(m), \mathbb{Z}_p)$ are given in the next section.

2.6. Cohomology of braid groups with other constant coefficients.
These cohomology groups were independently (and using different methods) calculated by F. Cohen [Cohen$_1$] and F. V. Vaĭnshteĭn [Vaĭn]. Here we present the results in the formulation of Vaĭnshteĭn. He used the cellular decomposition of the space $\widetilde{\mathbb{C}}^1(m)$ described in 2.1.

LEMMA ([Vaĭn]). *For a suitable choice of orientations for cells $e(m_1, \ldots, m_t)$ the differentials in the complex $\widetilde{\mathbb{C}}^1(m)$ are given by the formulas*

$$\partial e(m_1, \ldots, m_t) = \sum_{i=1}^{t-1} (-1)^i P_{m_i + m_{i+1}}^{m_i} \qquad (10)$$
$$\times e(m_1, \ldots, m_{i-1}, m_i + m_{i+1}, m_{i+2}, \ldots, m_t),$$

where $P_a^b = 0$ if both b and $a - b$ are odd, and $P_a^b = \binom{[a/2]}{[b/2]}$ otherwise. □

THEOREM ([Segal$_1$], [Cohen$_1$], [Vaĭn]). *For a prime $p \neq 2$ the ring $H^*(\mathrm{Br}, \mathbb{Z}_p)$ is the tensor product of the truncated polynomial algebras $\mathbb{Z}_p[x_i]/\{x_i^p\}$ generated by x_i, $i \geq 0$, $\dim x_i = 2p^{i+1} - 2$, and the exterior algebra generated by y_j, $j \geq 0$, $\dim y_j = 2p^j - 1$. The natural map $H^*(\mathrm{Br}, \mathbb{Z}_p) \to H^*(\mathrm{Br}(m), \mathbb{Z}_p)$ is an epimorphism, and its kernel is generated by those monomials $x_{r_1} \cdots x_{r_s} \cdot y_{l_1} \cdots y_{l_q}$ for which $2(p^{r_1+1} + \cdots + p^{r_s+1} + p^{l_1} + \cdots + p^{l_q}) > m$. The Bockstein operator β_p corresponding to the exact sequence $0 \to \mathbb{Z}_p \to \mathbb{Z}_{p^2} \to \mathbb{Z}_p \to 0$ is given by the formulas $\beta_p x_i = y_{i+1}$, $\beta_p y_i = 0$.* □

THEOREM (see [Cohen$_1$], [Vaĭn]). *For any $q \geq 2$ and m,*

$$H^q(\mathrm{Br}(m), \mathbb{Z}) = \bigoplus_p \beta_p H^{q-1}(\mathrm{Br}(m), \mathbb{Z}_p),$$

where the sum is taken over all prime p. In particular, the groups $H^q(\mathrm{Br}(m), \mathbb{Z})$ have no p^2-torsion. □

This implies the main results of the paper [Ar$_4$] (which initiated the study of the cohomology of braid groups):

COROLLARY (see [Ar$_4$]). 1. (*Finiteness theorem.*) *For any m all groups $H^q(\mathrm{Br}(m))$ are finite for $q \geq 2$ (and $H^0 = H^1 = \mathbb{Z}$).*
2. (*Stabilization theorem.*) $H^q(\mathrm{Br}(m)) \cong H^q(\mathrm{Br}(2q - 2))$ *for $m \geq 2q - 2$.*
3. *The lower stable groups $H^q(\mathrm{Br})$ are:*

$$H^0 = H^1 = \mathbb{Z}, \quad H^2 = 0, \quad H^3 = H^4 = \mathbb{Z}_2, \quad H^5 = H^6 = \mathbb{Z}_6.$$

§3. Homology of symmetric groups and configuration spaces

3.1. Configuration spaces.
Consider an infinite complete flag

$$\cdots \hookrightarrow \mathbb{R}^{n-1} \hookrightarrow \mathbb{R}^n \hookrightarrow \mathbb{R}^{n+1} \hookrightarrow \cdots. \qquad (11)$$

The imbeddings (11) define imbeddings of the corresponding configuration spaces

$$\cdots \hookrightarrow \mathbb{R}^n(m) \hookrightarrow \mathbb{R}^{n+1}(m) \hookrightarrow \cdots.$$

Passing to the limit we get the stable configuration space

$$\mathbb{R}^\infty(m) = \lim_{n \to \infty} \mathbb{R}^n(m).$$

LEMMA. *The space $\mathbb{R}^\infty(m)$ is a space of type $K(S(m), 1)$.*

In fact, it is easy to check that its universal covering $\mathbb{R}^\infty[m]$ is homotopically trivial (it is obtained by excising from the space $(\mathbb{R}^\infty)^m$ a finite number of subspaces of infinite codimension). □

The spaces $\mathbb{R}^n(m)$ can be stabilized with respect to m, not just n; compare 2.2. The corresponding diagram of imbeddings

$$
\begin{array}{ccccccc}
\downarrow & & \downarrow & & \downarrow & & \\
\longrightarrow & \mathbb{R}^n(m) & \longrightarrow & \mathbb{R}^{n+1}(m) & \longrightarrow & \cdots \mathbb{R}^\infty(m) & \\
\downarrow & & \downarrow & & \downarrow & & \\
\longrightarrow & \mathbb{R}^n(m+1) & \longrightarrow & \mathbb{R}^{n+1}(m+1) & \longrightarrow & \cdots \mathbb{R}^\infty(m+1) & \quad (12) \\
\downarrow & & \downarrow & & \downarrow & & \\
\vdots & & \vdots & & \vdots & & \\
\longrightarrow & \mathbb{R}^n(\infty) & \longrightarrow & \mathbb{R}^{n+1}(\infty) & \longrightarrow & \cdots \mathbb{R}^\infty(\infty) &
\end{array}
$$

is commutative, and the space in the lower right corner is of type $K(S(\infty), 1)$.

3.1.1. PROPOSITION. *All homomorphisms*

$$H^i(\mathbb{R}^{n+k}(m), \mathbb{Z}_2) \to H^i(\mathbb{R}^n(m), \mathbb{Z}_2), \qquad k = 1, 2, \ldots, \infty,$$

corresponding to the horizontal arrows of diagram (12) *are epimorphisms, and even isomorphisms for $i \leq n - 1$.*

This will be proved in 3.4.

If $n = 2$ then the space $\mathbb{R}^n(m)$ is a space of type $K(\mathrm{Br}(m), 1)$ considered in the preceding section.

On the other hand, the natural homomorphism $\mathrm{Br}(m) \to S(m)$ defines (up to homotopy) a map of classifying spaces

$$\mathrm{cl}\colon K(\mathrm{Br}(m), 1) \to K(S(m), 1).$$

3.1.2. PROPOSITION. *The map* cl *can be taken to be the imbedding $\mathbb{R}^2(m) \to \mathbb{R}^\infty(m)$ given by the obvious imbedding $\mathbb{R}^2 \to \mathbb{R}^\infty$.*

3.2. The Hopf algebra $H_*(S(\infty), \mathbb{Z}_2)$. Similarly to 2.4, the homology of $\mathbb{R}^n(\infty)$ has a Hopf algebra structure; this structure is compatible with the maps $\mathbb{R}^n(\infty) \to \mathbb{R}^{n+1}(\infty) \to \cdots$ and defines a limit structure on the groups $H_*(S(\infty), F) \sim H_*(\mathbb{R}^\infty(\infty), F)$.

Following [Hung$_1$] we describe the ring structure of the algebras $H_*(S(m), \mathbb{Z}_2)$.

Denote by J_+ the set of all finite sequences of nonnegative integers $l_0, l_1, \ldots, l_{s-1}$ with $l_0 \geq 1$. The length s of the sequence $K = (l_0, l_1, \ldots, l_{s-1}) \in J_+$ will be denoted by $|K|$.

THEOREM (see [Hung$_1$]). 1. *The Hopf algebra $H_*(S(\infty), \mathbb{Z}_2)$ is a free algebra with multiplicative generators D_K, where K runs over the set J_+, and $\dim D_K = l_0(2^s - 1) + l_1(2^s - 2) + \cdots + l_{s-1}(2^s - 2^{s-1})$ for $K = (l_0, l_1, \ldots, l_{s-1})$.*
2. *All obvious homomorphisms $H_*(S(m), \mathbb{Z}_2) \to H_*(S(m+1), \mathbb{Z}_2)$, and hence, also $i: H_*(S(m), \mathbb{Z}_2) \to H_*(S(\infty), \mathbb{Z}_2)$ are monomorphisms.*
3. *The additive basis of the subgroup $i(H_*(S(m), \mathbb{Z}_2)) \subset H_*(S(\infty), \mathbb{Z}_2)$ is formed by all monomials $D_{K_1} \cdot D_{K_2} \cdot \cdots \cdot D_{K_r}$ ($K_i = K_j$ is allowed) such that $2^{|K_1|} + \cdots + 2^{|K_r|} \leq m$.* □

In the subsequent sections 3.3 and 3.4 we will give a cellular realization of cohomology of symmetric groups and configuration spaces $\mathbb{R}^n(m)$.

3.3. Cellular decompositions of configuration spaces. Fix integers m, n. Let $\widetilde{\mathbb{R}}^n(m)$ be the one-point compactification of the space $\mathbb{R}^n(m)$. To each point of the space $\mathbb{R}^n(m)$ we associate a graph, and sets of points corresponding to different graphs will define a cellular decomposition of the space $\widetilde{\mathbb{R}}^n(m)$.

Define a linear function l on $\mathbb{R}^2$ and draw $n + 1$ parallel lines $L_j = l^{-1}(j)$, $j = 0, 1, \ldots, n$, with compatible orientations. Let $x_1, \ldots, x_n$ be a linear coordinate system in $\mathbb{R}^n$, $e_1, \ldots, e_n$ be the corresponding basis vectors, and π_j be the projection of the space $\mathbb{R}^n$ onto the j-dimensional plane $\{e_1, \ldots, e_j\}$ along $\{e_{j+1}, \ldots, e_n\}$. Let $\zeta \in \mathbb{R}^n(m)$ be a collection of m points $z_1, \ldots, z_m \in \mathbb{R}^n$. On each line L_j let us mark several points that are in one-to-one correspondence with the set $\pi_j(\zeta)$, such that the order of these points on the line L_j corresponds to the lexicographical order of points of the set $\pi_j(\zeta)$ defined by the coordinates $x_1, \ldots, x_j$. For each $j < n$ connect a pair of marked points on the lines L_j, L_{j+1} by an edge if there exists a point from the collection ζ that is sent to the points corresponding to these two by the maps π_j and π_{j+1}. In particular, the unique point on L_0 is connected with all points on L_1. Any graph constructed in such a way and considered up to a homeomorphism of the plane $\mathbb{R}^2$ preserving the function l and orientations of the lines L_j will be called a *standard (n, m)-tree*.

For example, let $n = 2$. Standard $(2, m)$-trees are in bijective correspondence with ordered decompositions of the number m: for example, to the $(2, 6)$-tree in Figure 9 corresponds the decomposition $(2, 3, 1)$. Therefore, the decomposition of the space $\mathbb{R}^2(m)$ into sets corresponding to different trees coincides with its decomposition into cells $e(m_1, \ldots, m_q)$ described in 2.2.

3.3.1. LEMMA. *For any n, m there exist the structure of a CW-complex on the space $\widetilde{\mathbb{R}}^n(m)$ with cells being sets of points corresponding to various (n, m)-trees and the added point.* □

Define the multiplicity of a vertex of a standard (n, m)-tree as the number

§3. HOMOLOGY OF SYMMETRIC GROUPS

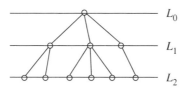

FIGURE 9

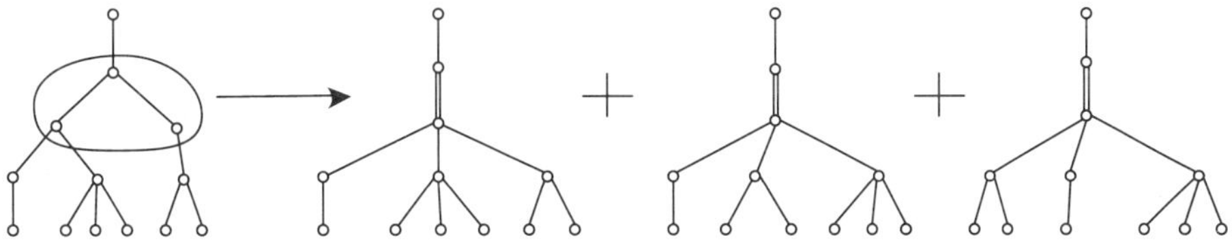

FIGURE 10

of vertices on the nth level subordinate to it. The reduced multiplicity of a vertex is its multiplicity minus 1.

3.3.2. LEMMA. *The codimension of the cell corresponding to a standard tree is equal to the sum of reduced multiplicities of all vertices of the corresponding tree except for the root vertex lying on L_0. The dimension of such a cell is equal to the number of edges in the corresponding tree.* □

Differentials in the cell complex $\widetilde{\mathbb{R}}^n(m)$ are defined as follows. Consider two arbitrary vertices of the tree that are neighbors on some line L_j, $j < n$, and are connected, by edges, with one vertex on L_{j-1}. Suppose that a and b edges, respectively, descend from these two vertices. To this pair of vertices we associate the sum of $\binom{a+b}{a}$ trees that coincide with the initial graph outside these two vertices and the part of the graph subordinate to them. Our two vertices are glued together into one point, and the $a + b$ branches that were connected to these two vertices are now connected to this one point and alternate with each other on the line L_{j+1} in any way preserving the order within each of these two groups of branches; see Figure 10. The boundary of a standard (n, m)-tree is equal to the sum of such chains taken over all pairs of vertices of the described type.

DEFINITION. The *depth* of a standard (n, m)-tree is the smallest number $j = 1, \ldots, n$ such that the intersection of the tree with the line L_j consists of m points.

3.3.3. LEMMA. *The coboundary operator does not increase the depth of a tree.*

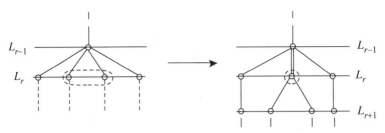

FIGURE 11

Indeed, let the tree have depth r. Any tree of depth $r+1$ geometrically incident to our tree and of codimension greater by 1 can be obtained from it in a unique way, and the corresponding cell is contained in the boundary of the initial one with multiplicity 2, see Figure 11. □

COROLLARY. *The complex of standard trees* mod 2 *splits into the sum of its subcomplexes spanned by trees of fixed depths.*

3.4. Stable cellular decomposition. For a pair of numbers $N > n$, we identify an (N, m)-tree with an (n, m)-tree if the former can be obtained from the latter by adding m trivial (directed vertically down) branches. These identifications preserve the codimension of cells and their incidence coefficients and, hence, make the following definition possible.

DEFINITION. A *stable m-tree* is the class of identified (n, m)-trees for various n. The stable complex of m-trees is the cochain complex over $\mathbb{Z}_2$ freely generated by stable m-trees with the dimension of a generator equal to the codimension of the stable tree.

PROPOSITION. *The group* $H^*(S(m), \mathbb{Z}_2)$ *is isomorphic to the cohomology group of the stable complex of m-trees.* □

PROOF OF PROPOSITION 3.1.1. Under the obvious imbedding $\mathbb{R}^n(m) \to \mathbb{R}^{n+k}(m)$ each cell in $\mathbb{R}^n(m)$ is regularly cut by the cell in $\mathbb{R}^{n+k}(m)$ identified with it. This operation defines a homomorphism of the complex of standard trees in $\mathbb{R}^{n+k}(m)$ into the similar complex for $\mathbb{R}^n(m)$; it is easy to see that this homomorphism agrees with the obvious homomorphism $H^*(\mathbb{R}^{n+k}(m), \mathbb{Z}_2) \to H^*(\mathbb{R}^n(m), \mathbb{Z}_2)$. Its kernel is generated by all trees of depth $> n$. Proposition 3.1.1 follows from the last corollary of 3.3 and the fact that all cells of depth $\geq n$ have codimension at least $n-1$. □

3.5. Cellular realization of the group $H^*(S(m), \mathbb{Z}_2)$. Suppose that $K = (l_0, l_1, \ldots, l_{s-1})$ is an arbitrary element of the set J_+ (see 3.2). Denote by Γ_K the stable 2^s-tree of the following type. On each line $L_0, \ldots, L_{l_0}$ it has exactly one point, and at the last of these points (on L_{l_0}) it branches into two isomorphic graphs each with one point on each of the lines $L_{l_0+1}, \ldots, L_{l_0+l_1}$,

§3. HOMOLOGY OF SYMMETRIC GROUPS

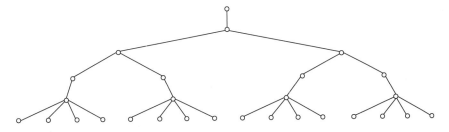

FIGURE 12

and at the last point each of these two graphs branches into two isomorphic graphs, etc. In particular, if $l_1 = 0$ then at the point lying on L_{l_0} the graph branches into four graphs, and if also $l_2 = 0$ then it branches into 8 graphs, etc. For example, in Figure 12 we have a 16-tree Γ_K, where $K = (1, 1, 2, 0)$. Note that the codimension of the cell corresponding to the tree Γ_K, $K = (l_0, l_1, \ldots, l_{s-1})$, is equal to $l_0(2^s - 1) + l_1(s^2 - 2) + \cdots + l_{s-1}(2^s - 2^{s-1})$ (cf. 3.2), and the depth of the tree Γ_K is equal to $\sum l_i + 1$. Denote the unique stable 1-tree by Γ_0.

Basis elements of the group $H^*(S(m), \mathbb{Z}_2)$ are constructed from the trees Γ_K using the following standard operations on trees. Let $\Gamma^1, \ldots, \Gamma^t$ be a collection of standard trees. Place them on the same plane so that they do not intersect and the roots lie in the correct order on the line L_0, and then act on the strip $l^{-1}[0, 1]$ by a transformation shrinking each line $l^{-1}(\varepsilon)$ by $1/\varepsilon$, so that all the roots get identified to one point. The resulting tree is denoted by $\{\Gamma^1, \ldots, \Gamma^t\}$. Obviously, the codimension of the cell corresponding to this tree is equal to the sum of codimensions of the cells corresponding to $\Gamma^1, \ldots, \Gamma^t$. For $\sigma \in S(t)$ denote by $\sigma(\Gamma^1, \ldots, \Gamma^t)$ the tree $\{\Gamma^{\sigma(1)}, \ldots, \Gamma^{\sigma(t)}\}$. Finally, let $\Omega(\Gamma^1, \ldots, \Gamma^t)$ be the chain that contains each of the graphs $\sigma(\Gamma^1, \ldots, \Gamma^t)$ exactly once. (For example, $\Omega(\Gamma^1, \Gamma^2)$ is $\{\Gamma^1, \Gamma^2\} + \{\Gamma^2, \Gamma^1\}$ if $\Gamma^1 \neq \Gamma^2$, and $\{\Gamma^1, \Gamma^2\}$ if $\Gamma^1 = \Gamma^2$.) For any collection of sequences $K_1, \ldots, K_t \in J_+$ such that $2^{|K_1|} + \cdots + 2^{|K_t|} \leq m$, denote by $[K_1, \ldots, K_t; m]$ the chain $\Omega(\Gamma_{K_1}, \ldots, \Gamma_{K_t}, \Gamma_0, \ldots, \Gamma_0)$ (where Γ_0 occurs $m - 2^{|K_1|} - \cdots - 2^{|K_t|}$ times).

THEOREM. *The additive basis in the group $H^*(S(m), \mathbb{Z}_2)$ is given by chains $[K_1, \ldots, K_t; m]$ corresponding to various collections $K_1, \ldots, K_t$ such that $2^{|K_1|} + \cdots + 2^{|K_t|} \leq m$. For each n, the additive basis in $H^*(\mathbb{R}^n(m), \mathbb{Z}_2)$ consists of such chains whose depth is at most n.*

PROOF. It can be checked directly that all the chains above are cycles, and for any stable m-tree Γ each graph in the chain $[K_1, \ldots, K_t; m]$ is contained in the boundary of Γ with zero coefficient. So all these cycles are independent. But the theorem of §3.2 implies that the number of these cycles in each dimension i coincides with the dimension of the group $H^i(S(m), \mathbb{Z}_2)$.

This proves the first statement of the theorem. The second one follows from the first and Lemma 3.3.3. □

The question of the description of the multiplication of our cocycles in $H^*(S(m), \mathbb{Z}_2)$ remains open. (The multiplicative structure of this ring was found in [Hung$_1$], and so the problem reduces to the question of how our generators are related to those used in [Hung$_1$].)

CONJECTURE. The nth power of each element in the algebra $H^*(\mathbb{R}^n(m), \mathbb{Z}_2)$ equals 0.

3.6. The dual realization of the Hopf algebra $H_*(S(\infty), \mathbb{Z}_2)$ and the homology of configuration spaces. To each index $K = (l_0, l_1, \ldots, l_{s-1}) \in J_+$ of depth n we associate a cycle in the space $\mathbb{R}^n(2^s)$; these cycles are multiplicative generators of the Hopf algebra $H_*(S(\infty), \mathbb{Z}_2)$ and of the "nonstable" Hopf algebras $H_*(\mathbb{R}^n(\infty), \mathbb{Z}_2)$; moreover, their suitably placed direct products form additive bases dual to the bases from 3.5 in all groups $H_*(\mathbb{R}^n(m), \mathbb{Z}_2)$. These cycles are direct generalizations of the cycle M from §8 of [Fuchs$_2$].

Let $K = (l_0, l_1, \ldots, l_{s-1}) \in J_+$. For each $d = 0, \ldots, s-1$ let $\tau_d = \sum_{i=0}^{d} l_i + 1$, so that τ_{s-1} is the depth of the tree Γ_K.

Fix a flag $\mathbb{R}^{\tau_0} \subset \mathbb{R}^{\tau_1} \subset \cdots \subset \mathbb{R}^{\tau_{s-1}}$ in the space $\mathbb{R}^{\tau_{s-1}}$, where each space $\mathbb{R}^{\tau_d}$ is the linear span of basis vectors $e_1, \ldots, e_{\tau_d}$. In the first space $\mathbb{R}^{\tau_0}$ of this flag consider a sphere of radius 1 and mark two antipodal points A_1 and A_2 on it. In the space $\mathbb{R}^{\tau_1}$ consider spheres of small radius ε with centers at these points and mark two antipodal points on each of them: A_{11}, A_{12} on the first and A_{21}, A_{22} on the second one. In $\mathbb{R}^{\tau_2}$ consider spheres of radius ε^2 with centers at each of these four points and mark a pair of antipodal points on each of them: $A_{111}, \ldots, A_{222}$. Finally, we will get 2^s distinct points in $\mathbb{R}^{\tau_{s-1}}$, that is a point of the space $\mathbb{R}^{\tau_{s-1}}(2^s)$. The set of all points of this space that can be obtained in such a way will be denoted by Δ_K. For an arbitrary collection $K_1, \ldots, K_t$ of elements of the set J_+ whose maximal depth does not exceed n, denote by $\Delta(K_1, \ldots, K_t)$ the submanifold in $\mathbb{R}^n(2^{|K_1|} + \cdots + 2^{|K_t|})$ diffeomorphic to the product of the manifolds $\Delta_{K_1}, \ldots, \Delta_{K_t}$ that consists of all collections of points in $\mathbb{R}^n$ in which the first $2^{|K_1|}$ points form a collection realizing a point of the set Δ_{K_1}, the next $2^{|K_2|}$ points are obtained from a collection $\zeta \in \Delta_{K_2}$ by translation along the vector $N \cdot \partial/\partial x_1$ (where the number N is sufficiently large), the next $2^{|K_3|}$ points are obtained from a collection $\xi \in \Delta_{K_3}$ by translation along $2N \cdot \partial/\partial x_1$, etc. Finally, for each number $m \geq 2^{|K_1|} + \cdots + 2^{|K_t|}$ define a cycle $\{K_1, \ldots, K_t; m\} \subset \mathbb{R}^n(m)$ diffeomorphic to $\Delta(K_1, \ldots, K_t)$ and consisting of all collections of m points obtained from collections in the set $\Delta(K_1, \ldots, K_t)$ by adding $m - 2^{|K_1|} - \cdots - 2^{|K_t|}$ points of the axis

x_1 with coordinates $x_1 = N \cdot m$, $N(m-1)$, $N(m-2)$, ... This manifold defines a cycle in $H_*(\mathbb{R}^n(m), \mathbb{Z}_2)$.

It is easy to see that the dimension of the compact cycle $\{K_1, \ldots, K_t; m\}$ is equal to the codimension of the closed cycle $[K_1, \ldots, K_t; m]$ defined in 3.5.

3.6.1. THEOREM. *Let $K_1, \ldots, K_t$ and $K'_1, \ldots, K'_q$ be two collections of elements of the set J_+ such that the maximal depth of indices K_i, K'_j is at most n, both numbers $2^{|K_1|} + \cdots + 2^{|K_t|}$ and $2^{|K'_1|} + \cdots + 2^{|K'_q|}$ do not exceed m, and the dimension of the compact cycle $\{K_1, \ldots, K_t; m\}$ is equal to the codimension of the closed cycle $[K'_1, \ldots, K'_q; m]$. Then the intersection index mod 2 of these cycles is 1 if and only if $t = q$ and the collection $K_1, \ldots, K_t$ can be obtained from $K'_1, \ldots, K'_q$ by a permutation of its elements K'_j.*

3.6.2. COROLLARY. *The Hopf algebra $H_*(\mathbb{R}^n(\infty), \mathbb{Z}_2)$ is a free algebra generated by cycles Δ_K overall $K \in J_+$ whose depth does not exceed n. In particular, the elements Δ_K (without any restriction on depth) can be taken as generators of the algebra $H_*(S(\infty), \mathbb{Z}_2)$ occurred in Theorem 3.2.*

3.6.3. COROLLARY. *For arbitrary n and m, an additive basis for the group $H_*(\mathbb{R}^n(m), \mathbb{Z}_2)$ is generated by all cycles $\{K_1, \ldots, K_t; m\}$ such that $2^{|K_1|} + \cdots + 2^{|K_t|} \le m$ and the depth of all Γ_{K_i} do not exceed n.*

This corollary follows immediately from Theorem 3.6.1 and the theorem of §3.5.

PROOF OF THEOREM 3.6.1. Let $t = q$, $K_1 = K'_1, \ldots, K_t = K'_q$. Then the required intersection point of the two cycles is the following collection of m points in $\mathbb{R}^n$. Let $K_1 = (l_0, l_1, \ldots, l_{s-1})$. Then the radius vectors of the first 2^s points of the required collection are

$$\pm e_{\tau_0} \pm \varepsilon e_{\tau_1} \pm \varepsilon^2 e_{\tau_2} \pm \cdots \pm \varepsilon^{s-1} e_{\tau_{s-1}}, \qquad (13)$$

for all possible combinations of signs. In a similar way we construct a collection of $2^{|K_2|}$ points using index K_2 and translate it along the axis x_1 by the length N, then translate the similar collection for K_3 by $2N$, and so on; the last $m - 2^{|K_1|} - \cdots - 2^{|K_t|}$ points are the last points from the definition of the cycle $\{K_1, \ldots, K_t; m\}$. It is easy to see that the union of the m points in $\mathbb{R}^n(m)$ so obtained is the point of the transversal intersection of this cycle with the cell $[\Gamma_{K_1}, \ldots, \Gamma_{K_t}, \Gamma_0, \ldots, \Gamma_0]$ of our cellular decomposition in $\widetilde{\mathbb{R}}^n(m)$, and it remains to prove that this cycle does not intersect any other cell of the same codimension and intersects our cell only at the above-mentioned point. First we consider the case $t = 1$. Let $K = (l_0, l_1, \ldots, l_{s-1})$; suppose $T = \dim \Delta_K \equiv l_0(2^s - 1) + l_1(2^s - 2) + \cdots + l_{s-1}(2^s - 2^{s-1})$.

LEMMA. *In the space $\mathbb{R}^n(2^{|K|})$ the cycle Δ_K does not intersect any closed cycle of the form $[K'_1, \ldots, K'_q; 2^{|K|}]$ of codimension $\ge T$, except for the* cycle $[K, 2^{|K|}]$, and has a unique intersection with this cycle at the point corresponding to the collection of $2^{|K|}$ points (13).

The general case of Theorem 3.6.1 follows directly from this lemma.

PROOF OF THE LEMMA. For each natural j denote by $d(j)$ the largest integer d such that $\tau_d \leq j+1$. Then the multiplicity of each vertex of the tree Γ_K lying on the line L_j is equal to $2^{|K|-d(j)}$. Consider an arbitrary point $x \in \Delta_K$ and the standard tree $\Gamma(x)$ corresponding to it. We show that the multiplicity of each vertex of this tree on the line L_j cannot exceed $2^{|K|-d(j)}$. In fact, let α be one of $2^{|K|}$ points of the collection x. All other points of this collection are classified by the degree of remoteness from the point α: the distance from $2^{|K|-1}$ of them to α is close to 2, from $2^{|K|-2}$ of them is close to 2ε, from $2^{|K|-3}$ of them is close to $2\varepsilon^2$, etc. Any point of the first of these groups (of cardinality $2^{|K|-1}$) differs from α by at least one of the first τ_0 coordinates because the vector of length ≈ 2 connecting them lies "almost in the plane $\mathbb{R}^{\tau_0}$"; similarly, points of the second group differ by one of the first $\mathbb{R}^{\tau_1}$ coordinates, etc. Hence the multiplicity of any vertex of the tree $\Gamma(x)$ on the line L_j does not exceed the multiplicity of any vertex of Γ_K on the same line. Using the fact that the number of vertices of any $(n, 2^{|K|})$-tree on any line L_j (taken with their multiplicities) is equal to $2^{|K|}$, and Lemma 3.3.2, we see that the cell corresponding to the tree $\Gamma(x)$ has codimension $\geq T$ if and only if $\Gamma(x) = \Gamma_K$. The fact that the intersection point of the cycles Δ_K and $[K, 2^{|K|}]$ is unique, is trivial, and the lemma is proved.

3.7. Homomorphism from the symmetric group to the orthogonal group. Similarly to 2.3 we can construct an m-dimensional vector bundle $T(m)$ over $\mathbb{R}^\infty(m) \sim K(S(m), 1)$ and over all spaces $\mathbb{R}^n(m)$, with fiber over a collection $\zeta \subset \mathbb{R}^n$ being the space of functions on its points. In the case $n = \infty$ this bundle coincides with the one induced by the trivial bundle over $BO(m)$ via the right map of (5).

THEOREM. *The ith Stiefel-Whitney class of the bundle $T(m)$ is equal to the sum of all cells corresponding to standard (n, m)-trees of depth 2 and codimension i.*

The proof is similar to the proof of Theorem 2.3. □

COROLLARY 1: *For any r, the rth degree of the $(m-1)$-dimensional Stiefel-Whitney class of the bundle $T(m)$ over $\mathbb{R}^n(m)$ is dual to the cell of the canonical cellular decomposition which corresponds to the standard (n, m)-tree with only one point on any line $L_1, \ldots, L_r$ and m points on L_{r+1} (that is, to the cell which consists of all collections $\{z_1, \ldots, z_m\}$ whose projections on the plane spanned by the orts $e_1, \ldots, e_r$ consist of only one point).*

Indeed, for $r = 1$ this cell is just the one from the previous theorem, and for any greater r the indicated cell is (the closure of) the intersection set of r manifolds in general position homological to the cell for $r = 1$.

COROLLARY 2. *If m is a power of 2, then this class $(w_{m-1}(T(M)))^r$ is nontrivial for any $r = 1, \ldots, n-1$.*

Indeed, if $m = 2^q$, then all trees from Corollary 1 with $r < n$ are m-trees of the form Γ_K with $K = (r, 0, \ldots, 0)$ (see 3.5) and are nontrivial by Theorem 3.6.1.

§4. Cohomology of braid groups and configuration spaces with coefficients in the sheaf $\pm\mathbb{Z}$

4.1. The sheaf $\pm\mathbb{Z}$ and representation $\pm\mathbb{Z}$. By the same symbol $\pm\mathbb{Z}$ we denote several different but closely related objects, namely,

a) the representation $S(m) \to \operatorname{Aut}(\mathbb{Z})$, where the odd permutations act by multiplication by -1;

b) the local system of groups on the space $\mathbb{R}^\infty(m) \sim K(S(m), 1)$ locally isomorphic to $\mathbb{Z}$ but nontrivial over any path corresponding to an odd permutation;

c) the local systems of groups on all spaces $\mathbb{R}^n(m)$ induced from the local system from b) by the obvious imbedding $\mathbb{R}^n \hookrightarrow \mathbb{R}^\infty$.

PROPOSITION. *$H^*(S(m), \pm\mathbb{Z}) \cong H^*(\mathbb{R}^\infty(m), \pm\mathbb{Z})$, where the symbols $\pm\mathbb{Z}$ are understood in the sense of a) and b), respectively.*

This follows from general facts about cohomology of groups (see, for example, [FF]).

Cohomology of configuration spaces $\mathbb{R}^2(m)$ with coefficients in the sheaf $\pm\mathbb{Z}$ appear in the computation of cohomology of Thom spaces of a braid group. More generally, let $E(n, m)$ be the total space of the vector bundle $T(m)$ over $\mathbb{R}^n(m)$ (see 3.7); let $E_0(n, m)$ be the space $E(n, m)$ with the zero section removed.

PROPOSITION.
$$H^i(\mathbb{R}^n(m), \pm\mathbb{Z}) \cong H^{i+m}(E(n, m), E_0(n, m); \mathbb{Z}). \quad \square$$

For odd n, $\pm\mathbb{Z}$ is the sheaf determined by the orientation of the space $\mathbb{R}^n(m)$: this orientation is violated for exactly those loops that induce odd permutations of the points of $\xi \in \mathbb{R}^n(m)$.

The main results of this section are the following two theorems (see [V$_3$]).

4.2. THEOREM. *The group $H^{m-1}(\operatorname{Br}(m), \pm\mathbb{Z})$ is trivial if m is not a power of a prime number, and is equal to $\mathbb{Z}_p$ if m is a power of a prime number p.*

4.3. THEOREM. *The natural homomorphisms*
$$H^i(S(m), \pm\mathbb{Z}) \to H^i(\mathbb{R}^{2k}(m), \pm\mathbb{Z}),$$
$$H^j(S(m), \mathbb{Z}) \to H^j(\mathbb{R}^{2k+1}(m), \mathbb{Z})$$

given by the imbeddings $\mathbb{R}^n \to \mathbb{R}^\infty$ are epimorphisms for all i, j, m, k, and isomorphisms for $i \leq 2k - 1$, $j \leq 2k$, respectively.

COROLLARY 1. *The homomorphism* $H^*(S(m), \pm\mathbb{Z}) \to H^*(\mathrm{Br}(m), \pm\mathbb{Z})$ *defined by the obvious map* $\mathrm{Br}(m) \to S(m)$ *is an epimorphism.*

Note that a similar statement is false in the case of integral coefficients: $H^1(\mathrm{Br}(m), \mathbb{Z}) = \mathbb{Z}$, but all groups $H^i(S(m), \mathbb{Z})$, $i \geq 1$, are finite.

COROLLARY 2. *All groups* $H^i(\mathbb{R}^{2k}(m), \pm\mathbb{Z})$, $H^i(\mathbb{R}^{2k+1}(m), \mathbb{Z})$ *for* $i \geq 1$ *are finite.*

PROOF OF COROLLARY 2. By the previous theorem it suffices to prove a similar result for the groups $H^i(S(m), \pm\mathbb{Z})$. Consider a line bundle E over $K(S(m), 1)$ with orientation violated over loops corresponding to odd permutations. Let E_0 be the space of the bundle E with the zero section removed.

LEMMA. 1. *For each* $i \geq 0$, $H^i(S(m), \pm\mathbb{Z}) \cong H^{i+1}(E, E_0; \mathbb{Z})$.
2. E_0 *is a space of type* $K(A(m), 1)$, *where* $A(m)$ *is the group of even-permutations.*

Statement 1 of the lemma is the Thom isomorphism for unoriented bundles; statement 2 is trivial. □

Hence the groups $H^i(E)$, $H^i(E_0)$ are finite for $i \geq 1$, since they coincide with the cohomology groups of finite groups. The required result now follows from the exact sequence of the pair (E, E_0). □

4.4. Cellular cohomology. To prove Theorems 4.2 and 4.3 we use once more the cellular decomposition of $\widetilde{\mathbb{R}}^n(m)$ constructed in 3.3. Define a cochain complex $\Xi(m, n)$ freely generated by pairs: (transversally oriented cell of this decomposition, basis section of the system $\pm\mathbb{Z}$ over it); change of the transversal orientation or of the basis section corresponds to multiplying the generator by -1; the dimension of such generator is equal to the codimension of the cell, and incidence coefficients are defined as follows. Let $e_1 \succ e_2$ be two cooriented cells of contiguous dimensions in $\mathbb{R}^n(m)$ equipped with basis sections of the system $\pm\mathbb{Z}$. At any point x of the cell e_2 fix a germ of a manifold of complementary dimension transversal to e_2. This manifold Λ intersects e_1 in a curve. To each local (near x) branch of this curve we associate two numbers α and β equal to ± 1 and defined as follows. The number α equals $+1$ or -1 depending on whether or not the fixed basis section of the system $\pm\mathbb{Z}$ over e_2 coincides with the section obtained from the fixed section over e_1 by following our branch. The number β equals $+1$ or -1 depending on whether or not the following two orientations of the manifold Λ coincide: the first one induced by the chosen coorientation of the cell e_2 at x, and the second given at every point of our branch by the frame tangent to Λ with the first vector directed along the branch towards x and others defining the chosen coorientation of the cell e_1.

§4. COHOMOLOGY OF BRAID GROUPS AND CONFIGURATION SPACES

Define the incidence coefficient of the generators of the complex $\Xi(m, n)$ corresponding to our cells (with given orientations and sections) as the sum of products $\alpha \cdot \beta$ over all local (close to x) branches of the curve $\Lambda \cap e_1$.

LEMMA. $H^*(\mathbb{R}^n(m), \pm \mathbb{Z}) \cong H^*(\Xi(m, n))$.

This is a version of the Poincaré duality theorem. □

Denote by $Y(m, n)$ the similar cell complex generated by the same cellular decomposition and computing the trivial integral homology of $\mathbb{R}^n(m)$: in the definition of its cells the choice of the section is omitted, and its incidence coefficients are defined as above but by the numbers β alone.

Let e be a cell of our decomposition in $\mathbb{R}^{n+j}(m)$ whose depth does not exceed n. Then under the natural imbedding $\mathbb{R}^n(m) \to \mathbb{R}^{n+j}(m)$ the transversal orientation of this cell induces the transversal orientation of the cell in $\mathbb{R}^n(m)$ stably equivalent to it. This allows us to define homomorphisms of complexes $\Xi(m, n+j) \to \Xi(m, n)$, $Y(m, n+j) \to Y(m, n)$; these homomorphisms agree with the maps in cohomology $H^*(\mathbb{R}^{n+j}(m), \pm \mathbb{Z}) \to H^*(\mathbb{R}^n(m), \pm \mathbb{Z})$, $H^*(\mathbb{R}^{n+j}(m), \mathbb{Z}) \to H^*(\mathbb{R}^n(m), \mathbb{Z})$, and their kernels are spanned y all cells of depth $\geq n+1$.

LEMMA (cf. 3.3). *The coboundary operator in the complex $\Xi(m, n)$ does not increase the depth of chains of even depth: the coboundary of any cell of depth $2k$ is the sum of cells of depth $2k$. In the complex $Y(m, n)$ the differential of any cell of depth $2k+1$ is the sum of cells of depth $2k+1$.*

COROLLARY. *The complex $\Xi(m, n)$ splits into the sum of $]n/2[$ * subcomplexes, the first spanned by cells of depth 1 and 2, the second by cells of depth 3 and 4, etc.*

The complex $Y(m, n)$ splits into the sum of subcomplexes, the first consisting of a single cell of depth 1, the second spanned by cells of depth 2 and 3, the third by cells of depth 4 and 5, etc.

This immediately implies Theorem 4.3.

For the proof of the lemma let us continue the proof of Lemma 3.3.3. Let e be a cell of depth $r < n$, e' be a cell of contiguous codimension and depth $r+1$, obtained by degeneration of e as in Figure 11; geometrically it occurs twice in the boundary of e. It is easy to verify that for any choice of transversal orientations of these cells and of basis sections of the system $\pm \mathbb{Z}$ over them the coefficients α from the definition of incidence coefficient corresponding to these two occurrences are opposite to each other, and the coefficients β coincide if and only if r is even. □

To prove Theorem 4.2 we consider the complex $\Xi(m, 2)$ in more detail; its cells will be denoted by $e(m_1, \ldots, m_t)$ as in §2. It is convenient to use the fact that the space $\Xi(m, 2)$ is oriented, and, instead of the transversal orientation of cells, to use their usual orientation.

---

*Editor's note. The notation $]a[$ means the smallest integer n such that $a \leq n$.

THEOREM. *For a suitable choice of basis sections of the system $\pm\mathbb{Z}$ over cells of the complex $\Xi(m, 2)$ and orientations of these cells, the differential in this complex is given by the formula*

$$\partial e(m_1, \ldots, m_t) = \sum_{i=1}^{t-1} (-1)^{i-1} \binom{m_i + m_{i+1}}{m_i} \qquad (14)$$
$$\times e(m_1, \ldots, m_{i-1}, m_i + m_{i+1}, m_{i+2}, \ldots, m_t).$$

Theorem 4.2 follows immediately from this formula applied to the case $t = 2$. In fact, the group $H^{m-1}(\mathrm{Br}(m), \pm\mathbb{Z})$ is generated by a single cell $e(m)$. By (14), this group is nontrivial if and only if all possible binomial coefficients $\binom{m}{l}$ $(0 < l < m)$ do not generate the group of integers. The latter happens if and only if m is a power of a prime number. □

For the proof of the theorem we make the sections of the local system $\pm\mathbb{Z}$ agree over different cells. For each cell $(m_1, \ldots, m_t)$ consider a path in $\mathbb{R}^2(m)$ that lies entirely in this cell, except for the end belonging to a cell of maximal codimension (m), in such a way that, if a point z_i from the collection $\{z_1, \ldots, z_m\}$ corresponding to the starting point of the path

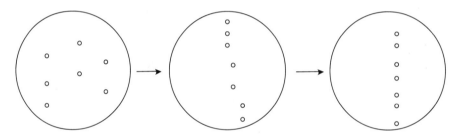

FIGURE 13

lies to the right of z_j, then the points $\tilde{z}_i, \tilde{z}_j$ obtained from them at the end of the path satisfy the condition $\mathrm{Im}\, \tilde{z}_i < \mathrm{Im}\, \tilde{z}_j$ (see Figure 13). Fix some basis section of the system $\pm\mathbb{Z}$ over cell (m) and extend it over the cell $(m_1, \ldots, m_t)$ by translation along such paths; obviously, the result does not depend on the choice of paths. Now we define orientation of cells. Rearrange points $z_1, \ldots, z_m$ of the collection belonging to this cell in such a way that for any $i < m$ either $\mathrm{Re}\, z_i < \mathrm{Re}\, z_{i+1}$ or $\mathrm{Re}\, z_i = \mathrm{Re}\, z_{i+1}$, $\mathrm{Im}\, z_i > \mathrm{Im}\, z_{i+1}$. The orientation of the cell $e(m_1, \ldots, m_t)$ is defined using the differential form $d(\mathrm{Re}$ of the first group of m_1 points$) \wedge d(\mathrm{Re}$ of the second group of m_2 points$) \wedge \cdots \wedge d(\mathrm{Re}$ of the group of m_t points$) \wedge \mathrm{Im}\, z_1 \wedge \cdots \wedge \mathrm{Im}\, z_n$. It is easy to check that for this choice of orientations and this agreement of sections we have formula (14). Note that each cell $e(m_1, \ldots, m_{i-1}, m_i + m_{i+1}, \ldots, m_t)$ is geometrically included in the boundary of the cell $e(m_1, \ldots, m_t)$ exactly $\binom{m_i + m_{i+1}}{m_i}$ times. All these inclusions make the same contribution to the incidence coefficient. □

4.5. More about the map of the braid group into the orthogonal group.
The generator of the group $H^{m-1}(\mathrm{Br}(m), \pm \mathbb{Z})$ can be described as follows.

The homomorphism $\mathrm{Br}(m) \to O(m)$ reduces to a homomorphism of $\mathrm{Br}(m)$ into the group $O(m-1)$ considered as a group of transformations of the plane $\{x | x_1 + \cdots + x_m = 0\}$. To this homomorphism corresponds a map $K(\mathrm{Br}(m), 1) \to BO(m-1)$, see (7). The space $BO(m-1)$ carries an orientation sheaf Or, which is a system of groups locally isomorphic to $\mathbb{Z}$ that follows the orientation of the universal vector bundle. It is easy to see that the system $\pm \mathbb{Z}$ on $K(S(m), 1)$ and, hence, also on $K(\mathrm{Br}(m), 1)$ is induced by this sheaf.

THEOREM. *Under the homomorphism*

$$H^{m-1}(BO(m-1), \mathrm{Or}) \to H^{m-1}(\mathrm{Br}(m), \pm \mathbb{Z})$$

the Euler class of the universal bundle over $BO(m-1)$ is mapped into the class of the cell $e(m)$.

This is proved exactly as the theorem from §2.3. □

§5. Cohomology of braid groups with coefficients in the Coxeter representation

The *Coxeter representation* of the group $S(m)$ (and hence of $\mathrm{Br}(m)$) is the representation $S(m) \to \mathrm{Aut}(\mathbb{Z}^m)$, which acts by permutations of basis vectors. The restriction of this representation to the sublattice $\{c | c_1 + \cdots + c_m = 0\} \subset \mathbb{Z}^m$ is called the *reduced Coxeter representation*. Notation for these representations is X_m and $\widetilde{X}_m$ respectively.

THEOREM. *The group $H^i(\mathrm{Br}(m), X_m)$ is trivial for $i > m-1$, isomorphic to $\mathbb{Z}$ for $i = 0$ and $i = m-1$, to $\mathbb{Z}^2$ for $i = 1$, and to $\mathbb{Z}^2 \oplus$ Torsion for $i = 2, \ldots, m-2$. The group $H^i(\mathrm{Br}(m), \widetilde{X}_m)$ is trivial for $i = 0$ and $i > m-1$, isomorphic to $\mathbb{Z}$ for $i = 1$ and $i = m-1$, and to $\mathbb{Z}^2 \oplus$ Torsion for $i = 2, \ldots, m-2$.*

PROOF. Proof is by induction on m. Denote by $\mathrm{Br}_1(m)$ the subgroup in $\mathrm{Br}(m)$ of index m consisting of braids such that the end of the mth string is again at the mth place. The representation X_m of the group $\mathrm{Br}(m)$ is induced by the trivial $\mathbb{Z}$-representation of $\mathrm{Br}_1(m)$. Therefore, by Shapiro's lemma (see [FF]), $H^*(\mathrm{Br}(m), X_m) \cong H^*(\mathrm{Br}_1(m), \mathbb{Z})$.

The base of induction: for $m = 2$, $\mathrm{Br}_1(m)$ is the group of colored braids on two strings. It is isomorphic to $\mathbb{Z}$, so its cohomology coincides with that of the circle.

Now let $m > 2$. Consider the homomorphism $l: \mathrm{Br}_1(m) \to \mathrm{Br}(m-1)$ obtained by erasing the mth string, and use the Hochschild-Serre spectral sequence of this homomorphism. By definition, the $E_2^{p,q}$ term of this sequence is the p-dimensional cohomology group of $\mathrm{Im}\, l = \mathrm{Br}(m-1)$ with coefficients in the natural representation of this group in the group $H^q(\operatorname{Ker} l)$. It is easy to see that $\operatorname{Ker} l$ is a free group on $m-1$ generators, and therefore $H^0(\operatorname{Ker} l) = \mathbb{Z}$, $H^1(\operatorname{Ker} l) = \mathbb{Z}^{m-1}$, $H^i(\operatorname{Ker} l) = 0$ for $i > 1$.

LEMMA. *The natural action of the group* $\operatorname{Br}(m-1)$ *on* $H^0(\operatorname{Ker} l)$ *is trivial, whereas the action on* $H^1(\operatorname{Ker} l)$ *is isomorphic to the Coxeter representation* X_{m-1}.

The assertion of the theorem about the group $H^*(\operatorname{Br}(m), X_m)$ follows immediately from this lemma, induction hypothesis, and the results on the group $H^*(\operatorname{Br}(m), \mathbb{Z})$ given at the end of §2.

To prove the lemma we use the following geometric realization of the Hochschild-Serre sequence. Realize the space $K(\operatorname{Br}(m-1), 1)$ as the configuration space $\mathbb{R}^2(m-1)$. Consider a bundle $\Pi \to \mathbb{R}^2(m-1)$ over this space with the fiber over $\zeta \in \mathbb{R}^2(m-1)$ being the plane $\mathbb{R}^2$ without $m-1$ points of the collection ζ. It is easy to see that Π is of type $K(\operatorname{Br}_1(m), 1)$, so that $H^*(\operatorname{Br}_1(m), \mathbb{Z}) \cong H^*(\Pi, \mathbb{Z})$, and our Hochschild-Serre sequence is just the Leray spectral sequence calculating the cohomology of the space Π and applied to the projection $\Pi \to \mathbb{R}^2(m-1)$. The basis in 1-cohomology of the fiber of this projection is given by $m-1$ small circles in $\mathbb{R}^2$ with centers at the points of the set ζ. When transported over loops in $\mathbb{R}^2(m)$, these circles are permuted in the same way as the ends of the corresponding strings. □

The statement about the groups $H^i(\operatorname{Br}(m), \widetilde{X}_m)$ follows from the exact cohomology sequence corresponding to the short sequence of coefficients

$$0 \to \widetilde{X}_m \to X_m \to \mathbb{Z} \to 0,$$

and the results on cohomology with coefficients in the modules X_m and $\mathbb{Z}$ above. □

CHAPTER II

Applications: Complexity of Algorithms, Superpositions of Algebraic Functions and Interpolation Theory

The topological complexity of a computational algorithm is the number of its branching nodes (operators IF). Smale [Smale$_4$] used the cohomology of the braid group to study the topological complexity of algorithms for finding approximations of the zeros of a complex polynomial. In particular, in [Smale$_4$] he found the lower bound of the minimal topological complexity of algorithms that compute with sufficient accuracy the roots of the polynomial $z^m + a_1 z^{m-1} + \cdots + a_m$:

$$\tau(m) \geq (\log_2 m)^{2/3}. \tag{1}$$

In §3 below we prove the two-sided estimate

$$m - 1 \geq \tau(m) \geq m - \min D_p(m), \tag{2}$$

where the minimum is taken over all prime p, and $D_p(m)$ is the sum of the digits in the expansion of m in base p; the estimate on the right-hand side of (2) is based on the results of §4 of Chapter I on the cohomology of the braid group with coefficients in the sheaf $\pm \mathbb{Z}$.

In particular, if m is a power of a prime number, the left and right estimates of (2) coincide, and the minimal number of branching nodes is equal to $m - 1$.

Moreover, in the latter case the number of branching nodes for the problem of computing one root is also $m - 1$.

The method of Smale is very general and allows one to estimate the topological complexity of any ill-posed computational problem using a topological parameter of a fibration associated to the problem, namely its genus, introduced and studied by A. S. Schwarz [Schwarz] (and rediscovered in [Smale$_4$]).

In §4 we apply this method to the problem (also raised by Smale) of the topological complexity of finding approximations to solutions of systems of polynomial equations in several variables. In most cases our estimates are also sharp up to order (the lower and upper estimates differ by a factor independent of the degree of the equations).

The study of cohomology of the braid group was initiated in [Ar$_4$] and [Fuchs$_2$] in relation to the problem of representability of algebraic functions by superpositions of functions in fewer variables. The use of the genus of a covering related to the algebraic function gives new obstructions to such a representation. This is discussed in §5 below.

In §6 we study the lowest possible dimensions of function spaces on manifolds, which interpolate any function at any k nodes of interpolation. The lower bound in this problem is formulated in the terms of characteristic classes of natural k-dimensional vector bundles over the corresponding configuration spaces (and is very close to the upper bound).

§1. The Schwarz genus

1.1. Definition and elementary properties. Let X, Y be normal Hausdorff spaces (for example, finite-dimensional manifolds or subsets of Euclidean spaces) and $f: X \to Y$ be a continuous map such that $Y = f(X)$.

DEFINITION. The *genus* of the map f is the minimal cardinality of an open cover of Y consisting of sets such that there exists a continuous section of the map f over each set. Notation: $g(f)$.

The following propositions directly follow from the definition:

1.1.1. PROPOSITION. *Let Y' be a subset of Y. Then the genus of the restriction of the map f to the set $X' = f^{-1}(Y')$ does not exceed the genus of f.* □

More generally, suppose there is a continuous map φ of a normal space Z into Y, and $f': X' \to Z$ is the map induced by it from the map $f: X \to Y$. (That is, the space X' consists of all pairs of the form $\{z \in Z$, point of the set $f^{-1}(\varphi(z))\}$, and the map f' sends such a pair to the point z.)

1.1.2. PROPOSITION. *The genus of the induced map $f': X' \to Z$ does not exceed the genus of f.* □

1.2. Genus and category. The genus of a map $X \to Y$ is a generalization of the category of the space Y.

DEFINITION (see [LSh]). The *category* of a topological space Y is the minimal cardinality of an open cover of Y consisting of sets contractible in Y.

Suppose Y is path connected. Consider the Serre fibration $S \to Y$ with the space consisting of all paths $[0, 1] \to Y$ sending 0 to the base point and projection that associates to each path its endpoint.

PROPOSITION ([Schwarz]). *The category of a path connected space Y coincides with the genus of its Serre fibration.* □

1.3. The estimate of the genus using the cohomological length of f. The map $f: X \to Y$ induces a homomorphism $f^*: H^*(Y, A) \to H^*(X, A)$ for any group of coefficients A. If A is a ring then the kernel $\operatorname{Ker} f^*$ of this homomorphism is a subring (and even an ideal) in $H^*(Y, A)$; the *cohomological length* of the map f is the length of this ring, which is the maximal number of its elements of positive dimension whose product is nonzero.

PROPOSITION (see [Schwarz], [Smale$_4$]). *The genus of the map $f: X \to Y$ is greater than its length.*

PROOF (see [Smale$_4$]). Suppose there exist elements $\gamma_1, \ldots, \gamma_k \in \operatorname{Ker} f^*$ such that $\gamma_1 \cdots \gamma_k \neq 0$, and an open cover $\{V_l\}$, $l = 1, \ldots, k$, of the space Y such that over each of its elements there is a section $\sigma_l: V_l \to X$ of the map f. For each $l = 1, \ldots, k$ consider the exact sequence of the pair

$$\cdots \to H^*(Y, V_l) \xrightarrow{i_l} H^*(Y) \xrightarrow{j_l} H^*(V_l) \to \cdots.$$

For each $\gamma_l \in \operatorname{Ker} f^*$ we have $j_l(\gamma_l) = \sigma_l^* f^*(\gamma_l) = 0$ by the definition of the section; by the exactness of the sequence this implies that there is an element $\alpha_l \in H^*(Y, V_l)$ such that $\gamma_l = i_l(\alpha_l)$. The cohomology product of cocycles $\alpha_1, \ldots, \alpha_k$ is an element of the ring $H^*(Y, \bigcup V_l) = 0$; since the multiplication in cohomology is natural, this product must be mapped to the class $\gamma_1 \cdots \gamma_k$ by the homomorphism $H^*(Y, \bigcup V_l) \to H^*(Y)$. Contradiction. □

Applying this theorem to the Serre fibration we get a theorem of Frolov-Èl'sgol'ts [FE]: the category of the space Y is greater than the length of the ring $H^*(Y)$.

1.4. Calculation of the genus of a fibration in terms of global sections. Let $f: X \to Y$ be a locally trivial fibration with typical fiber F. Let its *l*th *power* be a fibration whose fibers are the fiberwise joins of l copies of fibers of f, see Appendix 1. This fibration is denoted by f^{*l}, in particular, $f^{*1} \sim f$.

THEOREM (see [Schwarz]). *The genus of a locally trivial fibration $f: X \to Y$ is at most l if and only if its lth power has a continuous section.* □

In the case when the genus of f is finite there is a more general (but actually equivalent to the above) theorem.

THEOREM (see [Schwarz]). *If $g(f) < \infty$ then $g(f^{*l}) = [(1/l)(g(f)+l-1)]$ for all l.* □

COROLLARY. *If Y has the homotopy type of a k-dimensional CW-complex then the genus of any fibration over Y does not exceed $k + 1$.*

This follows directly from obstruction theory since the l-fold join of any space with itself is $(l - 2)$-connected.

1.5. Homological genus of a principal covering. Let $f: X \to Y$ be a principal G-covering (that is, a principal G-bundle with a discrete group G). Then (see Appendix 1) there exists a (unique up to homotopy) classifying map

$$\operatorname{cl}: Y \to K(G, 1) \tag{3}$$

such that the bundle f is induced from the universal G-bundle over $K(G, 1)$. Let A be an arbitrary G-module (or the corresponding local system on the space $K(G, 1)$, cf. I.4), and $\text{cl}^* A$ be the local system on Y induced from A by the map cl.

DEFINITION (see [Schwarz]). The *homological A-genus* of the principal G-bundle $f: X \to Y$ is the smallest number i such that the map

$$\text{cl}^*: H^j(K(G, 1), A) \to H^j(Y, \text{cl}^* A)$$

is trivial for all $j \geq i$. The homological A-genus of the bundle f is denoted by $h_A(f)$.

The *homological genus* $h(f)$ is the maximum of the numbers $h_A(f)$ over all G-modules A.

THEOREM (see [Schwarz]). *For any principal G-covering f, $g(f) \geq h(f)$.*
□

§2. Topological complexity of algorithms, and the genus of a fibration

2.1. Statement of problems. Let m be a positive integer and ε be a positive number. Denote by $\mathfrak{A}_m$ the space of all complex polynomials of the form

$$z^m + a_1 z^{m-1} + \cdots + a_{m-1} z + a_m \tag{4}$$

whose coefficients satisfy the condition $|a_i| \leq 1$. We consider computational algorithms that solve one of the following problems.

PROBLEM $P(m, \varepsilon)$: for each polynomial in $\mathfrak{A}_m$ given by the collection of its coefficients, compute all its roots with error at most ε.

PROBLEM $P_1(m, \varepsilon)$ differs from the problem $P(m, \varepsilon)$ in that only one root of the polynomial must be computed.

2.2. Algorithms. We use the definition of algorithm from [Smale$_4$]. Recall the definition applied to problem $P(m, \varepsilon)$.

An *algorithm* is a finite oriented tree (graph without cycles) with nodes of the following four types (see Figure 2 in the introduction).

1) The unique *input node* which accepts $2m$ real numbers $\text{Re}\, a_i$, $\text{Im}\, a_i$ (such that $(\text{Re}\, a_i)^2 + (\text{Im}\, a_i)^2 \leq 1$).

2) *Computing nodes*: at each of these certain real rational functions are computed from the input values and the values of similar functions computed at other computing nodes above the given one in the algorithm.

3) *Branching nodes*: here the value of one of the rational functions in $\text{Re}\, a$, $\text{Im}\, a$ computed earlier is compared to zero, and depending on whether this value is positive or not the control is passed to one of the two edges coming out of the node.

4) *Output nodes*: at each of them certain $2m$ rational functions in a computed earlier in the algorithm are declared to be the real and imaginary

parts of roots of the polynomial (4), denoted by $\operatorname{Re} z_i(a)$, $\operatorname{Im} z_i(a)$, and the execution of the program terminates.

(The absence of cycles in the graph implies that the number of output nodes exceeds the number of branching nodes by 1.)

Thus any collection of input values $a = (a_1, \ldots, a_m) \in \mathbb{C}^m$ defines an execution path of the algorithm and hence (if there is no division by zero along the way) one of the outputs and an ordered collection of complex numbers $z_1(a), \ldots, z_m(a)$. An algorithm *solves* the problem $P(m, \varepsilon)$ if for any collection of input values $a \in \mathfrak{A}_m$ the corresponding execution path does not pass through the division by zero and it is possible to order the roots ξ_i of the corresponding polynomial (4) so that $|z_i(a) - \xi_i(a)| \leq \varepsilon$ for each $i = 1, \ldots, m$.

Similarly, an algorithm *solving* the problem $P_1(m, \varepsilon)$ gives a pair of numbers that are declared to be the real and imaginary parts of the required root.

The *topological complexity of an algorithm* is the number of branching nodes in it. The *topological complexity of a problem* is the minimal topological complexity of algorithms solving this problem.

The topological complexities of the problems $P(m, \varepsilon)$ and $P_1(m, \varepsilon)$ will be denoted by $\tau(m, \varepsilon)$ and $\tau_1(m, \varepsilon)$. Obviously when ε decreases, these numbers do not decrease. Set

$$\tau(m) = \lim_{\varepsilon \to +0} \tau(m, \varepsilon), \qquad \tau_1(m) = \lim_{\varepsilon \to +0} \tau_1(m, \varepsilon).$$

2.3. Smale's theorem. Consider the *discriminant subset* Σ in $\mathfrak{A}_m$ consisting of polynomials with multiple roots. Over the set $\mathfrak{A}_m - \Sigma$ there is an $m!$-fold covering $f_m: M^m \to \mathfrak{A}_m - \Sigma$ with the fiber over a point a being various ordered collections of roots of the polynomial a. An exact solution (with $\varepsilon = 0$) of the problem $P(m, \varepsilon)$ with initial data a includes the choice of one of the $m!$ orderings of these roots.

PROPOSITION. *The space $\mathfrak{A}_m - \Sigma$ is a space of type $K(\operatorname{Br}(m), 1)$: its imbedding into the space of all nondiscriminant polynomials* (4) *is a homotopy equivalence. The space M^m is a space of type $K(I(m), 1)$, where $I(m)$ is the colored braid group.* □

SMALE'S THEOREM (see [Smale$_4$]). *For any m there is a number $\varepsilon_0 > 0$ such that for all $\varepsilon \in (0, \varepsilon_0]$ the topological complexity of the problem $P(m, \varepsilon)$ and the genus of the covering f_m are related by the inequality*

$$\tau(m, \varepsilon) \geq g(f_m) - 1. \tag{5}$$

Indeed, the covering and the system of sections used in the definition of genus are actually given by an algorithm. To each of its output nodes corresponds a semialgebraic set $W_i \subset \mathfrak{A}_m$ consisting of such input values that, upon being loaded at the root of the tree, we exit through exactly this output node. Let Σ_ε be the set of polynomials with a pair of roots at distance

$|\xi_i - \xi_j| \leq 2\varepsilon$. If ε is small then the covering $f_m \colon M^m \to \mathfrak{A}_m - \Sigma$ is equivalent to its restriction to $\mathfrak{A}_m - \Sigma_\varepsilon$, and, in particular, has the same genus. The algorithm defines an "ε-section" of the covering of M^m over each of the sets $W_i \cap (\mathfrak{A}_m - \Sigma_\varepsilon)$, that is, a map into M^m which differs from a section by ε. But by the definition of the set Σ_ε such ε-section can be deformed to a true section. Finally, this section over a semialgebraic set $W_i \cap (\mathfrak{A}_m - \Sigma_\varepsilon)$ can be extended to a section over a neighborhood of this set, and we get an open cover of $\mathfrak{A}_m - \Sigma_\varepsilon$ by sets corresponding to the output nodes of the algorithm, and over each of these open sets there is a section of the covering f_m. $\square$

REMARK. Proof of estimate (1) in [Smale$_4$] consists of this theorem, the estimate for the genus using the cohomological length of a map (see 1.3 above), and an estimate of this length for the covering f_m that follows from the results of D. B. Fuchs on the ring $H^*(\mathrm{Br}(m), \mathbb{Z}_2)$, see §I.2.

The method of Smale gives a similar estimate for the topological complexity of the problem $P_1(m, \varepsilon)$. Namely, together with the $m!$-fold covering f_m over the set $\mathfrak{A}_m - \Sigma$ we also have an m-fold covering $\varphi_m \colon N^m \to \mathfrak{A}_m - \Sigma$ whose space is the set of pairs (polynomial $a \in \mathfrak{A}_m - \Sigma$; one of the roots of a).

THEOREM. *For each m there is a number $\varepsilon_0 > 0$ such that for all $\varepsilon \in (0, \varepsilon_0]$*

$$\tau_1(m, \varepsilon) \geq g(\varphi_m) - 1. \tag{6}$$

This is proved by exactly the same argument as Smale's theorem. $\square$

2.4. The last two theorems can be viewed as special cases of "Smale's principle." Below is one of its versions.

Let $f \colon X \to Y$ be a smooth map of nonsingular (or at least stratified) analytic sets, $f(X) = Y$, and the problem is to find for each point $y \in Y$ a point $x \in X$ within a distance ε from one of the points in the set $f^{-1}(y)$. Excise from Y a proper submanifold so that over its complement the map f defines a locally trivial fibration. Then for sufficiently small ε the topological complexity of our problem is no less than the genus of this fibration minus 1.

§3. Estimates of the topological complexity of finding roots of polynomials in one variable

3.1. Statement of results.

3.1.1. THEOREM. *The topological complexity of the problem $P(m, \varepsilon)$ of finding all roots of a polynomial* (4) *with accuracy ε for sufficiently small ε and any prime p satisfies the estimate $\tau(m, \varepsilon) \geq m - D_p(m)$, where $D_p(m)$ is the same as in formula* (2).

COROLLARY. $\tau(m) \geq m - \log_2(m+1)$.

3.1.2. THEOREM. *For any $\varepsilon > 0$, $\tau(m, \varepsilon) \leq m - 1$.*

COROLLARY. *If m is a power of a prime number then $\tau(m, \varepsilon) = m - 1$.*

3.1.3. THEOREM. *If m is a power of a prime number then for sufficiently small $\varepsilon > 0$ the topological complexity of the problem $P_1(m, \varepsilon)$ of finding one of the roots of a polynomial* (4) *with accuracy ε equals $m - 1$.*

Denote by $\operatorname{prim}(m)$ the greatest divisor of m which is a power of a prime number.

3.1.4. THEOREM. *For any natural m and for sufficiently small $\varepsilon > 0$ the topological complexity of the problem $P_1(m, \varepsilon)$ is at least $\operatorname{prim}(m) - 1$.*

3.1.5. PROPOSITION. *For $m \to \infty$, $\operatorname{prim}(m) \gtrsim \ln m$ (i.e. for any $\delta > 0$ there is $m(\delta)$ such that $\operatorname{prim}(m) > (1 - \delta)\ln m$ for $m \geq m(\delta)$).*

PROBLEM. Is it true that the topological complexity $\tau_1(m)$ of the problem of finding one root monotonically depends on m? If this were true then Theorem 3.1.4 could be made much stronger by changing the lower estimate $\operatorname{prim}(m)$ to the largest integer not exceeding m that is a power of a prime. But I do not have any serious data that support this conjecture (contrary to the case of the problem of finding all roots, when the similar statement is true and not difficult to prove).

3.2. Proof of Theorem 3.1.1. Suppose the number m admits a decomposition into the sum of t powers of a prime number p: $m = m_1 + \cdots + m_t$, where $m_i = p^{k_i}$. Let a be an interior point of the polydisk $\mathfrak{A}_m$ corresponding to a polynomial that has exactly t geometrically distinct roots with multiplicities $m_1, \ldots, m_t$. Let $U \subset \mathfrak{A}_m$ be a small spherical neighborhood of a, and f_U be the restriction of the covering f_m to the preimage of the domain $U - \Sigma$.

THEOREM. *The homological $\pm\mathbb{Z}$-genus of the covering f_U is equal to $m + 1 - t$.*

Theorem 3.1.1 follows directly from this result, Smale's theorem, and Proposition 1.1.1 of this chapter.

PROOF OF THE THEOREM. The space $U - \Sigma$ is homotopy equivalent to the product

$$(\mathfrak{A}_{m_1} - \Sigma_1) \times \cdots \times (\mathfrak{A}_{m_t} - \Sigma_t), \qquad (7)$$

where $\mathfrak{A}_{m_i}$ is the standard unit polydisk in the space of polynomials of degree m_i with leading coefficient 1, and Σ_i is the set of polynomials with multiple roots. Over each of the spaces $\mathfrak{A}_{m_i} - \Sigma_i$ there is an $m_i!$-fold covering $f_{m_i}: M^{m_i} \to \mathfrak{A}_{m_i} - \Sigma_i$ constructed in §2. Denote by $f(i)$ the covering over the space (7) induced from f_{m_i} by projection onto the ith factor.

LEMMA. *Over $U - \Sigma$, the covering f_U is a disjoint union of $m!/(m_1! \cdots m_t!)$ copies of $m_1! \cdots m_t!$-fold coverings, each isomorphic to a product of the coverings f_i.*

Thus, under restriction to $U - \Sigma$ the group $S(m)$ of the covering f_m reduces to its subgroup $S(m_1) \times \cdots \times S(m_t)$, and the map $U - \Sigma \to K(S(m), 1)$ that classifies this covering factors as a composition

$$U - \Sigma \to K(S(m_1), 1) \times \cdots \times K(S(m_t), 1) \to K(S(m), 1).$$

The right arrow of this composition sends the local system $\pm \mathbb{Z}$ on $K(S(m), 1)$ to the tensor product of systems $\pm \mathbb{Z}$ on $K(S(m_i), 1)$. Now the theorem follows from Theorem 4.2 of Chapter I and the Künneth formula.

3.3. An algorithm of complexity $m - 1$: proof of Theorem 3.1.2. Now, for any given $\varepsilon > 0$ we give an explicit construction (modulo a realization of the Weierstrass approximation theorem) of an algorithm of topological complexity $m - 1$ that solves the problem $P(m, \varepsilon)$. The computing nodes of these algorithms contain only polynomial (not rational) functions. All these algorithms have the same topological structure (independent of ε), see Figure 14.

For the sake of convenience define the following metric in the polydisk $\mathfrak{A}_m$: the distance between two polynomials a and a' is

$$\min \max_i |\xi_i(a) - \xi_i(a')|,$$

the minimum being taken over all possible orderings of the roots $\xi_i(a)$, $\xi_i(a')$ of these polynomials. Split $\mathfrak{A}_m$ into sets $S_1, \ldots, S_m$, where S_t consists of polynomials such that the real parts of their roots take precisely t distinct values. (Hence the connected components of the set $S_t - \Sigma$ are exactly the intersections of the set $\mathfrak{A}_m - \Sigma \sim \mathbb{C}^1(m)$ with $(m + t)$-dimensional cells $e(m_1, \ldots, m_t)$ of the cellular decomposition from §I.2.) By the Weierstrass approximation theorem for any $t = 1, \ldots, m$ there exists a polynomial $\chi_t \colon \mathfrak{A}_m \to \mathbb{R}$ such that the domain $V_t = \{\chi_t \leq 0\}$ lies in the $(2^{-2t+1}\varepsilon)$-neighborhood of the set S_t and contains its $(2^{-2t}\varepsilon)$-neighborhood. (In particular, $V_m = \mathfrak{A}_m$.) Note that each point a of the set $V_t - V_{t-1}$ lies in the $(2^{-2t+1}\varepsilon)$-neighborhood of exactly one component of the set S_t. Let this component correspond to the cell $e(m_1, \ldots, m_t)$. Decompose the roots $\xi_1, \ldots, \xi_m$ of the polynomial a into groups of cardinality $m_1, \ldots, m_t$ according to the ordering of their real parts; in each group these real parts differ by no more than $2^{-2t+2}\varepsilon$. Associate to the point a the following $2m$ numbers: $\widetilde{\text{Re}}_1(a) = \cdots = \widetilde{\text{Re}}_{m_1}(a) =$ (arithmetic mean of the real parts of the roots from the first group), $\widetilde{\text{Re}}_{m_1+1} = \cdots = \widetilde{\text{Re}}_{m_1+m_2} =$ (arithmetic mean of the real parts of the roots from the second group), etc.; $\widetilde{\text{Im}}_1(a), \ldots, \widetilde{\text{Im}}_{m_1}(a)$ which are respectively the greatest, the next greatest, $\ldots$, the smallest of the numbers $\text{Im}(\xi_t)$, $t = 1, \ldots, m_1$; $\widetilde{\text{Im}}_{m_1+1}(a), \ldots, \widetilde{\text{Im}}_{m_1+m_2}(a)$ are the similarly ordered imaginary parts of the roots from the second group, and so on. Performing this operation for each point of the set $V_t - V_{t-1}$ we obtain $2m$

continuous functions on the set. Extend them to continuous functions on the whole set $\mathfrak{A}_m$ and, using the Weierstrass approximation theorem once more, approximate these functions by polynomials $\operatorname{Re}_1^t, \ldots, \operatorname{Re}_m^t, \operatorname{Im}_1^t, \ldots, \operatorname{Im}_m^t$ with accuracy $\varepsilon/2$.

Now we are ready to describe the required algorithm; see Figure 14. □

3.4. Proof of Theorem 3.1.3 is based on the following statement about the covering $\varphi_m: N^m \to \mathfrak{A}_m - \Sigma$ described at the end of 2.3.

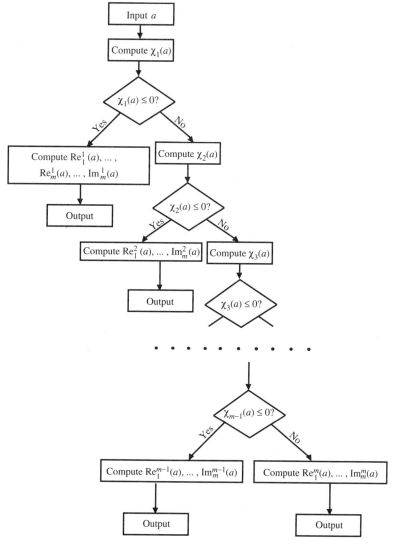

FIGURE 14

3.4.1. THEOREM (see [V_3]). *If m is a power of a prime then the genus of the covering φ_m is equal to m.*

To prove this we calculate an obstruction to the inequality $g(\varphi_m) < m$ for all m; for $m = p^k$ this obstruction is nontrivial.

Following 1.4 consider the bundle $\varphi^{*(m-1)} : \nabla^m \to \mathfrak{A}_m - \Sigma$ over $\mathfrak{A}_m - \Sigma$ with fiber over a being the join of $m - 1$ copies of the fiber of covering φ_m over a (in particular, this fiber is homotopy equivalent to the bouquet of $(m-1)^{m-1}$ copies of the $(m-2)$-dimensional sphere). From Theorem 1.4 and general obstruction theory it follows that the only obstruction to the inequality $g(\varphi_m) < m$ (or, equivalently, to the existence of a section of $\varphi_m^{*(m-1)}$) lies in the group $H^{m-1}(\mathfrak{A}_m - \Sigma, \Theta)$, where Θ is the local system of groups associated with $\varphi_m^{*(m-1)}$ whose fiber is the $(m-2)$th homotopy group of the fiber of $\varphi_m^{*(m-1)}$. By the formula for the homology of a join (see Appendix 1) and the Hurewicz theorem, Θ is the $(m-1)$th tensor power of a local system $\mathscr{H}$ whose fiber is the zero-dimensional homology group of the fiber of the covering φ_m modulo a point. (It is easy to see that the system $\mathscr{H}$ also has the following description: it is the local system on the space $K(\mathrm{Br}(m), 1)$ corresponding to the reduced Coxeter representation of the group $\mathrm{Br}(m)$; see §5 of Chapter I.)

Consider the cellular decomposition of the space $\mathbb{C}^1(m) \sim K(\mathrm{Br}(m), 1) \sim \mathfrak{A}_m - \Sigma$ used in Chapter I. The corresponding cochain complex which calculates the group $H^*(\mathfrak{A}_m - \Sigma, \mathscr{H}^{\otimes(m-1)})$ is nontrivial only in dimensions $i \le m-1$ and is isomorphic to $\mathscr{H}^{\otimes(m-1)}$ in dimension $m-1$. To choose one such isomorphism, let us order the fibers of the bundle φ_m over the unique cell $e(m)$ of codimension $m - 1$ (for example, in the order of decreasing imaginary values of points $\xi_1, \xi_2, \ldots, \xi_m$ in the corresponding collections $\xi \in \mathbb{C}^1(m)$). A basis in the sections of the system $\mathscr{H}$ over this cell is given by the elements $\xi_1 - \xi_2, \ldots, \xi_{m-1} - \xi_m$, and this determines a basis in sections of the system $\mathscr{H}^{\otimes(m-1)}$ over this cell. The group $H^{m-1}(\mathfrak{A}_m - \Sigma, \mathscr{H}^{\otimes(m-1)})$ is the quotient of the group $C^{m-1}(\Theta)$ of such sections of $\mathscr{H}^{\otimes(m-1)}$ modulo the group $\partial C^{m-2}(\Theta)$ generated by the boundary values of similar sections over various cells of the form $e(m', m - m')$.

3.4.2. THEOREM. *The obstruction to the existence of a continuous section of $\varphi_m^{*(m-1)}$ is equal to the class of the element*

$$(\xi_{m-1} - \xi_m) \otimes (\xi_{m-2} - \xi_{m-1}) \otimes \cdots \otimes (\xi_1 - \xi_2) \in \mathscr{H}^{\otimes(m-1)} \cong C^{m-1}(\Theta)$$

in the quotient group $C^{m-1}(\Theta)/\partial C^{m-2}(\Theta)$.

This immediately implies Theorem 3.4.1. In fact, under the obvious transformation of coefficients $\mathscr{H}^{\otimes(m-1)} \to \mathscr{H}^{\wedge(m-1)} \cong \pm \mathbb{Z}$ the above obstruction is mapped into a generator, while for $m = p^k$ all elements of the subgroup $\partial C^{m-2}(\Theta) \subset C^{m-1}(\Theta)$ are mapped to multiples of p (by formula (14) of Chapter I). □

§3. ROOTS OF POLYNOMIALS IN ONE VARIABLE

To prove Theorem 3.4.2 we will construct a section of $\varphi_m^{*(m-1)}$ over the complement of the cell $e(m)$. Represent the join of $m-1$ copies $A_1, \ldots, A_{m-1}$ of the m-point set A as an $(m-2)$-dimensional polyhedron whose vertices are all $m(m-1)$ points of their union and whose simplices are spanned by all collections of these points such that no one of these simplices has two vertices in the same set A_i. (In our case all A_i are copies of the fiber of the covering φ_m.)

First construct a section over the union of cells of the form $e(m_1, \ldots, 1)$, i.e. over the set of such collections $(z_1, \ldots, z_m) \subset \mathbb{C}^1$ that $\operatorname{Re} z_m > \operatorname{Re} z_i$ for any $i < m$. Namely, over each such point $\{z_1, \ldots, z_m\}$ the image of this section is in the vertex from A_1 corresponding to z_m. This section will be discontinuous near all cells of the form $e(m_1, \ldots, l)$, $l > 1$. Modify it so as to smooth these discontinuities near the union of the cells $e(m_1, \ldots, 2)$. A point of this union corresponds to a collection such that $\operatorname{Re} z_m = \operatorname{Re} z_{m-1} > \operatorname{Re} z_i$ for each $i < m-1$. For such a point consider the vertex of the fiber A_2 corresponding to the point z_m or z_{m-1} that has the larger imaginary part. The part of the one-dimensional skeleton of our bundle over this point consisting of segments connecting this point of A_2 to all the points of A_1 is, obviously, contractible. Therefore, we can deform the section chosen before in a small neighborhood of cells $e(m_1, \ldots, 2)$ so that it becomes continuous and its image lies in this part of the 1-skeleton. Proceed in the same manner: at the rth step we have a section over the union of cells of the form $e(m_1, \ldots, i)$, $i \leq r$, and the image of this section lies in the union of simplices whose vertices belong only to $A_1, \ldots, A_r$. Get rid of discontinuities of this section in neighborhoods of each point $\{z_1, \ldots, z_m\}$ of each cell of the form $e(m_1, \ldots, r+1)$ using a deformation whose image lies in the cone with the base being the union of $\{A_1, \ldots, A_r\}$ and the vertex being the point of A_{r+1} corresponding to z_i with the maximal imaginary part among the $r+1$ points with the maximal real part. Finally we get some section over the complement of the cell $e(m)$. Now consider a small $(m-1)$-dimensional disk transversal to the cell $e(m)$. Identify all the fibers of our bundle over it. Then the obtained section over its boundary defines an $(m-2)$-dimensional spheroid in the fiber of this bundle. We compute its value in the group $\pi_{m-2}(\text{fiber}) \cong \mathscr{H}^{\otimes(m-1)}$. Let our disk consist of collections $\zeta = \{z_1, \ldots, z_m\} \subset \mathbb{C}^1(m)$ such that $\operatorname{Im} z_j = -j\varepsilon$, and the collections of real parts of $z_1, \ldots, z_m$ run over the simplex in $\mathbb{R}^m$ spanned by the m points

$$u_i = \{x_i = \varepsilon, x_1 = \cdots = x_{i-1} = x_{i+1} = \cdots = x_m = -\varepsilon/(m-1)\},$$

$i = 1, \ldots, m$. Denote by Δ_i the face of this simplex spanned by the vertices $u_1, \ldots, \hat{u}_i, \ldots, u_m$. By induction hypothesis, before the last step of the construction of the section over the complement of $e(m)$, the obstruction to a retraction of the section over Δ_i into the set $\{A_1, \ldots, A_{m-2}\}$ is equal to

$$(\xi_{m-1} - \xi_m) \otimes \cdots \otimes (\xi_{i+1} - \xi_{i+2}) \otimes (\xi_{i-1} - \xi_{i+1}) \otimes (\xi_{i-2} - \xi_{i-1}) \otimes \cdots \otimes (\xi_1 - \xi_2).$$

At the last step almost all these obstructions are covered by cones consisting of segments connecting points of $\{A_1, \ldots, A_{m-2}\}$ to the first point in A_{m-1}, and only one, corresponding to Δ_1, to the second point. Orient these simplices Δ_i as elements of the boundary of our $(m-1)$-dimensional simplex. Then the sum of these cones in $\{A_1, \ldots, A_{m-1}\}$ (i.e. in the fiber of $\varphi_m^{*(m-1)}$) is equal to

$$\left[\sum_{i=2}^{m} (-1)^i (\xi_{m-1} - \xi_m) \otimes \cdots \otimes (\xi_{i+1} - \xi_{i+2}) \otimes (\xi_{i-1} - \xi_{i+1}) \otimes \cdots \otimes (\xi_1 - \xi_2) \right]$$
$$\otimes \xi_1 - (\xi_{m-1} - \xi_m) \otimes \cdots \otimes (\xi_2 - \xi_3) \otimes \xi_2$$
$$\equiv (\xi_{m-1} - \xi_m) \otimes \cdots \otimes (\xi_1 - \xi_2). \quad \square$$

3.5. Proof of Theorem 3.1.4. Let $m = q \cdot p^r$, p prime. In the space $\mathfrak{A}_m$ consider the set $\pi(m, p, \varepsilon)$ consisting of polynomials whose roots split into p^r groups with q roots (possibly multiple) in each, and the distance between the roots in each group is at most $\varepsilon/2$, while the distance between roots from different groups is at least 2ε. Over the set $\pi(m, p, \varepsilon)$ define a p^r-fold covering whose points are pairs of the form {polynomial $f \in \pi(m, p, \varepsilon)$; one of the groups of its roots}. Similarly to Smale's Theorem 2.3 one can prove that for sufficiently small ε the topological complexity of the problem $P_1(m, \varepsilon)$ is at least the genus of this covering minus 1. But the set $\pi(m, p, \varepsilon)$ together with this covering is homotopy equivalent to the covering φ_{p^r}, and it remains to apply Theorem 3.4.1.

3.6. Proof of Proposition 3.1.5. Let m be large. Consider its decomposition into prime factors. If among these factors there are numbers greater than $\ln m$ then the statement is proved. Suppose there are no such factors, and m is a product of prime numbers, each less than $\ln m$. By the asymptotic distribution law for prime numbers ($\pi(x) \sim x/\ln x$) the number of all such numbers is approximately $\ln m / \ln \ln m$. Hence the greatest of the powers of the numbers in the decomposition of m is asymptotically larger than the $(\ln m / \ln \ln m)$th root of m; this root is exactly $\ln m$, and the proposition is proved.

§4. Topological complexity of solving systems of equations in several variables

Here we generalize the results of the previous section to the case of several variables. There are several natural ways to generalize the set of admissible values $\mathfrak{A}_m$ considered above and the statements of problems $P(m, \varepsilon)$, $P_1(m, \varepsilon)$.

§4. SYSTEMS OF EQUATIONS IN SEVERAL VARIABLES

4.1. Formulation of problems; spaces of systems.

4.1.1. We consider computational problems of the following three classes.

A. Fix a topological space $\mathfrak{A}$ whose points are systems of k polynomials in n complex variables, a number $\varepsilon > 0$, and a domain $U \subset \mathbb{C}^n$. The problem is to construct an algorithm that for any system $F = (f_1, \ldots, f_k) \in \mathfrak{A}$ (given by the real and imaginary parts of its coefficients) computes all roots of the system in U with accuracy ε. This problem is formulated only for spaces $\mathfrak{A}$ consisting of systems for which it is known a priori that the number of their roots in U is finite and (taken with multiplicities) constant. An algorithm that solves this problem must accept an arbitrary system $F \in \mathfrak{A}$ at its input and produce a finite list of points $z_1, \ldots, z_{\#\mathfrak{A}} \in \mathbb{C}^n$ (given by the real and imaginary parts of their coordinates), where $\#\mathfrak{A}$ is the number of roots of the system F in U, and these roots $\xi_1, \xi_2, \ldots$ can be ordered so that for any i we have $|\xi_i - z_i| \leq \varepsilon$.

This problem is denoted by $P_T(\mathfrak{A}, U, \varepsilon)$, and its topological complexity by $\tau_T(\mathfrak{A}, U, \varepsilon)$.

B. Given a space $\mathfrak{A}$, number ε, and domain U as in problem A, construct an algorithm that for any system $F \in \mathfrak{A}$ computes, with accuracy ε, one root of this system in an ε-neighborhood of U or gives the correct statement that no roots exist in U. The corresponding notation is $P_1(\mathfrak{A}, U, \varepsilon)$ and $\tau_1(\mathfrak{A}, U, \varepsilon)$.

C. Problem $P_{\approx}(\mathfrak{A}, U, \varepsilon)$ consists in finding not points close to roots, but ε-roots, i.e. points satisfying the following condition:

DEFINITION. A point $z \in \mathbb{C}^n$ is said to be an *ε-root* of a system $F = (f_1, \ldots, f_k)$ if $|f_1(z)| < \varepsilon, \ldots, |f_k(z)| < \varepsilon$.

Given a system $F \in \mathfrak{A}$, an algorithm that solves problem $P_{\approx}(\mathfrak{A}, U, \varepsilon)$ must produce either its ε-root $z \in U$ or give the correct statement about the absence of $\varepsilon/2$-roots in U.

In the cases when the domain U coincides with the entire space $\mathbb{C}^n$ we omit the symbol $\mathbb{C}^n$ in the notation $P_\omega(\mathfrak{A}, U, \varepsilon)$, $\omega = T, 1$, or $\approx$.

From the point of view of applications Problem C is more natural than Problem B since in many important cases systems that have a root can lose this property under arbitrarily small perturbations (not picked up by a computer); if this does not happen because of certain restrictions on the space $\mathfrak{A}$ then Problems B and C are almost equivalent (each of them is reduced to a subproblem of the other by a change of ε).

Obviously, with ε decreasing the numbers $\tau_\omega(\mathfrak{A}, U, \varepsilon)$ do not decrease. Set $\tau_\omega(\mathfrak{A}, U) = \lim_{\varepsilon \to +0} \tau_\omega(\mathfrak{A}, U, \varepsilon)$.

In this section we study the behavior of limit topological complexities τ_ω, $\omega = T, 1, \approx$, for some natural sets $\mathfrak{A}$ of systems of equations with the growth of the degrees of polynomials in the systems.

4.1.2. *Basic spaces of systems of equations.* We will use the following standard notation. For a multi-index $\alpha = (\alpha_1, \ldots, \alpha_n) \in \mathbb{Z}_+^n$, $|\alpha|$ denotes the sum $\alpha_1 + \cdots + \alpha_n$; the coefficient of a polynomial f at the monomial $x^\alpha \equiv x_1^{\alpha_1} \cdot \cdots \cdot x_n^{\alpha_n}$ is denoted by f_α so that $f = \sum f_\alpha x^\alpha$.

For each collection of k natural numbers $D = (d_1, \ldots, d_k)$ denote by $H(D, n)$ the set of all collections of k homogeneous polynomials $\mathbb{C}^n \to \mathbb{C}$ of degrees $d_1, \ldots, d_k$.

We will consider sets $\mathfrak{A}$ of systems of polynomials belonging to one of the following two classes.

A. Fix $D = (d_1, \ldots, d_k) \in \mathbb{N}^k$. The space $\mathfrak{A}(D, n)$ consists of all systems of k polynomials $f_1, \ldots, f_k$, $f_i \colon \mathbb{C}^n \to \mathbb{C}$ (in general not homogeneous) with $\deg f_i \leq d_i$.

B. Fix a family of homogeneous polynomials $\Phi = (\varphi_1, \ldots, \varphi_k) \in H(D, n)$ and a positive number R. The space $\mathfrak{A}(\Phi, R)$ is defined to be the set of systems $F = (f_1, \ldots, f_k) \in \mathfrak{A}(D, n)$ such that the principal (of degree d_i) homogeneous parts of polynomials f_i coincide with φ_i, and the sum of the absolute values of coefficients of lower degree does not exceed R:

$$\sum_{|\alpha| < d_i} |f_{i,\alpha}| \leq R. \tag{8}$$

REMARKS. 1.
$$\dim_\mathbb{C} \mathfrak{A}(D, n) = \sum \binom{n + d_i}{n},$$
$$\dim_\mathbb{C} \mathfrak{A}(\Phi, R) = \sum \binom{n + d_i - 1}{n}.$$

2. Without loss of generality we can (and will) assume that the degrees d_i are ordered by the rule $d_1 \geq d_2 \geq \cdots$.

3. Among the above classes the problem $P_T(\mathfrak{A}, U, \varepsilon)$ can be considered only for $\mathfrak{A} = \mathfrak{A}(\Phi, R)$ such that the number k of polynomials in the system equals n, and the point 0 is an isolated solution of the system Φ: in other cases the number of solutions can be either infinite or nonconstant over the set $\mathfrak{A}$. Also the domain U must be so large compared to R that for any system $F \in \mathfrak{A}(\Phi, R)$ all its $d_1 \cdot \cdots \cdot d_n$ roots belong to this domain.

4.1.3. A computation algorithm which solves one of our problems must produce $2n$ (if it solves the problem P_1 or $P_\approx$) or $2n \cdot d_1 \cdot \cdots \cdot d_n$ (in the case of problem P_T) real numbers that are declared to be the real and imaginary parts of a root (ε-root, all roots); in the case of problems P_1, $P_\approx$ the algorithm can also produce a statement about the absence of roots in U. We say that an algorithm solves the problem $P_1(\mathfrak{A}, U, \varepsilon)$ (problem $P_\approx(\mathfrak{A}, U, \varepsilon)$, respectively) if for any collection of input data $F \in \mathfrak{A}$ in the corresponding execution of the algorithm there is no division by zero and the obtained point lies in the ε-neighborhood of a root of F in the ε-neighborhood of the domain U (is an ε-root, respectively) or the statement about the absence of roots ($\varepsilon/2$-roots) in U is true. An algorithm solves problem $P_T(\mathfrak{A}, U, \varepsilon)$ if the roots of the system F can be ordered so that the distance between the ith root and the ith computed root does not exceed ε for all i.

4.2. Statement of results.

4.2.1. *The topological complexity of the generic problem of finding all roots.* Let $k = n$ and $D = (d_1, \ldots, d_n)$. Define $W(D)$ to be the number of points $(a_1, \ldots, a_n)$ of the integral lattice $\mathbb{Z}^n$ satisfying the following inequalities: $0 \leq a_1 < [d_1/2]$, $0 \leq a_2 < d_2, \ldots, 0 \leq a_n < d_n$, $a_1 + \cdots + a_n < d_1$. For example, if all d_i are equal to a certain d, then $W(d, \ldots, d) = \binom{n+d-1}{n} - \binom{n+d-[d/2]-1}{n}$; for fixed n this number is $(1/n!)(1 - 2^{-n}) \cdot d^n(1 + O(1/d))$.

THEOREM. *Let $\Phi \in H(D, n)$ be a system of n homogeneous polynomials in n variables with an isolated solution at 0, and R be a positive number. Then*

(a) *The topological complexity of the problem of finding all roots of a system of equations $F \in \mathfrak{A}(\Phi, R)$ for any $\varepsilon > 0$ satisfies the inequality*

$$\tau_T(\mathfrak{A}(\Phi, R), \varepsilon) \leq \dim_\mathbb{R} \mathfrak{A}(\Phi, R); \tag{9}$$

(b) *If the system Φ is in general position(^{4}) in $H(D, n)$ then there is a positive number ε_0 such that for all $\varepsilon \in (0, \varepsilon_0]$*

$$\tau_T(\mathfrak{A}(\Phi, R), \varepsilon) \geq W(D). \tag{10}$$

In particular, this is true for the system $\Phi = (x_1^{d_1}, \ldots, x_n^{d_n})$.

The requirement of general position in part (b) of the theorem is probably unnecessary. Note that for $d_1 = \cdots = d_n = d$ and fixed n the leading terms of the upper and lower estimates (9), (10) differ by the multiplicative constant $2n/(1 - 2^{-n})$.

4.2.2. Now we consider problems of finding one root. We recall that for any natural number m we denote by $\operatorname{prim}(m)$ the greatest divisor of m which is a power of prime.

THEOREM. *For almost any system $\Phi \in H(d, n)$ the topological complexity of the problem $P_1(\mathfrak{A}(\Phi, R), \varepsilon)$ of finding one root is equal to 0 for all $\varepsilon > 0$ if the number k of equations in the system is less than n; if $k = n$ and ε is sufficiently small then for any $i = 1, \ldots, n$*

$$\tau_1(\mathfrak{A}(\Phi, R), \varepsilon) \geq \operatorname{prim}(d_i) - 1. \tag{11}$$

Here is a slight improvement of the last estimate.

(^{4}) Here and below the term "system in general position" or "almost every system" means "any system which does not belong to some proper algebraic subset in the space of systems."

PROPOSITION. *Let $k = n$ and suppose there is an even number d_j in the collection D. Then for any $i \neq j$ we can change d_i to $2d_i$ in the estimate* (11).

REMARK. Of course, under the assumptions of Theorem 4.2.1 the bound (9) is also an upper bound for the problem of finding one root.

Now let $\mathfrak{A} = \mathfrak{A}(D, n)$.

4.2.3. THEOREM. *For any $D = (d_1, \ldots, d_k) \in \mathbb{N}^k$ and $n \in \mathbb{N}$ the topological complexity of the problem $P_1(\mathfrak{A}(D, n), U, \varepsilon)$ satisfies the following relations*:

(a) *If U is a bounded semialgebraic domain in $\mathbb{C}^n$ then the numbers $\tau_1(\mathfrak{A}(D, n), U, \varepsilon)$ are bounded from above by a constant $c(D, n, U)$ independent of ε;*

(b) *If V_r^{2n} is a ball of radius r in the space $\mathbb{C}^n$ then for all $\varepsilon > 0$, $\tau_1(\mathfrak{A}(D, n), V_r^{2n}, \varepsilon) \geq \tau_1(\mathfrak{A}(D, n-1), V_r^{2n-2}, \varepsilon)$;*

(c) *For any fixed $k \leq n$ and nonempty domain U we have the asymptotic estimate*

$$\tau_1(\mathfrak{A}(D, n), U) \gtrsim (d/k)^k (1 + O(1/d))$$

for $d_1 = \cdots = d_k = d \to \infty$.

The last statement follows directly from Theorem 4.2.4 below. In order to state the theorem we introduce more notation. For each natural m define $pp(m)$ to be the largest number which is at most m and is a power of prime. It is well known that $\lim_{m \to \infty} pp(m)/m = 1$ and, moreover, $m - pp(m) = o(m^{38/61+\alpha})$ for any $\alpha > 0$; see [Ing].

For all $d = 0, 1, \ldots;$ $k = 1, 2, \ldots$ define the numbers $M(d, k)$, $a(d, k)$ by the recurrence rule $M(d, 1) = d$, $a(d, 1) = 1$,

$$a(d, k) = [M(d - [d/k], k-1)/a(d, 1) \cdots \cdot a(d, k-1)],$$
$$M(d, k) = [d/k] \cdot a(d, 1) \cdots \cdot a(d, k). \tag{12}$$

In particular, for a fixed k and $d \to \infty$,

$$M(d, k) = (d/k)^k (1 + O(1/d)),$$
$$a(d, k) = d(k-1)^{k-2}/k^{k-1}(1 + O(1/d)).$$

4.2.4. THEOREM. *If $k \leq n$ and the degrees $d_1, \ldots, d_k$ satisfy the conditions*

$$d_2 \geq a(d_1, 2), \ldots, d_k \geq a(d_1, k), \tag{13}$$

then for any nonempty domain $U \subset \mathbb{C}^n$ there is an $\varepsilon_0 > 0$ such that

$$\tau_1(\mathfrak{A}(D, n), U, \varepsilon) \geq pp(M(d_1, k)) - 1 \tag{14}$$

for all $\varepsilon \in (0, \varepsilon_0]$.

EXAMPLE. Let $k = n$ be fixed, and $d_1 = \cdots = d_n = d \to \infty$. Then the condition (13) is satisfied, and the leading terms of the asymptotics of the lower bound (14) differ from the real dimension of the space $\mathfrak{A}(D, n)$ by the factor $(n-1)!/2n^n$.

There is an estimate similar to (14) (and slightly improving it) in the case when the degrees d_i do not satisfy the conditions of (13). In order to state it we introduce the following notation. Let $D - l$ be the sequence $(d_1 - l, d_2, \ldots, d_k)$. Set $\Lambda_1(D) = d_1$. Define a numerical sequence $a_j(D)$, $t_j(D)$, $\Lambda_j(D)$ $(j = 2, \ldots, k)$ recursively by the following rules. Suppose that for some $j = 1, \ldots, k$ we know the numbers $a_2(D), \ldots, a_j(D)$ and $\Lambda_j(D - l)$, $l \geq 0$. Define nonnegative numbers $a_{j+1}(D)$, $t_{j+1}(D)$ by the following conditions:

(a) $a_{j+1}(D) \leq d_{j+1}$;
(b) $\Lambda_j(D - t_{j+1}(D)) \geq a_2(D) \cdot \ldots \cdot a_j(D) - 1$;
(c) The number

$$(t_{j+1}(D) + 1)a_2(D) \cdot \ldots \cdot a_{j+1}(D) + \Lambda_j(D - t_{j+1}(D) - 1) \qquad (15)$$

takes the greatest possible value over all $a_j(D)$, $t_j(D)$ satisfying conditions (a) and (b).

This value (15) is denoted by $\Lambda_{j+1}(D)$.

4.2.5. THEOREM. *For any collection of natural numbers $D = (d_1, \ldots, d_n)$ and any nonempty domain $U \subset \mathbb{C}^n$*

$$\tau_1(\mathfrak{A}(D, n), U, \varepsilon) \geq \operatorname{pp}(\Lambda_k(D)) - 1 \qquad (16)$$

for sufficiently small $\varepsilon > 0$.

REMARK. The bound (16) slightly improves (in the second term of the asymptotics in d_1) the bound (14) even in the case when the restrictions (13) are satisfied. For example, if $k = 2$ then

$$a_2(D) = \min(d_2, [(d_1 + 3)/2]), \qquad t_2 = d_1 + 1 - a_2(D),$$
$$\Lambda_2(D) = (a_2(D) - 1)(t_2(D) + 1) + d_1.$$

But it is more convenient to use estimate (14).

4.2.6. *Problem of finding an ε-root.*

THEOREM. (A) *Everywhere in the statements of Theorems 4.2.2–4.2.5 we can substitute the problem $P_\approx(\cdot, \varepsilon)$ of finding an ε-root (and the corresponding topological complexity $\tau_\approx(\cdot, \varepsilon)$) for the problem $P_1(\cdot, \varepsilon)$ of finding one root with accuracy ε (and $\tau_1(\cdot, \varepsilon)$).*

(B) *If the set of systems of equations $\mathfrak{A}$ has the form $\mathfrak{A}(\Phi, R)$ or is a bounded domain in $\mathfrak{A}(D, n)$ then for any $\varepsilon > 0$ and any bounded domain $U \subset \mathbb{C}^n$ the topological complexity of this problem satisfies the inequality $\tau_\approx(\mathfrak{A}, U, \varepsilon) \leq \dim_\mathbb{R} \mathfrak{A} + 1$. In the case when $\mathfrak{A} = \mathfrak{A}(\Phi, R)$ and the system of equations Φ has an isolated solution at 0 we can remove the requirement that U be bounded.*

For example, if $d_1 = \cdots = d_n \to \infty$ and U is bounded then our upper and lower bounds for $\tau_{\approx}(\mathfrak{A}(D, n), U, \varepsilon)$ differ by the factor
$$(n - 1)!/2n^n(1 + O(1/d_1)).$$

4.3. Proof of the lower estimates.

4.3.1. *Estimates in the problem of finding all roots.* Fix a system $\Phi = (\varphi_1, \ldots, \varphi_n)$ of homogeneous polynomials $\varphi_j: \mathbb{C}^n \to \mathbb{C}$, $\deg \varphi_j = d_j$, and let 0 be its unique solution in $\mathbb{C}^n$; let R be a positive number. Then each system $F = (f_1, \ldots, f_n) \in \mathfrak{A}(\Phi, R)$ has exactly $d_1 \cdots d_n$ roots in $\mathbb{C}^n$ counted with their multiplicities. Select a subset Σ in $\mathfrak{A}(\Phi, R)$ consisting of all systems with multiple roots. This is an algebraic subset of codimension 1, and its complement will be denoted by $B(\Phi, R)$. Over this complement there is a $(d_1 \cdots d_n)!$-fold covering $S_{\Phi, R}: \tilde{B}(\Phi, R) \to B(\Phi, R)$ whose points are pairs of the form (point $F \in \mathfrak{A}(\Phi, R)$, ordered collection of all roots of the system F). Recall the notation $g(\cdot)$ for the genus of a covering; see §1.

LEMMA 1. *For sufficiently small $\varepsilon > 0$ the topological complexity of the problem of finding all roots satisfies the inequality*
$$\tau_T(\mathfrak{A}(\Phi, R), \varepsilon) \geq g(S_{\Phi, R}) - 1.$$

This lemma is proved similarly to Smale's theorem from 2.3.

DEFINITION. A point $z \in \mathbb{C}^n$ is a *simple double root* of the system $F = (f_1, \ldots, f_n)$ if

a) $f_1(z) = \cdots = f_n(z) = 0$;

b) Some $n - 1$ divisors $\{f_i\}$ are nonsingular in a neighborhood of z and are in general position so that their intersection in that neighborhood is a smooth curve of multiplicity 1;

c) The restriction to this curve of the remaining function f_i has a Morse singularity at the point z.

By a local diffeomorphism and reordering of the functions f_i in a neighborhood of such point the system F can be transformed to the form $f_1 = x_1^2 + \tilde{f}(x_2, \ldots, x_n)$, $f_2 = x_2, \ldots, f_n = x_n$.

LEMMA 2. 1) *Let the system $F \in \mathfrak{A}(\Phi, R)$ have a simple double root at z, and let Σ_z be the local (near F) component of the set Σ consisting of systems $\tilde{F}$ with multiple roots near z. Then for $\tilde{F} \in \Sigma_z$ sufficiently close to F all these multiple roots are simple double roots, and the component Σ_z is a smooth divisor near F;*

2) *If at least one of the numbers d_i is greater than 1 then systems with one simple double root form a dense set in Σ;*

3) *A path in $B(\Phi, R)$ going around Σ along a small circle centered at a point F that has a unique simple double root permutes the two roots obtained as a result of the decay of this double root and preserves all other roots of the system.*

Proof is elementary.

§4. SYSTEMS OF EQUATIONS IN SEVERAL VARIABLES

DEFINITION. A point $F \in \mathfrak{A}(\Phi, R)$ is a *standard m-point* of the discriminant Σ if the system F has m distinct simple double roots, and in a neighborhood of F the corresponding m local components Σ_{z_i} of the divisor Σ are in general position (i.e. the pair $(\mathfrak{A}(\Phi, R)$, union of these components) is locally biholomorphically equivalent to the product of a linear space and the pair $(\mathbb{C}^m$, union of all coordinate hyperplanes in $\mathbb{C}^m)$.

Note that in this definition we do not require the absence of other multiple roots of the system F; in particular, a standard m-point can simultaneously be a standard $(m+1)$-point.

LEMMA 3. *If a discriminant* $\Sigma \subset \mathfrak{A}(\Phi, R)$ *has standard m-points then* $g(S_{\Phi, R}) > m$.

EXAMPLE. Let $K = n = 1$, $\varphi_1 = x^d$. Then Lemma 3 and Smale's principle imply that $\tau(d) > [d/2]$. In fact we have the estimate (2) which is almost twice as strong.

PROOF OF LEMMA 3. Let F be a standard m-point of the discriminant Σ, V be a small ball in $\mathfrak{A}(\Phi, R)$ centered at the point F, and σ be the union of m local components of the set Σ at F corresponding to simple double roots. The set $V - \sigma$ is homotopy equivalent to the m-dimensional torus which is the distinguished boundary of the standard polydisk in $\mathbb{C}^m$; in particular, the ring $H^*(V - \sigma, \mathbb{Z}_2)$ is the exterior algebra on m generators, and the length of the subring

$$\bigoplus_{i \geq 1} H^i(V - \sigma, \mathbb{Z}_2) \tag{17}$$

is equal to m (see definition of the length of a ring in 1.3). Let $S_{\Phi, R, V}$ be the restriction of the projection of the covering $S_{\Phi, R}$ to the set $S_{\Phi, R}^{-1}(V - \Sigma)$.

LEMMA 4. (A) *The map*

$$\mathrm{in}^i \colon H^i(V - \sigma, A) \to H^i(V - \Sigma, A)$$

induced by inclusion is a monomorphism for any i and any ring A.

(B) *If* $i \geq 1$ *and* $A = \mathbb{Z}_2$ *then the image of this map belongs to* $\operatorname{Ker} S^*_{\Phi, R, V}$, *i.e.* $S^*_{\Phi, R, V} \circ \mathrm{in}^i \equiv 0$.

Lemma 3 follows directly from this lemma, Proposition 1.1.1 and the proposition from 1.3.

PROOF OF LEMMA 4. Instead of part (A) we prove the dual statement: the map

$$\mathrm{in}_i \colon H_i(V - \Sigma, A) \to H_i(V - \sigma, A)$$

induced by the imbedding is an epimorphism. Choose local coordinates $x_1, \ldots, x_N$ in V so that the components $\sigma_1, \ldots, \sigma_m$ of the set σ are given by the equations $x_1 = 0, \ldots, x_m = 0$. Since $V \not\subset \Sigma$, there is a standard m-point $\widetilde{F} \in \sigma_1 \cap \cdots \cap \sigma_m$ in V such that the m-dimensional complex plane $X = \{x \mid x_{m+1} \equiv x_{m+1}(\widetilde{F}), \ldots, x_N \equiv x_N(\widetilde{F})\}$ does not belong to Σ. We show that the generator of the group $H_m(V - \sigma, A) \cong A$ can be represented by an m-dimensional torus in $V - \Sigma$ lying in the plane X.

Consider the set of $\varkappa_1 \in \mathbb{C}^1$ such that the plane $\{x_1 \equiv \varkappa_1\} \cap X$ belongs to $\Sigma \cap X$. Since $X \not\subset \Sigma$, this set is discrete and in a neighborhood of the point 0 consists of only this point. Draw a circle $S^1_{(1)}$ in the line $\mathbb{C}^1 = \mathbb{C}^1(x_1)$ that does not intersect this set and bounds only the one point 0 of this set. For any point $\varkappa_1$ of this circle consider the set of $\varkappa_2$ on the line $\mathbb{C}^1(x_2)$ such that the plane $\{x_1 \equiv \varkappa_1, x_2 \equiv \varkappa_2\} \cap X$ belongs to Σ. By the choice of $\varkappa_1$, for a sufficiently small $\delta > 0$ the circle with radius δ does not intersect this set and bounds the unique point 0 of this set, and the same is true for all nearby values of $\varkappa_1$. By the compactness of the circle $S^1_{(1)}$ it is possible to choose a common value of δ satisfying these conditions for all points of $S^1_{(1)}$. The circle of radius δ in $\mathbb{C}^1(x_2)$ will be denoted by $S^1_{(2)}$. Consider the torus $S^1_{(1)} \times S^1_{(2)}$ in the plane $\mathbb{C}^1(x_1) \times \mathbb{C}^1(x_2)$. Similar arguments enable us to construct a torus in the space $X \cong \mathbb{C}^1(x_1) \times \cdots \times \mathbb{C}^1(x_m)$ that does not intersect Σ. This torus in $X - \Sigma$ realizes the generator of the group $H_m(V - \sigma)$, and its various coordinate subtori form a complete collection of generators of the groups $H_i(V - \sigma)$ for all $i < m$. This proves part (A) of the lemma.

It suffices to prove part (B) for the multiplicative generators of the ring (17), i.e. for the linking indices mod 2 with the components of the set σ. But by part 3) of Lemma 2 any circle in $S^{-1}_{\Phi, R}(V - \Sigma)$ projects onto a circle whose linking indices with these components are even, and Lemma 4 is proved.

REMARK. Let F be a standard m-point of the discriminant, and assume that F has only m simple and no other multiple roots. In general, we cannot guarantee that the map $H^*(B(\Phi, R), \mathbb{Z}_2) \to H^*(V - \Sigma, \mathbb{Z}_2)$ given by the imbedding is a monomorphism; however, all symmetric polynomials in standard one-dimensional generators of the latter ring lie in the image of this map. Indeed, following [Fuchs$_2$] consider a vector $d_1 \cdots d_n$-dimensional bundle over $B(\Phi, R)$ whose fiber over F is the space of real functions on the set of roots of F. Under our homomorphism the ith Stiefel-Whitney class of this bundle is mapped precisely to the ith basic symmetric polynomial in linking indices with the components of the set $\Sigma \cap V$, cf. Chapter IV, §4.

LEMMA 5. *For any $D \in \mathbb{N}^n$ and $m \in \mathbb{N}$ the set of homogeneous systems $\Phi = (\varphi_1, \ldots, \varphi_n) \in H(D, n)$ such that the discriminant $\Sigma \subset \mathfrak{A}(\Phi, R)$ has standard m-points is open in the Zariski topology.*

This follows directly from definitions, part 1 of Lemma 2 and the Tarski-Seidenberg theorem.

To prove assertion (b) of Theorem 4.2.1 it remains to prove that for $m = W(D)$ this set is nonempty.

4.3.1.1. THEOREM. *Let $\Phi = (x_1^{d_1}, \ldots, x_n^{d_n})$. Then for any $R > 0$ the* *discriminant* $\Sigma \subset \mathfrak{A}(\Phi, R)$ *contains a standard* $W(D)$*-point.*

PROOF. For $i = 2, \ldots, n$ set $f_i = x_i(x_i - t) \cdot (x_i - 2t) \cdots (x_i - (d_i - 1)t)$. For d_1 even set $f_1 = x_1^2 \cdot (x_1 - t)^2 \cdots (x_1 - ([d_1/2] - 1)t)^2$, and for d_1 odd set f_1 equal to the product of this polynomial by $x_1 + t$. If $t > 0$ is sufficiently small then the collection $F = (f_1, \ldots, f_n)$ belongs to $\mathfrak{A}(\Phi, R)$. It has $[d_1/2] \cdot d_2 \cdots d_n$ double points. We prove that it is possible to choose $W(D)$ points among them in such a way that F satisfies the conditions in the definition of a standard $W(D)$-point. Consider a plane Λ in $\mathfrak{A}(\Phi, \infty)$ consisting of systems $\tilde{F} = (\tilde{f}_1, \ldots, \tilde{f}_n)$ with $\tilde{f}_i = f_i$ for $i \geq 2$ and an arbitrary $\tilde{f}_1$. This plane is naturally equivalent to the space $\mathfrak{A}(d_1 - 1, n)$ of polynomials of degree $\leq d_1 - 1$. By the implicit function theorem it suffices to prove that in the intersection with Λ some $W(D)$ of the local components of Σ at F are already in general position.

LEMMA 6. *Let* z *be a simple double point of the system* F. *Then in a neighborhood of the point* F *the local component* $\Sigma_z \cap \Lambda$ *of the set* $\Sigma \cap \Lambda$ *consisting of systems with a double point near* z *is nonsingular and is tangent to the hyperplane consisting of systems* $\tilde{F} = (\tilde{f}_1, f_2, \ldots, f_n)$ *such that* $\tilde{f}_1(z) = 0$.

This lemma can easily be checked.

Now consider the space $C(F)$ of complex-valued functions on the set of all double roots of the system F. There is an obvious homomorphism

$$\text{value:} \; \mathfrak{A}(d_1 - 1, n) \to C(F) \tag{18}$$

that assigns to each polynomial the collection of its values at all the double points of F. Now Theorem 4.3.1.1 follows immediately from Lemma 6 and the following assertion.

LEMMA 7. *The rank of the map* (18) *is at least* $W(D)$.

To prove this consider, instead of the space $\mathfrak{A}(d_1 - 1, n)$, the space $\text{Pol}([d_1/2], d_2, \ldots, d_n)$ consisting of linear combinations of monomials $x_1^{\alpha_1} \cdots x_n^{\alpha_n}$ with $\alpha_1 < [d_1/2], \alpha_2 < d_2, \ldots, \alpha_n < d_n$. Note that the set of multi-indices α satisfying these relations is homothetic to the set of double points of the system F and, in particular, has the same cardinality. The next lemma can easily be proved.

LEMMA 8. *The map* value: $\text{Pol}([d_1/2], d_2, \ldots, d_n) \to C(F)$ *is an isomorphism.*

Now Lemma 7 follows from the fact that $W(D)$ is just the dimension of the intersection of $\mathfrak{A}(d_1 - 1, n)$ with $\text{Pol}([d_1/2], d_2, \ldots, d_n)$. The proof of Theorem 4.3.1.1 and part (b) of Theorem 4.2.1 is complete.

4.3.2. *Lower estimates for algorithms that compute one root for the systems of the space* $\mathfrak{A}(D, n)$. Proofs of Theorems 4.2.4 and 4.2.5 are based on the following generalization of inequality (6) (see 2.3). Let an arbitrary collection of natural numbers $D = (d_1, \ldots, d_n)$ and a domain $U \subset \mathbb{C}^n$ be given. Let Γ be a semialgebraic subset in the space $\mathfrak{A}(D, n)$ such that all systems $F \in \Gamma$ have the same number $\#\Gamma$ of roots in the ε-neighborhood of U, and all these roots lie in U and are distinct. Let θ_Γ be a $\#\Gamma$-fold covering over Γ whose points are pairs of the form (point $F \in \Gamma$, one of the roots of the system F in U).

PROPOSITION. *The topological complexity of the problem of approximate (with accuracy ε) computation of one root of systems of the set Γ for sufficiently small $\varepsilon > 0$ satisfies the inequality*

$$\tau_1(\Gamma, U, \varepsilon) \geq g(\theta_\Gamma) - 1. \tag{19}$$

By Proposition 1.1.1 we can replace $\tau_1(\Gamma, U, \varepsilon)$ in inequality (19) by $\tau_1(\mathfrak{A}(D, n), U, \varepsilon)$. The inequality (19) is proved in the same way as Smale's theorem. We will reduce Theorems 4.2.4 and 4.2.5 to the inequality in the case $k = n$. The case $k < n$ reduces to this one if one considers a subset in $\mathfrak{A}(D, n)$ consisting of systems depending only on the first k coordinates; the same method proves part (b) of Theorem 4.2.3.

It remains to exhibit, for any collection D and any nonempty domain U, a set $\Gamma \subset \mathfrak{A}(D, n)$ satisfying the above conditions and such that the genus of the covering θ_Γ is greater than the right-hand sides of (14) and (16) (here we can, without any loss of generality, assume that U contains the origin in $\mathbb{C}^n$). To do this we use Theorem 3.4.1 of §3, which describes the genus of a covering corresponding to the universal polynomial (4) in one variable.

Let $a_2, \ldots, a_n$ be a collection of natural numbers satisfying $a_2 \leq d_2, \ldots, a_n \leq d_n$. Consider the curve in $\mathbb{C}^n$ given by the conditions

$$x_2 = x_1^{a_2}, \ldots, x_n = x_{n-1}^{a_n}. \tag{20}$$

This is a smooth curve, and one can take the restriction of x_1 as the coordinate in it. Any monomial $x^\alpha = x_1^{\alpha_1} \cdot \ldots \cdot x_n^{\alpha_n}$ restricted to this curve equals x_1 in the power

$$\alpha_1 + \alpha_2 a_2 + \cdots + \alpha_n a_2 a_3 \cdot \ldots \cdot a_n. \tag{21}$$

Let us now consider the set of all exponents of monomials of degree $\leq d_1$, i.e. the set

$$\{\alpha \in \mathbb{Z}^n \mid \alpha_1 \geq 0, \ldots, \alpha_n \geq 0, |\alpha| \leq d_1\}. \tag{22}$$

Each vector α satisfying these conditions defines an integer $w(\alpha)$ by formula (21). Suppose that all integers from 0 to N are the numbers $w(\alpha)$ for some α from the set (22). For each integer $l \in [0, N]$ choose some α such that $l = w(\alpha)$ and denote it by $\alpha(l)$. Consider the subset $\widetilde{\Gamma} \subset \mathfrak{A}(D, n)$ consisting of systems $(f_1, \ldots, f_n)$ such that $f_2 = x_2 - x_1^{a_2}, \ldots, f_n = x_n - x_{n-1}^{a_n}$, and f_1 is a linear combination of monomials x^α such that $\alpha = \alpha(l)$ for some $l \in [0, N]$, with the coefficient of the monomial $x^{\alpha(N)}$ equal to 1. Finally, let Γ be the subset of $\tilde{\Gamma}$ obtained by removing the discriminant. Then the covering θ_Γ (see (19)) is equivalent to the covering φ_N defined at the end of 2.3. In particular, if N is a power of a prime then the genus of this covering equals N. Now Theorems 4.2.4 and 4.2.5 (case $k = n$) follow directly from the next lemma.

LEMMA. (A) Let a_j, $j = 2, \ldots, n$, be the number $a_j(D)$ defined before the statement of Theorem 4.2.5. Then each number $l = 0, 1, \ldots, \Lambda_n(D)$ is equal to the value of the expression (21) for some α satisfying (22).

(B) Let a collection $D = (d_1, \ldots, d_n)$ satisfy the conditions (13), where the numbers $a(d_1, j)$, $j = 2, \ldots, n$, are defined by (12). Then part (A) of this lemma is true for any number $l = 0, 1, \ldots, M(d_1, n)$.

Part (B) will be proved by induction on n. Set $A_s = a(d_1, 2) \cdots a(d_1, s)$. Suppose that for any $q = 1, 2, \ldots, n$ any number $l = 0, \ldots, M(q, n-1)$ can be realized as a sum $\alpha_1 + A_2\alpha_2 + \cdots + A_{n-1}\alpha_{n-1}$ for some multi-index $\alpha = (\alpha_1, \ldots, \alpha_{n-1})$. Then all numbers of the set $\{0, 1, \ldots, M(d, n-1)$, $A_n + \{0, 1, \ldots, M(d-1, n-1)\}, 2A_n + \{0, \ldots, M(d-2, n-1)\}, \ldots$, $(d-1)A_n + \{0, \ldots, M(1, n-1)\}, dA_n\}$ can be realized by similar sums with an additional term $A_n\alpha_n$ and the restriction $|\alpha| \le d$. A break in the set of these numbers can occur only when

$$M(d-r, n-1) < A_n - 1, \tag{23}$$

and in this case the greatest number in this set that can be reached without breaks in the sequence of natural numbers is $\tilde{r}A_n + M(d-\tilde{r}, n-1)$, where $\tilde{r}$ is the smallest r satisfying (23). It is easy to verify that in our case this $\tilde{r}$ is at least $[d_1/n]$, and hence the number $M(d_1, n) = A_n[d_1/n]$ satisfies part (B) of the lemma. Part (A) is verified similarly: the condition (b) for the numbers $a_j(D)$ and $t_j(D)$ given before the statement of Theorem 4.2.5 means exactly that there is no break in the chain $0, 1, \ldots, \Lambda_{j+1}(D)$.

4.3.3. *Proof of Theorem* 4.2.2 *(case* $k = n$). For $n = 1$ this theorem coincides with Theorem 3.1.4, proved in 3.5. For arbitrary n it is deduced in the same manner from the following lemma.

LEMMA 1. *Let $D = (d_1, \ldots, d_n)$ be a collection of natural numbers with $d_1 = qp^r$, U be an arbitrary neighborhood of 0 in $\mathbb{C}^n$. Then for almost any collection of homogeneous polynomials $\Phi \in H(D, n)$ and any number $R > 0$ there is a compact subset Γ in $\mathfrak{A}(\Phi, R)$ such that*

(a) *the number of roots of any system $F \in \Gamma$ counted with their multiplicities equals $d_1 \cdots d_n$;*

(b) *all these roots are in U;*

(c) *these roots for all $F \in \Gamma$ can be simultaneously divided into p^r groups, $q \cdot d_2 \cdots d_n$ roots in each, so that each multiple root is entirely in one group, and this decomposition is continuously preserved in passing to neighboring systems $F' \in \Gamma$;*

(d) *the set Γ is homotopy equivalent to the space of polynomials of type* (1) *with $d = p^r$ distinct roots*;

(e) *the p^r-fold covering over Γ whose points are the pairs ($F \in \Gamma$, one of the groups of roots of the system F) is homotopy equivalent to the covering φ_{p^r} of Theorem* 3.1.3.

PROOF. First note that we can consider $U = \mathbb{C}^n$. Indeed, suppose that a set $\Gamma \subset \mathfrak{A}(\Phi, R)$ satisfies the conditions of the lemma with $U = \mathbb{C}^n$. Consider an action of the group $\mathbb{R}_+$ on Γ defined by $t: (f_1, \ldots, f_n) \to (t^{-d_1} f_1(tx), \ldots, t^{-d_n} f_n(tx))$. For $t > 1$ this action sends $\mathfrak{A}(\Phi, R)$ intoitself, and for any $F \in \mathfrak{A}(\Phi, R)$ the collection of roots of the system $t(F)$ is obtained from that of F by a homothety with coefficient $1/t$. In particular, for sufficiently large t all roots of the system F can be mapped into U. By compactness of Γ we can choose this number t the same for all $F \in \Gamma$ and replace Γ by $t(\Gamma)$. Now Lemma 1 follows from Lemmas 2 and 3 below.

LEMMA 2. *In the space $H(D, n)$ there is a nonempty Zariski open subset such that if the assertion of Lemma 1 (with $U = \mathbb{C}^n$) is true for one point Φ of this subset then it is also true for any other point of the subset.*

LEMMA 3. *Any nonempty Zariski open subset in $H(D, n)$ contains systems Φ satisfying the assertions of Lemma 1.*

PROOF OF LEMMA 2. In the space $\mathfrak{A}(D, n)$ consider a region $\mathfrak{A}(D, n, R)$ determined by conditions (8), i.e. the region consisting of systems from the sets $\mathfrak{A}(\Phi, R)$ for suitable $\Phi \in H(D, n)$. This region splits naturally into the product $H(D, n) \times \mathfrak{A}(0, R)$, where 0 is the zero system. In the space $\mathfrak{A}(D, n, R) \times \mathbb{C}^n$ consider the set of pairs (system F, point x) such that x is a root of F. Consider any semialgebraic Whitney stratification of this set, and for every stratum mark the set of critical points of its projection to the space $H(D, n)$. By the lemmas of Sard and of Tarski-Seidenberg, the union of such sets over all strata is contained in a proper algebraic subset of the space $H(D, n)$. By the Thom isotopy theorem (see [Ph], [AVGL$_1$]), the complement of this subset satisfies the conditions of Lemma 2.

PROOF OF LEMMA 3. First let $\Phi = (x_1^{d_1}, \ldots, x_n^{d_n})$. In this case take Γ to be the set of systems of the form $(f_1(x), x_2^{d_2}, \ldots, x_n^{d_n})$, where f_1 depends only on the coordinate x_1 and its values range over the set $\pi(d, p, \varepsilon)$ considered in the proof of Theorem 3.1.4 in 3.5. Let Γ_δ be a δ-neighborhood of the set Γ in the space $\mathfrak{A}(D, n)$. If δ is sufficiently small then for any system Φ' close to Φ in the space $H(D, n)$, the set $\mathfrak{A}(\Phi', R) \cap \Gamma_\delta$ satisfies Lemma 1. But any Zariski open subset in $H(D, n)$ intersects any neighborhood (in the usual topology) of the point Φ, and Lemma 3 is proved; this completes the proof of Theorem 4.2.2.

To prove Proposition 4.2.2 it is sufficient to consider the case $d_1 = q \cdot 2^r$, d_2 even. The proof is based on a lemma similar to Lemma 1 with just one change: instead of 2^r groups the set of roots is divided into 2^{r+1} groups of $q \cdot d_2 \cdots d_n/2$ roots in each. In the proof of the analogue of Lemma 3 we must consider, in the space $\mathfrak{A}(\Phi, R)$, the set of systems of the form $(f_1(x_1, x_2), (-x_1 + x_2^2 + ax_2)^{d_2/2}, x_3^{d_3}, \ldots, x_n^{d_n})$ for the same Φ. The last $n-1$ functions in this system define a smooth curve with coordinate x_2. Changing the parameter a and coefficients of degree $< d_1$ in f_1 we can make the polynomial f_1 restricted to this curve equal to any given polynomial of degree $2d_1$ in x_2 with leading coefficient 1. Now the proposition is reduced to Theorem 3.1.4 as before.

4.3.4. The lower estimates for algorithms computing ε-roots follow from the proofs in 4.3.2 and 4.3.3, since in all the cases considered there ε-roots can lie only in sufficiently small neighborhoods of true roots. This proves the lower estimates in Theorem 4.2.6.

4.4. Upper estimates.

4.4.1. *Proof of part* (a) *of Theorem* 4.2.1.

LEMMA 1. *Let A be a compact subset of the N-dimensional Euclidean space. Then for any open cover $\{U_\alpha\}$ of A, $A = \bigcup U_\alpha$, there is a cover of A by $N+1$ open sets such that the cover of A by the connected components of the closures of these sets is a refinement of the cover $\{U_\alpha\}$.*

PROOF. Let $1 \gg t > 0$. Consider the function $S = \sin(x_1/t) + \cdots + \sin(x_n/t)$ on $\mathbb{R}^n$. Let V_0 be the t^2-neighborhood of the set of minimum points of this function. For any $i = 1, \ldots, N$ define the set $\widetilde{V}_i$ to be the union of lower separatrices of the field $\operatorname{grad} S$ for all singular points of signature $(n-i, i)$, and define the set V_i inductively as the t^{i+2}-neighborhood of the set $\widetilde{V}_i \setminus (V_0 \cup \cdots \cup V_{i-1})$. For sufficiently small t the connected components of the sets $\overline{V}_i \cap A$ form a refinement of any given open cover of A, and Lemma 1 is proved.

Now suppose that the assumptions of Theorem 4.2.1 are satisfied, F is an arbitrary point of the set $\mathfrak{A}(\Phi, R)$, and $\xi_1(F), \xi_2(F), \ldots$, are all the roots of the system F in $\mathbb{C}^n$. Then for any system F' from some neighborhood $\Omega(F)$ of the system F in $\mathfrak{A}(\Phi, R)$ these points form a collection of roots computed with accuracy $\varepsilon/2$. Construct a cover of the space $\mathfrak{A}(\Phi, R)$ by $N+1$ open sets $V_0, \ldots, V_N$ (where $N = \dim_\mathbb{R} \mathfrak{A}(\Phi, R)$) satisfying Lemma 1 with respect to the cover of $\mathfrak{A}(\Phi, R)$ by the regions $\Omega(F)$. For any $i = 0, 1, \ldots, N$ define an ordered collection of $d_1 \cdots d_n$ functions $r_{i,j}: V_i \to \mathbb{C}^n$ that coincides with the collection $(\xi_1(F), \xi_2(F), \ldots)$ on any connected component of the set V_i, where F is an arbitrary point in $\mathfrak{A}(\Phi, R)$ such that this connected component belongs to $\Omega(F)$. Using the Weierstrass approximation theorem, for any $j = 1, \ldots, d_1 \cdots d_n$ we construct a real polynomial map $\pi_{i,j}: \mathfrak{A}(\Phi, R) \to \mathbb{C}^n$ that approximates the function $r_{i,j}$ everywhere on V_i with accuracy $\varepsilon/2$.

Using the Weierstrass theorem again, we construct N polynomial functions $\chi_0, \ldots, \chi_{N-1} : \mathfrak{A}(\Phi, R) \to \mathbb{R}$ such that for any $i = 0, \ldots, N-1$ the set $\{F \mid \chi_i(F) < 0\}$ belongs to V_i, and $\{F \mid \chi_0(F) \geq 0, \ldots, \chi_{N-1}(F) \geq 0\}$ belongs to V_N. Now the desired algorithm can be described as follows (cf. 3.3).

Input (coefficients of the system F are entered). Compute the function $\chi_0(F)$. $\chi_0(F) < 0$? If yes then compute the collection of polynomials $\pi_{0,j}(F)$ and send them to output. If no then compute $\chi_1(F)$. $\chi_1(F) < 0$? If yes then compute $\pi_{1,j}(F)$ and send to output. If no then pass to $\chi_2(F)$, etc. The topological complexity of this algorithm is N, and Theorem 4.2.1 is proved completely.

4.4.2. Part (b) of Theorem 4.2.6 is proved similarly. Here we use the fact that if the system F has an ε-root at a point $z \in U$ or has no $\varepsilon/2$-roots in U then the same is true for all systems in a neighborhood of the point F. The only difference is that we have to use one more open set V_{N+1} in our cover of the space $\mathfrak{A}$ by domains V_i, namely, the union of all connected components of the previously constructed domains $V_0, \ldots, V_N$ in which we have to give the message of absence of $\varepsilon/2$-roots instead of a numerical answer.

In the case of the problem of finding one root with accuracy ε the situation is very different: it can happen that a system from $\mathfrak{A}(D, n)$ that has a root at z can be transformed by an arbitrarily small perturbation to a system with roots far from z or even without any roots in $\mathbb{C}^n$. The situation is saved by the fact that the existence of roots of a system is determined by a semialgebraic condition on its coefficients; this makes the arguments below possible.

4.4.3. *Proof of part* (a) *of Theorem* 4.2.3. Let $\|F\|^2$ be the sum of squares of the real and complex parts of the coefficients of a system F, and $\mathfrak{A}^1$ be the subset in $\mathfrak{A}(D, n)$ consisting of all systems F with $\|F\| \leq 1$. Consider the set of all pairs (F, x) in the space $\mathfrak{A}^1 \times \overline{U}$ such that $F(x) = 0$. This is a compact real semialgebraic set. According to [Loj$_2$], there exist real algebraic Whitney stratifications of this set and of the set $\mathfrak{A}^1$ such that the preimage under the natural projection of each stratum of the second stratification is the union of several strata of the first stratification, and the restriction of the projection to this preimage is a locally trivial smooth fibration.

Let $\sigma \subset \mathfrak{A}^1$ be an arbitrary stratum of the second stratification, and $\bar\sigma$ be its closure. For any system $F \in \bar\sigma$ fix any of its roots ξ belonging to $\overline{U}$ (if it exists) or the statement about its absence. Then there exists a neighborhood $W(F)$ of the system F in $\mathfrak{A}^1$ such that any system $F' \in \bar\sigma \cap W(F)$ has a root at distance $< \varepsilon$ from ξ (possibly not in $\overline{U}$) in the first case and no roots in U in the second case. The domains $W(F)$ form a cover of the set $\bar\sigma$; apply Lemma 1 to it and get some cover of $\bar\sigma$ by $N+1$ sets $V_0(\sigma), \ldots, V_N(\sigma)$. As in 4.4.2 we define one more set $V_{N+1}(\sigma)$ consisting of connected components of closures of the domains $V_i(\sigma)$ such that for F in these components we have to announce the absence of roots.

The desired algorithm can be described as follows. First replace F by $F^1 \equiv 2F/(\|F\|^2 + 1)$. Then in a finite (and independent on ε) number of steps determine whether F^1 belongs to the first stratum of our stratification of the set $\mathfrak{A}$. If it does then apply the algorithm from 4.1–4.2 correspondingto the cover of this stratum by the domains $V_i(\sigma)$. If not, then determine whether F^1 belongs to the second stratum, etc.

4.4.4. *Proof of Theorem 4.2.2 for $k < n$.* If $k < n$ and $\Phi = (\varphi_i, \dots, \varphi_k)$ is a system in general position in $H(D, n)$ then the Jacobian $\{\partial \varphi_i / \partial x_j\}$ has rank k at any nonzero point $x \in \mathbb{C}^n$ with $\Phi(x) = 0$. Hence, there exists a ball $T \subset \mathbb{C}^n$ centered at 0 and large enough so that the same is true for any system $F \in \mathfrak{A}(\Phi, R)$ everywhere outside T, and the set $F^{-1}(0)$ is transversal to the boundary of T. Consider the fibration over $\mathfrak{A}(\Phi, R)$ with fiber $F^{-1}(0) \setminus T$ over the point F. It follows from the implicit function theorem and the Thom isotopy theorem that this fibration is locally trivial. But $\mathfrak{A}(\Phi, R)$ is a contractible set; hence this fibration is trivializable and, in particular, has a section. The proof is completed by yet another application of the Weierstrass theorem.

4.5. Two problems. Let an algebraic subset Σ in the space $\mathbb{C}^n$ (or in any other complex manifold) be given. Which elements of the group $H^*(\mathbb{C}^n - \Sigma)$ are stable with respect to adding new components to Σ, i.e. which elements have the property that for any algebraic subset $\Delta \subset \mathbb{C}^n$ the image of these elements under the natural homomorphism $H^*(\mathbb{C}^n - \Sigma) \to H^*(\mathbb{C}^n - (\Sigma \cup \Delta))$ is not 0? For example, the assertion (A) of Lemma 4 of 4.3.1 guarantees that if Σ is a collection of hyperplanes through the origin with normal intersections then the entire group $H^*(\mathbb{C}^n - \Sigma)$ consists of stable elements.

Let S be a subset in the lattice $\mathbb{Z}_+^n$, $P(S)$ be the set of all polynomials in $\mathbb{C}^n$ that are linear combinations of the monomials x^α, $\alpha \in S$, and $C(S)$ be the space of complex-valued functions on the set S. Is it true that the map value: $P(S) \to S$ is an isomorphism? Lemma 8 of 4.3.1 says that this is true if S is the intersection of $\mathbb{Z}_+^n$ with the parallelepiped whose sides are parallel to the axes.

§5. Obstructions to representing algebraic functions by superpositions

The study of the cohomology of the braid group was initiated in [Ar$_4$], [Ar$_6$] in connection with the problem of representability of algebraic functions by superpositions of algebraic functions in fewer variables. The results of D. B. Fuchs about the rings $H^*(\text{Br}(m), \mathbb{Z}_2)$ allowedV. I. Arnol'd to prove the following theorem (see [Ar$_6$], [Br$_4$]).

THEOREM. *The universal algebraic function* (4) *in m variables $a_1, \dots, a_m$ cannot be represented as a complete superposition of algebraic functions in l*

variables if $l < m - D_2(m)$ (where $D_2(m)$ is the number of units in the binary expansion of m).

Definitions of all notions used in this statement are recalled in the following sections 5.1–5.2.

The results about the groups $H^*(\mathrm{Br}(m), \pm\mathbb{Z})$ and the genus of the covering $f_m \to \mathbb{C}^m - \Sigma$ presented above enable us to slightly improve this result.

THEOREM (see [V_3]). *The algebraic function* (4) *is not representable by a complete superposition of algebraic functions in l variables if $l < m - D_p(m)$ for some prime p (and, more generally, if $l < g(f_m) - 1$, where f_m is the canonical $m!$-fold covering over the complement of the discriminant; see 2.3).*

The last theorem itself has only methodological importance since a stronger result has been obtained by V. Ya. Lin ([Lin$_1$], [Lin$_2$]): in the statements of the previous theorems the numbers $m - D_2(m)$ ($m - D_p(m)$ and $g(f_m) - 1$, respectively) can always be replaced by $m - 1$. The method used in the proof of these theorems gives, for example, similar estimates for other algebraic functions.

5.1. Algebraic functions. An *entire m-valued algebraic function* in k complex variables $a_1, \ldots, a_k$ is a polynomial in variables $z, a_1, \ldots, a_k$ of the form

$$z^m + p_1(a)z^{m-1} + \cdots + p_{m-1}(a)z + p_m(a), \qquad (24)$$

where $p_i(a) = p_i(a_1, \ldots, a_k)$ are polynomials. Its values at the point $a = (a_1, \ldots, a_k)$ are various roots of the polynomial (24) considered as a polynomial in z with coefficients $p_i(a)$. The *universal* entire m-valued algebraic function is the polynomial (4); in this case $k = m$. We say that an algebraic function Φ of the form (24) depends on l variables if there is a similar m-valued algebraic function $\Phi': \mathbb{C}^l \to \mathbb{C}$ and a polynomial map $F: \mathbb{C}^k \to \mathbb{C}^l$ such that $\Phi = \Phi' \circ F$. For example, any m-valued function (24) depends on m variables: here it suffices to take $\Phi': \mathbb{C}^m \to \mathbb{C}$ to be the universal function (4), and the map $F: \mathbb{C}^k \to \mathbb{C}^m$ is given by polynomials p_i from (24). (Moreover, it is easy to see that any m-valued function (24) is reduced to a function in $m - 1$ variables which can be obtained from (4) by removing the monomial $a_1 z^{m-1}$.)

5.2. Superpositions of algebraic functions. Consider the space $\mathbb{C}^{k+r}$ with linear coordinates $a_1, \ldots, a_k, z_1, \ldots, z_r$. For any $i = 1, \ldots, r$ consider an m_i-valued algebraic function $\Phi_i: \mathbb{C}^{k+i-1} \to \mathbb{C}^1$ that depends on the coordinates $a_1, \ldots, a_k, z_1, \ldots, z_{i-1}$:

$$\Phi_i = z_i^{m_i} + p_{1,i} z_i^{m_i - 1} + \cdots + p_{m_i, i}, \qquad (25)$$

where all $p_{l,i}$ are polynomials in $a, z_1, \ldots, z_{i-1}$. We will consider Φ_i as a polynomial in all variables $a_1, \ldots, a_k, z_1, \ldots, z_r$ which does not depend on $z_{i+1}, \ldots, z_r$. Then the polynomials $\Phi_1, \ldots, \Phi_r$ define r divisors in $\mathbb{C}^{k+r}$. Under the projection onto the plane $\mathbb{C}^k = \{a, z \mid z = 0\}$ the k-dimensional algebraic subset $\Delta \subset \mathbb{C}^{k+r}$ determined by these divisors becomes the space of an $m_1 \cdots m_r$-fold ramified covering over this plane. A *complete superposition* of the functions Φ_i is an $M = m_1 \cdots m_r$-valued function on $\mathbb{C}^k$ which associates to each value $a = (a_1, \ldots, a_k)$ all (possibly multiple) values of the coordinate z_r of points of the set Δ that project to a.

It is easy to see that a superposition of algebraic functions is algebraic, i.e. is given by a suitable equation of the form

$$z_r^M + \pi_1(a) a_r^{M-1} + \cdots + \pi_M(a) = 0. \tag{26}$$

An algebraic function $z_r(a_1, \ldots, a_r)$ is called an (incomplete) *superposition* of the functions $\Phi_1, \ldots, \Phi_r$ if the corresponding polynomial (24) is a divisor of the polynomial (26).

The algebraic function (24) decomposes into a (complete) superposition of functions in l variables if its representation in the form of a (complete) superposition contains only functions Φ_i that depend on $\leq l$ variables.

We have defined all the notions used in the statement of the theorems from the introduction to this section. Now we begin their proof.

The problem of representability of an algebraic function in the form of an incomplete superposition is much more difficult; see [Hilbert], [Wiman], [Khovanskii].

5.3. ARNOL'D'S LEMMA. *If the universal algebraic function* (4) *is decomposed into a complete superposition of algebraic functions* $\Phi_1, \ldots, \Phi_r$ *then all* Φ_i *except one are single-valued.*

PROOF (see [Ar$_6$]). Let again Δ be a subset of $\mathbb{C}^{m+r}$ defined by equations $\Phi_1 = \cdots = \Phi_r = 0$.

The ramified covering $\Delta \to \mathbb{C}^m$ is the composition of r maps:

$$\Delta \equiv \Delta_r \to \Delta_{r-1} \to \cdots \to \Delta_1 \to \Delta_0 \equiv \mathbb{C}^m, \tag{27}$$

where Δ_j is the subset of $\mathbb{C}^{m+j-1}$ given by the conditions $\Phi_1 = \cdots = \Phi_j = 0$, and the map $\Delta_{j+1} \to \Delta_j$ is the projection $(a, z_1, \ldots, z_j) \to (a, z_1, \ldots, z_{j-1})$. Let Δ'_j be the set of points in Δ_j whose projection into $\mathbb{C}^m$ belongs to the set $\mathbb{C}^m - \Sigma$. Then the restrictions of the chain (27) to the sets Δ'_j define a sequence of nonramified coverings

$$\Delta' \equiv \Delta'_r \to \cdots \to \Delta'_1 \to \Delta'_0 \equiv \mathbb{C}^m - \Sigma.$$

The composite projection $\Delta'_r \to \mathbb{C}^m - \Sigma$ is equivalent to the m-fold covering $\varphi_m : \mathbb{N}^m \to \mathbb{C}^m - \Sigma$ considered in 2.3; in particular, the image of the monodromy group of this covering is the symmetric group $S(m)$. But it is easy to show that a covering with this monodromy group cannot be decomposed into nontrivial composition of coverings (see [Ar$_6$]); the lemma is proved.

5.4. COROLLARY. *Suppose the universal m-valued function* (4) *decomposes into a complete superposition of algebraic functions in l variables. Then there is an l-dimensional topological space* L, *an m-fold covering* $\lambda \to L$, *and a map* $\mathbb{C}^m - \Sigma \to L$ *such that the covering* $\Delta' \to \mathbb{C}^m - \Sigma$ *is equivalent to the one induced by this map from the covering* λ.

PROOF. Suppose the universal function Φ of the form (4) is decomposed into a complete superposition of functions depending on l variables. Then it follows from Lemma 5.3 that there exists a polynomial map $P: \mathbb{C}^m \to \mathbb{C}^l$, an entire algebraic m-valued function $\Psi: \mathbb{C}^l \to \mathbb{C}^1$ (i.e. a polynomial $t^m + \varkappa_1(b)t^{m-1} + \cdots + \varkappa_m(b)$, $b \in \mathbb{C}^l$), and a polynomial $Q: \mathbb{C}^{l+1} \to \mathbb{C}$ such that $\Phi(a) = Q(\Psi(P(a)), a)$. Let $\mathbb{C}^l - \sigma$ be a subset in $\mathbb{C}^l$ consisting of points b such that all m values of the function $\Psi(b)$ are distinct. By construction, $P(\mathbb{C}^m - \Sigma) \subset \mathbb{C}^l - \sigma$ and the m-fold covering $\Delta' \to \mathbb{C}^m - \Sigma$ is equivalent to the covering induced by the map $P: \mathbb{C}^m - \Sigma \to \mathbb{C}^l - \sigma$ from the similar covering over $\mathbb{C}^l - \sigma$ whose points are the pairs ($b \in \mathbb{C}^l - \sigma$; one of the values of the function $\Psi(b)$). But $\mathbb{C}^l - \sigma$ is an l-dimensional Stein manifold, and in particular, has the homotopy type of an l-dimensional CW-complex. This proves Corollary 5.4.

5.5. To each m-fold covering corresponds the associated principal $S(m)$-bundle, and hence the classifying map of its base into the space $K(S(m), 1)$; see Appendix 1.

In our case we get a diagram of maps

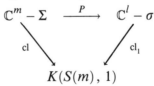

This diagram is homotopy commutative since the defining coverings P are induced from each other. In particular, for each $S(m)$-module A we get a commutative diagram

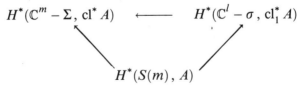

Now let $l < m - D_2(m)$. Then commutativity of this diagram contradicts the fact that $H^{m-D_2(m)}(\mathbb{C}^m - \Sigma, \mathbb{Z}_2) \neq 0$ and the left vertical arrow is an epimorphism (see Chapter I, §2.3). This proves the first theorem from the introduction to this section.

To prove the inequality $l \geq m - D_p(m)$ (see the second theorem) we have to take $A = \pm \mathbb{Z}$ and restrict the previous arguments to the domain $U \subset \mathbb{C}^m$ considered in 3.2.

Finally, we prove the second theorem in the homotopy formulation: $l \geq g(f_m) - 1$. Consider the $m!$-fold covering over $\mathbb{C}^l - \sigma$ associated with our m-fold covering (its fiber over the point b consists of various orderings of the values of the function $\Psi(b)$). The covering f_m is induced from this one by the map P. Now the desired conclusion follows from the fact that the genus of the induced covering is at most the genus of the inducing covering (Proposition 1.1.2), and the genus of any covering over an l-dimensional base is at most $l + 1$ (this is the last assertion of 1.4).

§6. On the function spaces interpolating at any k points

Let M be a topological space, and let L be a finite dimensional vector subspace of the space of continuous real functions on M. Interpolational properties of the functions of the space L depend on the choice of the nodes of interpolation. We investigate here the dimensions of such spaces L which can interpolate any function on M at any k points.

DEFINITION 1. The space L is called *k-interpolating* if any (continuous) real function on M can be interpolated at any k points of M by an appropriate function from the space L.

In other words, the space L is k-interpolating if for any k different points $z_1, \ldots, z_k$ of M and any numbers $a_1, \ldots, a_k$ there is a function $F \in L$ such that $F(z_i) = a_i$ for any i.

Denote by $I(M, k)$ the lowest dimension of k-interpolating function spaces on M. The problem is to find the numbers $I(M, k)$ for all M and k.

EXAMPLES. If $M = R^1$, then $I(M, k) = k$ by Newton's interpolation theorem: for a k-interpolating space we can take the space of all polynomials of degree $\leq k - 1$.

If $M = S^1$ and k is odd, then $I(M, k) = k$: a k-interpolating space is provided by the trigonometric polynomials of degree $\leq [k/2]$.

THEOREM 1. *If $M = S^1$ and k is even, then $I(M, k) = k + 1$.*

Indeed, the upper estimate is provided again by the trigonometric polynomials, the lower bound will be proved later.

Now, let M be the plane $\mathbb{R}^2$.

THEOREM 2. *The number $I(\mathbb{R}^2, k)$ satisfies the inequality*

$$2k - d(k) \leq I(\mathbb{R}^2, k) \leq 2k - 1, \tag{28}$$

where $d(k) \equiv D_2(k)$ is the number of units in the binary representation of k. In particular, if k is a power of 2, then $I(\mathbb{R}^2, k) = 2k - 1$.

Again the upper estimate is straightforward: the estimate can be realized by the space L spanned by the functions 1, $\mathrm{Re}(z^i)$, $\mathrm{Im}(z^i)$, $i = 1, \ldots, k-1$, where z is a complex coordinate on the plane.

It is an intriguing fact, that in the first case, when the two bounds from theorem 2 do not coincide, it is the complicated lower bound that is realistic and not the easy upper bound: indeed, for $k = 3$, the four dimensional function space spanned by the functions 1, x, y, and x^2+y^2 is 3-interpolating on the plane.

THEOREM 3. *For any n and any n-dimensional manifold M,*

$$I(M,k) \leq k(n+1). \tag{29}$$

CONJECTURE 1. *If n is a power of 2, then*

$$I(\mathbb{R}^n, k) \geq k + (n-1)(k - d(k)). \tag{30}$$

This conjecture is closely connected to some problems concerning the multiplicative structure in the cohomology of configuration spaces (see Conjecture 2 below).

DEFINITION 2. A space L of functions $M \to \mathbb{R}$ is called *k-distinguishing* if the linear hull of L and the function equal identically to 1 is a k-interpolating space.

This notion has a pictural geometrical interpretation: the functions $f_1, \ldots, f_N$ form a basis of a k-distinguishing space iff the map $M \to \mathbb{R}^N$ given by them maps any k different points of M into the vertices of some $(k-1)$-dimensional simplex in $\mathbb{R}^N$. In particular, the calculation of the lowest possible dimension $D(M, k)$ of k-distinguishing spaces includes (for $k = 2$) the problem about the imbeddings of manifolds into Euclidean spaces. On the other hand, $I(M, k) = D(M \cup \{\text{a point}\}, k)$.

THEOREM 4. *Theorems 1, 2, and 3 remain valid if we replace the numbers $I(M, k)$ by $D(M, k) + 1$ in their statements.*

To prove the lower bounds in Theorems 1 and 2, consider the *configuration space* $M(k)$; that is, the space of all subsets of cardinality k in M. Let $T(M, k)$ be the k-dimensional vector bundle over $M(k)$ whose fiber over a point $\{z_1, \ldots, z_k\}$ is the space of real-valued functions on the set of these k points $z_1, \ldots, z_k$ (cf. §I.3).

For any N-dimensional function space L on M, consider the trivial N-dimensional vector bundle $\{L\} \equiv L \times M(k)$ over $M(k)$. There is an obvious restriction homomorphism $\mathrm{Restr} : \{L\} \to T(M, k)$: over any points $\{z_1, \ldots, z_k\} \in M(k)$, to any function $f \in L$ there corresponds its restriction to the set of these k points.

Now we are in position to prove the lower bound in Theorem 1. Indeed, if we have $I(S^1, k) = k$, then the restriction homomorphism is an isomorphism of vector bundles. But it is easy to calculate that for even k the bundle $T(S^1, k)$ is nonorientable; in particular, it cannot be isomorphic to the trivial bundle.

§6. ON THE FUNCTION SPACES INTERPOLATING AT ANY k POINTS

In the general case, for an arbitrary manifold M and a function space L, if L is k-interpolating, then the restriction homomorphism is an epimorphism; in particular, the trivial bundle $\{L\}$ contains a subbundle isomorphic to $T(M, k)$. The basic facts about Stiefel-Whitney characteristic classes (see [MS] now imply the following theorem.

THEOREM 5. *For any M and k,*

$$I(M, k) \geq k + \deg(w^{-1}(T(M, k))) \tag{31}$$

where $w(E)$ is the total Stiefel-Whitney class of the vector bundle E, $w^{-1}(E)$ is its inverse in the multiplicative group of the cohomology ring of the space $M(k)$ with coefficients in $\mathbb{Z}_2$, and $\deg$ is the highest dimension of nontrivial homogeneous parts of an element of a graded ring.

For instance, let M be the plane $\mathbb{R}^2$. Then it follows from the calculations of Fuchs (see §I.2) that $\deg(w(T(M, k))) = k - d(k)$ and that the operation of taking the inverse element in the multiplicative group of the ring $H^*(Br(k), \mathbb{Z}_2)$ is identical. This implies the lower bound in (28).

Theorem 3 follows immediately from Thom's multijet transversality theorem (see Appendix 4). Indeed, denote by $F(M, k)$ the space of ordered subsets of cardinality of k in M. A set of functions $f_1, \ldots, f_N$ on M maps $F(M, k)$ into the space of $k \times N$ matrices. The set of matrices of nonmaximal rank in this space has codimension $N - k + 1$, and hence, if $N - k + 1 > k \times \dim M$, then the image of $F(M, k)$ under the map defined by a generic collection $\{f_1, \ldots, f_N\}$ does not meet this set.

REMARK. The estimate from Theorem 3 is nonexact already for $k = 2$: it follows from Whitney's imbedding theorem that $I(M^n, 2) \leq 2n - 1$.

CONJECTURE 2. *If n is a power of 2, then the nth degree of any element of positive dimension in $H^*(B(\mathbb{R}^n, k), \mathbb{Z}_2)$ equals 0. In particular, $(w(T(\mathbb{R}^n, k)))^n = 1$.*

Conjecture 1 follows immediately from this one and from the fact that if k is a power of a prime, then the class $(w_{k-1}(T(\mathbb{R}^n, k)))^{n-1}$ is nontrivial, see Corollary 2 in subsection I.3.7.

CHAPTER III

Topology of Spaces of Real Functions without Complicated Singularities

An important characteristic of a manifold is the topological structure of the space of smooth functions on it (or, more generally, its maps into $\mathbb{R}^n$) without too complicated critical points, that is, points in the closure of some stratum in the space of singularities. The systematic study of such spaces was motivated by the work of Smale and Cerf in differential topology, see [Smale$_{1,3}$], [Cerf$_1$], [Milnor$_2$]; for their results and applications see [CL], [Serg], [Wagoner], [Volodin], [Hatcher], [HW], [Sharko], [Igusa$_{1-3}$], [Cerf$_2$], [Gromov], [GZ], [Ar$_{18}$].

The present chapter contains computations of cohomology rings of such spaces. We prove, in particular, the following reduction theorem: if the codimension of the discriminant (that is, the set of functions with prohibited singularities) in the function space is at least 2 then instead of spaces of admissible functions we can consider spaces of admissible sections of jet bundles.

In particular, the space $\mathscr{F} - \Sigma_k$ of functions $\mathbb{R}^1 \to \mathbb{R}^1$ without zeros of multiplicity k ($k \geq 3$) and identically one outside some compact set is homology (and for $k \geq 4$ even homotopy) equivalent to the loop space ΩS^{k-1}.

We also compute the cohomology rings of spaces of real polynomials

$$x^d + a_1 x^{d-1} + \cdots + a_{d-1} x + a_d,$$

without roots of multiplicity k for any k and d. With d increasing, these cohomology rings (and homotopy groups) get stabilized to those of $\mathscr{F} - \Sigma_k$.

The additive structure of these cohomology rings (and the fundamental group of such spaces for $k = 3$) were found by V. I. Arnol'd [AVGL$_1$], [Ar$_{18}$] using a different method.

§1. Statements of reduction theorems

1.0. Let M be a smooth m-dimensional manifold with boundary. By $J^k(M, \mathbb{R}^n)$ we denote the space of k-jets of smooth maps $M \to \mathbb{R}^n$; see Appendix 4.

Let $\mathfrak{A}$ be an arbitrary subset in $J^k(\mathbb{R}^m, \mathbb{R}^n)$ invariant under the natural action of the diffeomorphism group of the space $\mathbb{R}^m$. Then for any m-dimensional M there is an invariantly defined subset $\mathfrak{A}(M) \subset J^k(M, \mathbb{R}^n)$: the choice of a local chart $\mathbb{R}^m \to M$ identifies $J^k(\mathbb{R}^m, \mathbb{R}^n)$ with a domain in $J^k(M, \mathbb{R}^n)$, and the set $\mathfrak{A}(M)$ in this domain is the image of the set $\mathfrak{A}$ under this identification.

Let $\varphi: M \to \mathbb{R}^n$ be a smooth map without singular points of type $\mathfrak{A}$ near ∂M.

NOTATION. $A(M, \mathfrak{A}, \varphi)$ is the space of all smooth maps $M \to \mathbb{R}^n$ equal to φ in some neighborhood of ∂M and such that their k-jets do not belong to $\mathfrak{A}(M)$ at any point. $B(M, \mathfrak{A}, \varphi)$ is the space of smooth sections of the evident bundle $(J^k(M, \mathbb{R}^n) - \mathfrak{A}(M)) \to M$ that are equal to the k-jet extension of φ in a neighborhood of ∂M. In the case when M has no boundary these spaces do not depend on φ and are denoted by $A(M, \mathfrak{A}), B(M, \mathfrak{A})$. There is an obvious imbedding

$$j^k: A(M, \mathfrak{A}, \varphi) \to B(M, \mathfrak{A}, \varphi), \tag{1}$$

given by the k-jet extensions of the maps.

Supply these spaces with the Whitney topology (see Appendix 4). Imbedding (1) is continuous in this topology.

1.1. The first main theorem. *Let the manifold M be compact, and the set $\mathfrak{A} \subset J^k(\mathbb{R}^m, \mathbb{R}^n)$ be semialgebraic, closed, and of codimension at least $m+2$. Then the imbedding (1) defines an isomorphism of cohomology rings.*

Proof is given in §4 of this chapter.

EXAMPLE. Let $n = 1$. In the space of k-jets ($k \geq 3$) of functions on M there are, in particular, the following closed subsets invariant under diffeomorphisms and under adding a constant:

(A) The set of codimension m consisting of all jets with a singularity.

(B) The set of codimension $m+1$ that is the closure of the set of singularities of type A_2. Near singular points of this type a function can be written (in suitable coordinates) in the form

$$x_1^3 \pm x_2^2 \pm \cdots \pm x_m^2 + c. \tag{2}$$

Topologists call this singularity "birth-death," since it appears during standard surgery when two Morse points of contiguous indices merge together and disappear; in the coordinates (2) this surgery is given, for example, by the one-parameter family

$$f_t = f - tx_1;$$

see Figure 15.

(C) The set $\overline{A}_3$ of codimension $m+2$. It is the closure of the set A_3 consisting of singularities that can be written (in suitable coordinates) in the form

$$\pm x_1^4 \pm x_2^2 \pm \cdots \pm x_m^2 + c.$$

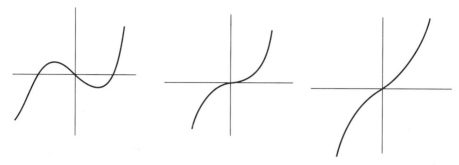

FIGURE 15

(Of course, there are other subsets of those codimensions: for example, from the class of all singularities we can take the closure of the set of Morse singularities of some fixed index, etc.)

The first main theorem implies that for any compact manifold with boundary M and any function φ without singularities near ∂M the spaces $A(M, \overline{A}_3, \varphi)$ and $B(M, \overline{A}_3, \varphi)$ are homology equivalent.

1.2. The second main theorem. *If in the conditions of the first main theorem* $\operatorname{codim} \mathfrak{A} > m + 2$, *then the imbedding* (1) *is a weak homotopy equivalence.*

1.3. This theorem follows immediately from Theorem 1.1 and the following Whitehead's theorem.

THEOREM. *Let X, Y be two simply-connected topological spaces, and f a continuous map from X to Y. Then for any $q > 1$ the following statements are equivalent*:

1) *The homomorphism $f_*: \pi_r(X) \to \pi_r(Y)$ is an isomorphism for $r < q$ and an epimorphism for $r = q$*;

2) *The homomorphism $f_*: H_r(X) \to H_r(Y)$ is an isomorphism for $r < q$ and an epimorphism for $r = q$.*

For example, the imbedding (1) satisfies the conditions of this theorem if the conditions of Theorem 1.1 are satisfied and $\operatorname{codim} \mathfrak{A} \geq m + 3$: indeed, in this case both spaces in (1) are simply connected.

PROBLEM. Is it true that the additional restriction in Theorem 1.2 is superfluous, i.e. under the assumption of Theorem 1.1 the imbedding (1) is always a weak homotopy equivalence? A homotopy equivalence? Note that the analogous statement is false in a very similar situation considered in Chapter IV.

1.4. The topology of spaces in (1) does not depend on φ.

PROPOSITION. *Under the assumptions of Theorem 1.1, for any pair of maps $\varphi, \psi: M \to \mathbb{R}^n$ without singularities of type $\mathfrak{A}$ near ∂M there are homotopy equivalences $A(M, \mathfrak{A}, \varphi) \sim A(M, \mathfrak{A}, \psi)$, $B(M, \mathfrak{A}, \varphi) \sim B(M, \mathfrak{A}, \psi)$ commuting with the imbeddings* (1).

PROOF. In a closed neighborhood U of the boundary ∂M define a *collar*, i.e. a diffeomorphism $U \sim \partial M \times [0, 1]$ such that $\partial M \subset V$ corresponds identically to $\partial M \times \{0\} \subset \partial M \times [0, 1]$.

LEMMA. *If the maps φ, ψ have no singular points of type $\mathfrak{A}$ in M then there exists a smooth map $\{\varphi, \psi\}: \partial M \times [-2, 1] \to \mathbb{R}$ without singularities of type $\mathfrak{A}$ and equal to φ in $\partial M \times [0, 1] \sim U$ and to the function ψ_{-2} in $\partial M \times [-2, -1]$ given by $\psi_{-2}(u, t) = \psi(u, t+2)$.*

This is a direct consequence of the Whitney extension theorem and the Thom transversality theorem.

Fix a smooth diffeomorphism $\lambda: [-2, 1] \to [0, 1]$ such that $\lambda(t) = t + 2$ for $t \in [-2, -5/3]$, and $\lambda(t) = t$ for $t \in [2/3, 1]$. This diffeomorphism determines a diffeomorphism $\tilde{\lambda}: M \times [-2, 1] \to \partial M \times [0, 1]$.

For any function $f: M \to \mathbb{R}^n$ equal φ near ∂M define a function $f_{\{\varphi, \psi\}}$ equal f outside the neighborhood $U \sim \partial M \times [0, 1]$ and to $\{\varphi, \psi\} \circ \tilde{\lambda}^{-1}$ in this neighborhood.

By construction, this function is equal to ψ near ∂M, and if f has no singularities of type $\mathfrak{A}$ then neither does $f_{\{\varphi, \psi\}}$. Hence we get a map $A(M, \mathfrak{A}, \varphi) \to A(M, \mathfrak{A}, \psi)$. Similarly we construct a map $B(M, \mathfrak{A}, \varphi) \to B(M, \mathfrak{A}, \psi)$. It is easy to check that the compositions of these maps with the ones in the opposite direction $f \to f_{\{\varphi, \psi\}} \to (f_{\{\varphi, \psi\}})_{\{\psi, \varphi\}}$ are homotopic to identities, and Proposition 1.4 is proved.

1.5. Singularities of functions on one-dimensional manifolds. Any $\mathrm{Diff}(\mathbb{R}^1)$-invariant semialgebraic closed irreducible subset in $J^r(\mathbb{R}^1, \mathbb{R}^1)$ is given either by the condition $f'(x) = \cdots = f^{(l)} = 0$ or by the condition $f(x) = \alpha$, $f'(x) = \cdots = f^{(l)}(x) = 0$. Denote the first of these subsets by $\mathfrak{A}^l$, and the other by $\mathfrak{A}^l_\alpha$.

PROPOSITION. *For any one-dimensional compact manifold M, any numbers $l \geq 2$ and α, and arbitrary functions $\varphi, \eta: M \to \mathbb{R}$ without singularities on ∂M of types $\mathfrak{A}^{l+1}$ and $\mathfrak{A}^l_\alpha$, respectively, there are homotopy equivalences*

$$A(M, \mathfrak{A}^{l+1}, \varphi) \sim A(M, \mathfrak{A}^l_\alpha, \eta),$$
$$B(M, \mathfrak{A}^{l+1}, \varphi) \sim B(M, \mathfrak{A}^l_\alpha, \eta).$$

Indeed, for $\alpha = 0$ this homotopy equivalence is given by taking the derivative, and the homotopy equivalence for all classes $A(M, \mathfrak{A}^l_\alpha, \eta)$ (or $B(M, \mathfrak{A}^l_\alpha, \eta)$) for various α follows from Proposition 1.4.

§2. Spaces of functions on one-dimensional manifolds without zeros of multiplicity three

2.1. Notation. $\mathscr{F}$ is the space of smooth (C^∞) functions $\mathbb{R}^1 \to \mathbb{R}^1$ that are equal to 1 outside some compact set. Φ is the space of smooth

§2. FUNCTIONS WITHOUT ZEROS OF MULTIPLICITY THREE

functions $S^1 \to \mathbb{R}^1$. Σ_k is the set of smooth functions on $\mathbb{R}^1$ or S^1 with at least one root of multiplicity k. $B(k) \sim B([0,1], \mathfrak{A}_0^{k-1}, 1)$ is the space of smooth sections of the l-jet bundle $J^l(\mathbb{R}^1, \mathbb{R}^1)$ $(l \geq k-1)$ coinciding with the l-jet extension of the function equal to 1 outside some compact set and disjoint from the set in $J^l(\mathbb{R}^1, \mathbb{R}^1)$ defined by the condition $f(x) = \cdots = f^{(k-1)}(x) = 0$.

From the last proposition of §1 follows

COROLLARY. *All spaces $A(M, \mathfrak{A}^k, \varphi)$, $A(M, \mathfrak{A}_\alpha^{k-1}, \eta)$ $(k \geq 3)$ are homotopy equivalent to $\mathscr{F} - \Sigma_k$ or $\Phi - \Sigma_k$ depending on whether M is a segment or a circle. Similarly, all spaces $B([0,1], \mathfrak{A}^k, \varphi)$, $B([0,1], \mathfrak{A}_\alpha^{k-1}, \eta)$ are homotopy equivalent to $B(k)$.*

Obviously, $B(k)$ is the loop space of the set $\mathbb{R}^k - \{0\}$ with base point $(1, 0, \ldots, 0)$; in particular, $B(k)$ is homotopy equivalent to $\Omega(S^{k-1})$. Similarly, the space $B(S^1, \mathfrak{A}_0^{k-1})$ is the space of free loops, i.e. smooth maps $S^1 \to \mathbb{R}^k - \{0\}$.

In particular, $\pi_1(B(3)) = \pi_2(B(3)) = \mathbb{Z}$ (see Appendix 3).

2.2. Statements. This section contains proofs of the following two theorems.

THEOREM 1 (see [Ar$_{18}$]). $\pi_1(\mathscr{F} - \Sigma_3) = \mathbb{Z}$, $\pi_1(\Phi - \Sigma_3) = \mathbb{Z}$.

THEOREM 2. *The maps*
$$j_*^2 \colon \pi_i(\mathscr{F} - \Sigma_3) \to \pi_i(B(3)), \quad \pi_i(\Phi - \Sigma_3) \to \pi_i(B(S^1, \mathfrak{A}_0^2)),$$
given by 2-jet extensions of functions are isomorphisms for $i = 1$ and epimorphisms for $i = 2$.

2.3. Proof of Theorem 1. Choose the function 1 (the identity) to be the base point of the space $\mathscr{F}$ (and all spaces $\mathscr{F} - \Sigma_k$). Without loss of generality, we will call any continuous map $\mathbb{R}^r \to X$ that sends the complement of some ball in $\mathbb{R}^r$ to the base point an r-dimensional spheroid of the space X. In particular, a 1-spheroid of the space $\mathscr{F} - \Sigma_3$ is a one-parameter family of smooth functions $f_\lambda(x)$, $\lambda \in \mathbb{R}^1$, $x \in \mathbb{R}^1$, such that $f_\lambda(x) = 1$ for sufficiently large values of $|x| + |\lambda|$, and for no value of λ the function f_λ has zeros of multiplicity ≥ 3.

Define a function $F \colon \mathbb{R}^2 \to \mathbb{R}^1$ by $F(x, \lambda) \equiv f_\lambda(x)$.

Without loss of generality we can consider our spheroids to be smooth (i.e., f_λ depends smoothly on λ) and nondegenerate (i.e. $\operatorname{grad} F(x, \lambda)$ and $F(x, \lambda)$ do not vanish simultaneously). Indeed, by the Weierstrass theorem and the Thom transversality theorem, any spheroid of the space $\mathscr{F} - \Sigma_3$ can be approximated by a smooth and nondegenerate one. Define the curve $\Gamma(F) \subset \mathbb{R}^2$ to be the set of zeros of the function F (see Figure 16).

LEMMA 1 (see [Ar$_{18}$]). *A smooth nondegenerate 1-spheroid of the space $\mathscr{F}$ is disjoint from the set Σ_3 if and only if the curve $\Gamma(F)$ has no inflection point with horizontal tangent (i.e., no inflection points at which the tangent line coincides with some line $\lambda = $ const).* □

The group $\pi_1(\mathscr{F} - \Sigma_3)$ is in one-to-one correspondence with the set of plane curves without inflection points with horizontal tangents, considered up to some "3-cobordism" equivalence; see [Ar$_{18}$]. Namely, consider the product of $\mathbb{R}^2 = \mathbb{R}^1_x \times \mathbb{R}^1_\lambda$ by $[0, 1]$. Let p be the projection of it onto $[0, 1]$. A *3-cobordism* is a smooth compact surface $K \subset \mathbb{R}^2 \times [0, 1]$ transversal to the planes $\mathbb{R}^2 \times 0$, $\mathbb{R}^2 \times 1$ such that its projection to $[0, 1]$ is a Morse function, and

 a) for any $s \in [0, 1]$ the curve $\Gamma_s = K \cap p^{-1}(s)$ has no nonsingular inflection points where the tangent is parallel to the x axis;

 b) if $(x, \lambda, s) \in K$ is a saddle critical point of the projection (hence, near this point the curve $K \cap p^{-1}(s)$ is a pair of intersecting smooth curves) then the tangents to these curves at this point are also not parallel to the x axis.

Two curves $\Gamma, \Gamma' \subset \mathbb{R}^2$ are 3-cobordant if there exist a 3-cobordism K with $\Gamma_0 = \Gamma, \Gamma_1 = \Gamma'$.

Being 3-cobordant is an equivalence relation on the set of closed curves with no horizontal inflection points. There is a group structure on the set of these cobordism classes: addition is given by the disjoint sum operation, and inverses are formed by reflection with respect to the x axis.

THEOREM 3 (see [Ar$_{18}$]). *The group $\pi_1(\mathscr{F} - \Sigma_3)$ is isomorphic to the group of classes of smooth closed 3-cobordant curves without horizontal inflection points.* □

Here to each closed curve correspond functions $F(x, \lambda)$ (or families $f_\lambda(x)$ tautologically associated to them) that are identically 1 outside some sufficiently large disk and 0 on this curve.

LEMMA 2. *Any curve without horizontal inflection points is 3-cobordant either to a sum of several curves depicted in Figure 16 ("kidneys") or curves symmetric to them with respect to the x axis ("antikidneys"). The sum of a kidney and antikidney is 3-cobordant to zero.*

Proof consists of the direct surgery, see [Ar$_{18}$]. □

This implies that the group $\pi_1(\mathscr{F} - \Sigma_3)$ is cyclic. We prove that it is free.

To each closed curve without vertical inflection points corresponds its index (see [Ar$_{18}$]) defined as the difference between the number of points with a horizontal tangent where the restriction of the function λ to the region bounded by the curve has a local maximum and the number of similar points of local minima. For example, the index of a kidney is 1. The index of a curve is invariant under all elementary 3-cobordisms, and hence determines an epimorphism of the 3-cobordism group onto $\mathbb{Z}$. This completes the proof

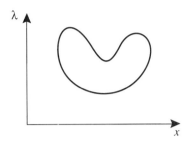

FIGURE 16

of Theorem 1 for the spaces $\mathscr{F} - \Sigma_3$. The case of functions on a circle is dealt with similarly. In particular, the generator of the group of 3-cobordisms of curves on the cylinder $S^1 \times \mathbb{R}^1$ is the same "kidney" whose inflection points do not touch the circles $S^1 \times (\cdot)$ of this cylinder. □

2.4. Proof of Theorem 2. First we prove that the map $j_*^2 \colon \pi_2(\mathscr{F} - \Sigma_3) \to \pi_2(B(3))$ is an epimorphism. As before, consider a 2-spheroid in the space $\mathscr{F}$ as a family $f_\lambda(X)$, $\lambda \in \mathbb{R}^2$, $x \in \mathbb{R}^1$, with $f_\lambda(x) = 1$ for sufficiently large $|\lambda| + |x|$. In the space $\mathbb{R}^3 = \mathbb{R}_\lambda^2 \times \mathbb{R}_x^1$ consider a surface Γ on which the function $F(x, \lambda) \equiv f_\lambda(x)$ vanishes. If the spheroid $\{f_\lambda\}$ is smooth and nondegenerate (which we can still assume) then Γ is a smooth surface in $\mathbb{R}^3$. Project it to the space $\mathbb{R}_\lambda^2$ along the axis $\mathbb{R}_x^1$.

LEMMA 3. *A smooth nondegenerate 2-spheroid in $\mathscr{F}$ is disjoint from the set Σ_3 if and only if the projection of the corresponding surface Γ to $\mathbb{R}_\lambda^2$ has only foldings but no Whitney cusps or more complicated singularities.* □

Now we construct an invariant for such 2-spheroids.

On the surface Γ consider the critical set (fold) of its projection to $\mathbb{R}_\lambda^2$. This is a smooth closed curve, that is a finite collection of circles imbedded in $\mathbb{R}^3$. This collection can be divided naturally into two subcollections, positive and negative, consisting of points (λ, x) such that x is respectively a point of minimum (maximum) of the function f_λ. Moreover, the fold curve has a natural orientation. In fact, in each of its points there is a nondegenerate frame in $\mathbb{R}^3$ whose first vector is $\partial/\partial x$, the second is $\operatorname{grad} F$, and the third is any vector tangent to the fold. Let us assume the direction of this vector to be chosen in such a way that the frame is positive. This rule determines the compatible directions at all points of every component of the fold, and hence, orients it.

DEFINITION. The *Hopf invariant* of a smooth 2-spheroid $\{f_\lambda\}$, $\lambda \in \mathbb{R}^2$, is the linking index in $\mathbb{R}^3$ of the positive and negative subsets of its fold, with the natural orientations.

LEMMA 4. *The Hopf invariant is an invariant of 2-spheroids in the group $\pi_2(\mathscr{F} - \Sigma_3)$.*

LEMMA 5. *Under the 2-jet imbedding $\mathscr{F} - \Sigma_3 \to B(3)$ the 2-spheroid $\{f_\lambda\} \subset \mathscr{F} - \Sigma_3$ with Hopf invariant χ is mapped to the χ-fold generator of the group $\pi_2(B(3)) \sim \mathbb{Z}$.*

LEMMA 6. *There is a family f_λ with Hopf invariant one.*

These three lemmas imply immediately the second part of Theorem 2. Lemma 4 follows from Lemma 5.

PROOF OF LEMMA 5. Associate to a 2-spheroid $\{f_\lambda\} \subset \mathscr{F} - \Sigma_3$ the map $X \colon \mathbb{R}^3 \to \mathbb{R}^3 - \{0\} \sim S^2$ given by the formula $X(\lambda, x) = (f_\lambda(x), f'_\lambda(x), f''_\lambda(x))$ (the derivatives are always taken with respect to x). Then the complement of some ball is mapped to the point $(1, 0, 0)$, so that X is a 3-spheroid in $\mathbb{R}^3 - \{0\} \sim S^2$. The Hopf invariant of this spheroid coincides with the one for the family $\{f_\lambda\}$. In fact, by definition it is equal to the linking index in $\mathbb{R}^3$ of the preimages of any pair of rays in $\mathbb{R}^3 - \{0\}$, and these rays can be taken to be the two halves of the line $\{f = 0, f' = 0\}$.

It remains to remark that our construction assigning a 3-spheroid in $\mathbb{R}^3 - \{0\}$ to each 2-spheroid in $\mathscr{F} - \Sigma_3$ is the composition of the imbedding j^l and the construction realizing the isomorphism $\pi_2(\Omega S^2) \cong \pi_3(S^2)$, see Appendix 3. □

The family $\{f_\lambda\}$ used in the proof of Lemma 6 is shown in Figure 17. Namely, the curve in this figure is the set of critical points of the projection of the desired surface Γ to the plane $\mathbb{R}^2_\lambda$ (or, equivalently, the set of those values of λ for which the function f_λ has a double root). The Roman numerals in the components of the complement of the curve show the number of roots of f_λ for λ from these components (the letter Z means zero). On each smooth segment of the curve (far from intersections) a number i is shown signifying that in passing through this segment, from the component with the larger number of roots, the ith and $i+1$th roots (counting from left to right) of the functions merge together and disappear. Here the positive and negative parts of the fold Γ correspond to the lines with odd (respectively even) numbers associated to them. It is easy to see that the linking index of these parts is equal to ± 1 depending on the choice of orientation in $\mathbb{R}^3$, and the lemma is proved.

The assertion of Theorem 2 that $\pi_1(\mathscr{F} - \Sigma_3) \to \pi_1(B(3))$ is an isomorphism is proved similarly: the 1-spheroid $\{f_\lambda\}$, $\lambda \in \mathbb{R}^1$, determines a 2-spheroid in $\mathbb{R}^3 - \{0\}$; the degree of the 2-spheroid corresponding to the generator of the group $\pi_1(\mathscr{F} - \Sigma_3)$ equals 1.

Indeed, the degree of a 2-spheroid in $\mathbb{R}^3 - \{0\}$ can be computed as the number of its positive intersections with the semiaxis $\{f = f' = 0, f'' > 0\}$ (with respect to given orientations of the spheroid and the semiaxis) minus the number of negative intersections. Such intersection points are precisely

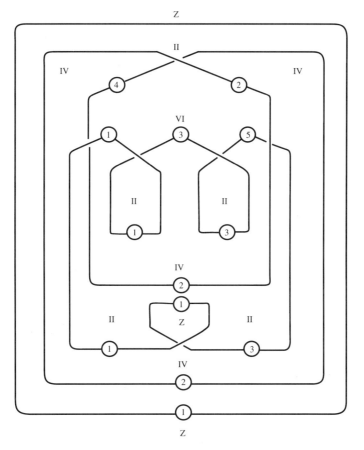

FIGURE 17

the points of the curve Γ used in the definition of its index, and the local intersection indices corresponding to the points of local minimum and maximum of the function λ are opposite to each other.

Finally, we prove the assertions of Theorem 2 about the space $\Phi - \Sigma_3$ of functions without triple zeros on the circle.

In the space Φ there is a subspace consisting of functions that are identically 1 near the base point of S^1. This space is identified in the obvious way with the space $\mathscr{F}$, and we obtain the maps $I: \pi_i(\mathscr{F} - \Sigma_k) \to \pi_i(\Phi - \Sigma_k)$ for all i, k. The maps $\pi_i(B(k)) \to \pi_i(B(S^1, \mathfrak{A}_0^{k-1}))$ are constructed similarly. It is easy to see that these maps are part of the commutative diagram

$$\begin{array}{ccccc}
\pi_i(\mathscr{F} - \Sigma_k) & \xrightarrow{j_*} & \pi_i(B(k)) & \simeq & \pi_i(\Omega S^{k-1}) \\
\downarrow I & & \downarrow I & & \downarrow \\
\pi_i(\Phi - \Sigma_k) & \xrightarrow{j_*} & \pi_i(B(S^1, \mathfrak{A}_0^{k-1})) & \simeq & \pi_i(\Omega_f S^{k-1})
\end{array} \qquad (3)$$

where the right isomorphisms are described in 2.1, and the right vertical arrow is the left homomorphism from the natural exact sequence

$$0 \to \pi_i(\Omega S^{k-1}) \to \pi_i(\Omega_f S^{k-1}) \to \pi_i(S^{k-1}) \to 0 \qquad (4)$$

(see Appendix 3). Let $k = 3$. Then from this sequence for $i = 1$ we see that the right (and middle) vertical arrows in (3) are isomorphisms. But by the above description of the generators of the groups $\pi_1(\mathscr{F} - \Sigma_3)$, $\pi_1(\Phi - \Sigma_3)$ the left vertical arrow is also an isomorphism, and the desired conclusion about the groups π_1 follows from commutativity of the diagram. Now let $i = 2$. It follows from sequence (4) that $\pi_2(B(S^1, \mathfrak{A}_0^2)) \cong \mathbb{Z} \oplus \mathbb{Z}$, and one of the generators of this group is obtained from the generator of the group $\pi_2(\mathscr{F} - \Sigma_3)$ shown in Figure 17 by a composition of arrows I and j_* taken in arbitrary order. In particular, this generator lies in the image of the group $\pi_2(\Phi - \Sigma_3)$ under the lower homomorphism j_*. The other generator in $\pi_2(B(S^1, \mathfrak{A}_0^2))$ is covered by a 2-spheroid in $\Phi - \Sigma_3$, which we will now describe.

Fix an arbitrary point $\alpha \in S^1$ and consider the unit sphere S^2 in the fiber of the bundle $J^2(S^1, \mathbb{R}^1)$ over this point. Let us assume that we have succeeded in constructing a map of this sphere into the space $\Phi - \Sigma_3$ such that the function f_λ corresponding to the point $\lambda \in S^2 \subset J_\alpha^2(S^1, \mathbb{R}^1)$ has a 2-jet at α equal to this point λ. It follows from sequence (4) that the corresponding element of the group $\pi_2(B(S^1, \mathfrak{A}_0^2))$ generates the quotient of this group by the subgroup $I(\pi_2(B(3)))$; hence, together with the element constructed before, it generates the whole group.

Let us now construct such a 2-spheroid. It can be realized in the space of all trigonometric polynomials of degree one, i.e. in the set of functions on S^1 of the form

$$a + b \cos t + c \sin t. \qquad (5)$$

Indeed, to each point $\lambda \in J_\alpha^2(S^1, \mathbb{R}^1)$ corresponds a unique triple of numbers a, b, c such that the 2-jet of the polynomial (5) at the point α equals λ; on the other hand, such a polynomial has a root of multiplicity ≥ 3 on S^1 if and only if $a = b = c = 0$.

Theorem 2 of this section is proved.

§3. Cohomology of spaces of polynomials without multiple roots

Denote by P_d the space of polynomials

$$x^d + a_1 x^{d-1} + \cdots + a_{d-1} x + a_d \qquad (6)$$

with real coefficients a_i.

§3. POLYNOMIALS WITHOUT MULTIPLE ROOTS

In this section we compute the rings $H^*(P_d - \Sigma_k)$ for all $d \geq k \geq 3$. (For $d \to \infty$ these rings and the homotopy groups of spaces $P_d - \Sigma_k$ stabilize to those of the space $\mathscr{F} - \Sigma_k$.) Here for the first time (and in its simplest version) appears the spectral sequence that is the main tool in the proof of Theorem 1.1 and all further computations in this chapter and in Chapters IV and V.

3.1. Statement of results.

THEOREM 1. *Let $k \geq 3$. Then the group $H^i(P_d - \Sigma_k)$ is isomorphic to $\mathbb{Z}$ if i is divisible by $k - 2$ and $i/(k - 2) \leq d/k$, and is trivial for all other i. For a suitable choice of generators $e_a \in H^{a(k-2)}(P_d - \Sigma_k)$ multiplication in the cohomology of the space $P_d - \Sigma_k$ is given by the following rules. If k is even then $e_a e_b = \binom{a+b}{a} e_{a+b}$; if k is odd then $e_1 e_1 = 0$, $e_1 e_{2a} = e_{2a+1}$, $e_{2a} e_{2b} = \binom{a+b}{a} e_{2a+2b}$ (here we assume that $e_c \equiv 0$ if $c > d/k$). In particular, e_1 is the class defined by the linking index with the whole stratum $\Sigma_k \subset P_d$ endowed with a suitable orientation at all nonsingular points.*

REMARK. 1. The groups $H^*(P_d - \Sigma_k)$ have been computed by V. I. Arnol'd; see [Ar$_{18}$]. Almost all of the results of this chapter were obtained in solving his problem on the multiplicative structure of the rings $H^*(P_d - \Sigma_k)$ and thinking over the answer.

2. The constants $e_a e_b / e_{a+b}$ in Theorem 1 corresponding to the case of odd k are precisely the coefficients P_{a+b}^a from 2.6 of Chapter I used in the description of the integral cohomology of the braid group.

We now give an explicit realization of the generators of the ring $H^*(P_d - \Sigma_k)$. The set Σ_k is a $(d - k + 1)$-dimensional algebraic subset in P_d. We shall see that it can be represented as the image of the space $\mathbb{R}^{d-k+1}$ under a proper map into P_d; in particular, it has the integral (noncompact) fundamental cycle $[\Sigma_k] \in \overline{H}_{d-k+1}(\Sigma_k)$. This cycle determines a local orientation of the set Σ_k near any of its nonsingular points. Fix an arbitrary orientation of the space P_d. Then the class $e_1 \in H^{k-2}(P_d - \Sigma_k)$ is defined as the Alexander dual of Σ_k; i.e. it is defined by the linking indices with Σ_k in the contractible space P_d. Now consider an arbitrary nonsingular point of the maximal (of multiplicity $[d/k]$) self-intersection of the stratum Σ_k: it corresponds to a polynomial f with $[d/k]$ pairwise distinct roots of multiplicity k. Let V be a small spherical neighborhood of this point. The set $V - \Sigma_k$ is diffeomorphic to the product of $[d/k]$ spheres S^{k-2} and the linear space of dimension $d - (k-2)[d/k]$; the set $\Sigma_k \cap V$ consists of $[d/k]$ smooth $(d - k + 1)$-dimensional components in the general position. Each of these components corresponds to one of the multiplicity k roots of the polynomial f; let their order correspond to the decreasing order of these roots. The orientation of these components is induced from the global orientation of the stratum $\Sigma_k \subset P_d$ defined by the fundamental cycle $[\Sigma_k]$. The algebra

$H^*(V - \Sigma_k)$ is isomorphic to the product (tensor or exterior depending on k being even or odd) of $[d/k]$ algebras $H^*(S^{k-2})$; the multiplicative generators $\alpha_1, \ldots, \alpha_{[d/k]} \in H^{k-2}(V - \Sigma_k)$ of this algebra are the linking indices with the components of the set $V \cap \Sigma_k$.

The rth symmetric polynomial $\sigma_r \in H^{r(k-2)}(V - \Sigma_k)$ is the sum of monomials $\alpha_{i_1} \cdots \alpha_{i_r}$ over all k-tuples of natural numbers $i_1 < \cdots < i_r \leq [d/k]$.

THEOREM 2. *For $k \geq 3$ the map $H^*(P_d - \Sigma_k) \to H^*(V - \Sigma_k)$ given by the imbedding $V \to P_d$ is a monomorphism and sends the generator e_r to the rth symmetric polynomial in α_i.*

Theorem 1 follows directly from this result. Moreover, Theorem 2 immediately determines all cohomology operations for the cohomology of the space $P_d - \Sigma_k$ with coefficients in any abelian group.

REMARK. Theorems 1, 2 also make sense for $k = 2$. In this case the space $P_d - \Sigma_k$ consists of $[d/k] + 1$ contractible components so that $H^i(P_d - \Sigma_k) = 0$ for $i > 0$. However, it is possible to define a grading and compatible multiplication on the groups $H^0(P_d - \Sigma_2)$ and $H^0(V - \Sigma_2)$ so that if we replace the dimensional grading and the standard cohomology multiplication by these ones, the assertion of the theorems remains valid. For example, this grading in the group $H^0(P_d - \Sigma_2)$ is dual to the grading in $H_0(P_d - \Sigma_2)$ such that the set of polynomials (6) with $d - 2l$ real roots has degree l.

Let $c > d$. Then near any nonsingular point of the set $\Sigma_d \subset P_c$, the set $P_c - \Sigma_k$ is diffeomorphic to the product of $P_d - \Sigma_k$ with the linear space $\mathbb{R}^{c-d}$; any such diffeomorphism determines an imbedding

$$I_{d,c}: P_d - \Sigma_k \to P_c - \Sigma_k.$$

THEOREM 3. *For $k \geq 3$ any map $I_{d,c}$ induces an isomorphism of homology groups in dimensions $\leq ([d/k] + 1)(k - 2) - 1$, and, if $k > 3$, an isomorphism of homotopy groups in dimensions $\leq ([d/k] + 1)(k - 2) - 2$, and an epimorphism of homotopy groups in dimension $([d/k] + 1)(k - 2) - 1$. In particular, for $[d/k] = [c/k]$ the map $I_{d,c}$ is a homotopy equivalence.*

THEOREM 3'. *For any d there exists an imbedding $I_d: P_d \to \mathcal{F}$ such that for any $k \geq 3$ the set $\Sigma_k \cap P_d$, and only this set, is mapped into $\Sigma_k \cap \mathcal{F}$. The corresponding map $H_i(P_d - \Sigma_k) \to H_i(\mathcal{F} - \Sigma_k)$ is an isomorphism for $i \leq ([d/k] + 1)(k - 2) - 1$, and, if $k > 3$, a similar map of homotopy groups π_i is an isomorphism for $i \leq ([d/k] + 1)(k - 2) - 2$ and an epimorphism for $i = ([d/k] + 1)(k - 2) - 1$.*

(The last epimorphism is not always an isomorphism: for example, for $d = 6, 7, 8$ the space $P_d - \Sigma_k$ has the homotopy type of the bouquet $S^1 \vee S^2$, so that $\pi_2(P_d - \Sigma_k) = \mathbb{Z}^\mathbb{Z}$ (Yu. G. Makhlin, 1990).)

The imbedding I_d can be defined, for example, as follows. Fix an arbitrary cut-off function $\chi\colon \mathbb{R}^1 \to [0, 1]$ such that $\chi(x) = 0$ for $|x| \leq 1$, $\chi(x) = 1$ for $|x| \geq 2$, and $\partial\chi/\partial x \neq 0$ for $1 < |x| < 2$. Define a function $\mu\colon P_d \to [1, \infty)$ equal to the sum of absolute values of the coefficients of the polynomial (6). If d is even, then for any such polynomial f define the function $I_d(f)$ by the condition $I_d(f)(x) \equiv f(x) + (1 - f(x))\cdot\chi(x/\mu(f))$; if d is odd, define the imbedding I_d to be the composition $I_{d+1} \circ I_{d,d+1}$ of the already constructed imbeddings.

3.2. Two preliminary remarks about the spaces $P_d - \Sigma_k$. Together with P_d consider an affine subspace $P'_d \subset P_d$ consisting of polynomials (6) with the zero coefficient a_1.

PROPOSITION 1 (cf. [Ar$_4$]). *The space $P'_d - \Sigma_k$ is a deformation retract of $P_d - \Sigma_k$; in particular, these spaces are homotopy equivalent.*

Indeed, the space P_d fibers into curves diffeomorphic to lines and containing with each polynomial $f(x)$ all polynomials $f(x - a)$, $a \in \mathbb{R}$. Each such curve intersects the set P'_d at one point and either lies in the set Σ_k entirely, or does not intersect it at all.

PROPOSITION 2. *For any ball $U \subset P_d$ centered at 0 the set $U - \Sigma_k$ is a deformation retract of $P_d - \Sigma_k$.*

This deformation is performed along the orbits of the one-parameter group $\mathbb{R}_+$ acting on P_d by the $f_t(x) = t^{-d} f(tx)$.

3.3. The spectral sequence for computing the groups $H^*(P_d - \Sigma_k)$.

NOTATION. We shall always denote by $\overline{H}_*(X)$ the homology group of X with closed support (or, equivalently, the homology of the one-point compactification of X reduced modulo the added point).

By the Alexander duality theorem, for $i \geq 1$ we have

$$H^i(P_d - \Sigma_k) \cong \overline{H}_{d-i-1}(\Sigma_k). \tag{7}$$

We construct a filtered space whose closed homology can be immediately computed and coincides with that of Σ_k.

Let Σ_k^r be the set of polynomials from Σ_k with at least r geometrically distinct roots of multiplicity $\geq k$. These sets have singularities; in particular $\Sigma_k^{r+1} \subset \operatorname{sing}\Sigma_k^r$. We construct a resolution of these singularities. Denote by $\mathbb{R}(r)$ the set of unordered collections of r distinct points of the line $\mathbb{R} = \mathbb{R}^1$. Let $\widetilde{\Sigma}_k^r$ be the set of all pairs of the form

$$\text{(collection } \{t_1, \ldots, t_r\} \in \mathbb{R}(r); \text{ polynomial } f \in P_d \text{ with} \tag{8}$$
$$\text{roots of multiplicity } \geq k \text{ at all points } t_1, \ldots, t_r).$$

The set $\widetilde{\Sigma}_k^r$ is endowed with the obvious structure of a smooth affine $(d - kr)$-bundle over $\mathbb{R}(r)$, in particular, with the structure of a smooth manifold.

(Indeed, the condition, that the polynomial f has roots of given multiplicity at the prescribed points, is a system of linear conditions on the coefficients of f.) Forgetting the first element of the pair (8), we obtain the projection of $\widetilde{\Sigma}_k^r$ onto Σ_k^r; it is a diffeomorphism over the Zariski open subset in Σ_k^r consisting of polynomials with exactly r roots of multiplicity exactly k.

Let us fix a smooth regular imbedding of the line $\mathbb{R}^1$ into the space $\mathbb{R}^N$ of sufficiently large dimension in such a way that images of no $[d/k]$ distinct points lie in the same $([d/k] - 2)$-dimensional affine plane in $\mathbb{R}^N$. (For example, the Veronese imbedding in $\mathbb{R}^{[d/k]-1}$ given by $x_1(t) = t$, $x_2(t) = t^2, \ldots, x_{[d/k]-1}(t) = t^{[d/k]-1}$, possesses this property.)

For each $r = 1, \ldots, [d/k]$ and any point of the set $\widetilde{\Sigma}_k^r$ (i.e. a collection $(\{t_1, \ldots, t_r\}, f)$) we consider an $(r-1)$-dimensional open simplex in the space $P_d \times \mathbb{R}^N$ whose projection to P_d is the point f, and the projection to $\mathbb{R}^N$ is an open simplex spanned by the images of the points $t_1, \ldots, t_r \in \mathbb{R}^1$ under our imbedding $\mathbb{R}^1 \to \mathbb{R}^N$. The union of such simplices over all points of the manifold $\widetilde{\Sigma}_k^r$ will be denoted by $\{\Sigma_k, r\}$; it is a locally trivial (and hence, trivializable) bundle over $\widetilde{\Sigma}_k^r$ with the simplex as a fiber. Define the set $\{\Sigma_k\} \subset P_d \times \mathbb{R}^N$ as the union of sets $\{\Sigma_k, r\}$ over all $r = 1, \ldots, [d/k]$. This set has a natural increasing filtration: its term $F_\tau\{\Sigma_k\}$ is the union of sets $\{\Sigma_k, r\}$ over all $r \leq \tau$.

Denote by π the map $\{\Sigma_k\} \to \Sigma_k$ given by the evident projection

$$P_d \times \mathbb{R}^N \to P_d; \qquad (9)$$

see Figure 18.

LEMMA 1. *The map π is proper and can be continuously extended to a map $\bar{\pi}\colon \overline{\{\Sigma_k\}} \to \overline{\Sigma_k}$ of the one-point compactifications of $\{\Sigma_k\}$ and Σ_k. The corresponding map of the closed homology groups*

$$\overline{H}_*(\{\Sigma_k\}) \to \overline{H}_*(\Sigma_k) \qquad (10)$$

is an isomorphism.

REMARK. In fact, $\bar{\pi}$ is even a homotopy equivalence, but the proof of this is more difficult, and the result will not be needed.

PROOF OF THE LEMMA. The first statement of the lemma is clear. Let us prove the second. By a theorem from [Loj$_1$], [Loj$_2$] semialgebraic sets can be triangulated; therefore, there exists a cellular decomposition of the set $\overline{\Sigma}_k$ such that over each open cell the projection $\bar{\pi}$ is a trivializable bundle with simplex as a fiber. Filter the space $\overline{\Sigma}_k$ by the skeletons of this decomposition, and the space $\overline{\{\Sigma_k\}}$ by the preimages of these skeletons under the map $\bar{\pi}$. Let $\mathscr{E}_{p,q}^r$ be the spectral sequence converging to the homology of the space $\overline{\{\Sigma_k\}}$ and generated by the filtration. By construction of this filtration, the

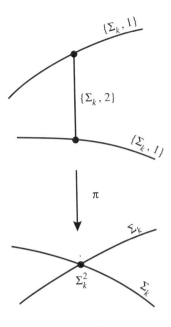

FIGURE 18

term $\mathscr{E}^1_{p,q}$ vanishes for $q > 0$, and the complex $\{\mathscr{E}^1_{*,0}; d_1\}$ is isomorphic to the cellular differential complex of the cellular decomposition of $\overline{\Sigma_k}$, the isomorphism being given by the projection $\overline{\pi}$. □

It remains to compute the group $\overline{H}_*(\{\Sigma_k\})$. Here we use the spectral sequence $E^r_{p,q}$ generated by the natural filtration $F_1\{\Sigma_k\} \subset \cdots \subset F_{[d/k]}\{\Sigma_k\} \equiv \{\Sigma_k\}$. Its term $E^1_{p,q}$ is the group $\overline{H}_{p+q}(\{\Sigma_k, p\})$. But the manifold $\{\Sigma_k, p\}$ is homeomorphic to the ball of dimension $\dim \Sigma^p_k + (p-1) = d - p(k-2) - 1$, and hence $E^1_{p,q} = \mathbb{Z}$ if $q = d - p(k-1) - 1$, $q \geq 0$, and is trivial for all other p, q; see Figure 19 (next page). It is clear that the spectral sequence degenerates at this term and gives the additive part of Theorem 1.

3.4. Proof of Theorem 2. As a cycle with closed support in $\{\Sigma_k\}$ corresponding to the generator of the group $E^1_{p,d-p(k-1)-1}$ of our spectral sequence, we can take the cycle in $F_p\{\Sigma_k\}$ consisting of the cell $\{\Sigma_k, p\}$ and surface $L \subset F_{p-1}\{\Sigma_k\}$ spanning the cycle $\partial(\{\Sigma_k, p\})$. (That this cycle is homologous to zero follows from our spectral sequence.)

Let $f \in P_d$ be a point of a regular multiplicity p self-intersection of the stratum Σ_k; i.e. f is a polynomial with exactly p roots of multiplicity k and without roots of greater multiplicity. Let V be a small spherical neighborhood of f in P_d, and $(S^{k-2})^p$ be an imbedded smooth cycle in $V - \Sigma_k$ realizing the generator of the group $H_{p(k-2)}(V - \Sigma_k)$.

The projection of the cycle $\{\Sigma_k, p\} + L \subset F_p\{\Sigma_k\}$ to Σ_k given by the

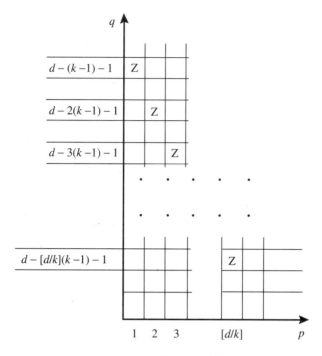

FIGURE 19

projection $\pi\colon \{\Sigma_k\} \to \Sigma_k$ determines an element of the group

$$H_{d-p(k-2)-1}(\Sigma_k, \Sigma_k \cap \partial V).$$

PROPOSITION. *The linking index of the element $\pi_*(\{\Sigma_k, p\} + L)$ with the cycle $(S^{k-2})^p$ is equal to ± 1.*

Theorem 2 follows directly from this proposition, applied near the point of $[d/k]$-fold self-intersection of Σ_k to the cycle $(S^{k-2})^{[d/k]}$ and to all its $\binom{[d/k]}{r}$ coordinate subcycles of the form $S^{k-2}_{i_1} \times \cdots \times S^{k-2}_{i_r}$, $1 \le i_1 < \cdots < i_r \le [d/k]$.

Let us prove this proposition.

LEMMA 2. *The group $H_{d-p(k-2)-1}(F_{p-1}\{\Sigma_k\} \cap \pi^{-1}(V), \pi^{-1}(\partial V))$ is trivial.*

LEMMA 3. *There exists a chain $L' \subset F_{p-1}\{\Sigma_k\} \cap \pi^{-1}(V)$ such that the chain $\{\Sigma_k, p\} \cap \pi^{-1}(V) + L'$ is a cycle modulo $\pi^{-1}(\partial V)$, and the projection of the latter chain to Σ_k has linking index one with the cycle $(S^{k-2})^p$.*

The proposition follows immediately from these two lemmas. Indeed, $\{\Sigma_k, p\} + L = \{\Sigma_k, p\} + L' - L' + L$. Both sums $\{\Sigma_k, p\} + L'$ and $-L + L'$ are relative cycles in $\{\Sigma_k\} \cap \pi^{-1}(V)$ modulo $\pi^{-1}(\partial V)$; by Lemmas 2 and 3 the intersection indices of their projections with $(S^{k-2})^p$ are equal to 1 and 0.

§3. POLYNOMIALS WITHOUT MULTIPLE ROOTS   93

To prove the lemmas, we introduce coordinates in V in such a way that the first local component of the set Σ_k is given by the condition $z_1 = \cdots = z_{k-1} = 0$, the second one by the condition $z_k = \cdots = z_{2k-2} = 0$, etc., and the last $d - p(k-1)$ coordinates are coordinates on the set Σ_k^p. Consider a vector field in V which is the Euler vector field with respect to the coordinates $z_1, \ldots, z_{p(k-1)}$, i.e. equal to $z_1 \partial/\partial z_1 + \cdots + z_{p(k-1)} \partial/\partial z_{p(k-1)}$. This field preserves the set Σ_k and all its strata Σ_k^t, and lifts by the projection π to a vector field on $\{\Sigma_k\} \cap \pi^{-1}(V)$. This last vector field pushes the set $F_{p-1}\{\Sigma_k\} \cap \pi^{-1}(V) \equiv (\{\Sigma_k\} - \{\Sigma_k, p\}) \cap \pi^{-1}(V)$ out onto its intersection with $\pi^{-1}(\partial V)$, and Lemma 2 is proved. Consider a $d - p(k-1)$-dimensional set in V given by conditions $z_1 \geq 0$, $z_2 = \cdots = z_{k-1} = 0$, $z_k \geq 0$, $z_{k+1} = \cdots = z_{2k-2} = 0$, $z_{2k-1} \geq 0, \ldots, \cdots = z_{p(k-1)} = 0$; the last $d - p(k-1)$ coordinates are arbitrary. The boundary of this set lies in $\Sigma_k \cap \partial V$. Consider the complete preimage of this boundary in $\pi^{-1}(\Sigma_k \cap V)$. It contains the set $\{\Sigma_k, p\}$; their difference is the desired frame L'.

Theorem 2 is proved completely, together with Theorem 1.

3.5. Proof of Theorem 3. First of all, we consider the spectral sequence from 3.3 in its cohomology form: for this rename the term $E_{p,q}^r$ of the sequence by $E_r^{-p, d-q-1}$. Also set $E_r^{0,0} \equiv \mathbb{Z}$, $E_r^{0,q} \equiv 0$ for $q > 0$. The resulting sequence is shown in Figure 20 (next page); by Alexander duality it converges to the cohomology of $P_d - \Sigma_k$. If $c > d$, then for $p \geq -[d/k]$ all terms $E_r^{p,q}$, $r \geq 1$, of spectral sequences corresponding to $P_d - \Sigma_k$ and to $P_c - \Sigma_k$ are isomorphic; we show that this isomorphism agrees with the imbedding $P_d \to P_c$.

Let f be a regular point of the stratum $\Sigma_d \subset P_c$ (i.e. the polynomial f has one root of multiplicity d and has no other multiple roots). Then in some neighborhood $U \subset P_c$ of f the pair (U, Σ_k) is diffeomorphic to the product of the space $\mathbb{R}^{c-d}$ and the pair (P_d, Σ_k). This diffeomorphism defines an isomorphism $H^*(P_d - \Sigma_k) \sim H^*(U - \Sigma_k)$. It remains to compare $H^*(U - \Sigma_k)$ and $H^*(P_c - \Sigma_k)$.

The group $H^i(U - \Sigma_k)$ is the Alexander dual of $\overline{H}_{c-i-1}(\Sigma_k \cap U)$; the latter group is isomorphic to the group $\overline{H}_{c-i-1}(\{\Sigma_k\} \cap \pi^{-1}(U))$, which can be computed, similarly to $\overline{H}_*(\{\Sigma_k\})$, using the spectral sequence corresponding to the natural filtration of the set $\{\Sigma_k\}$. The imbedding $U \to P_c$ defines a homomorphism of these spectral sequences

$$E_r^{p,q}(\{\Sigma_k\}) \to E_r^{p,q}(\{\Sigma_k\} \cap \pi^{-1}(U)).$$

For $p \geq -[d/k]$ this homomorphism of the terms $E_1^{p,q}$ is an isomorphism: this follows from the fact that for any $t = 1, 2, \ldots, [d/k]$ both sets

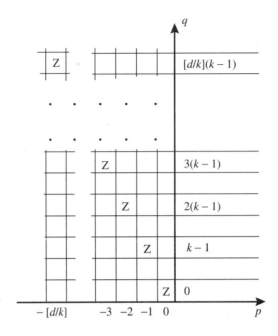

FIGURE 20

$\{\Sigma_k, t\}$ and $\{\Sigma_k, t\} \cap \pi^{-1}(U)$ are homeomorphic to disks of the same dimension. Since for both spectral sequences $E_1 = E_\infty$, and all their nonzero terms of total dimension $p + q \leq ([d/k] + 1)(k - 2) - 1$ lie in the columns with $p \geq -[d/k]$ and coincide, the limit homomorphism

$$\overline{H}_{c-i-1}(\{\Sigma_k\}) \to \overline{H}_{c-i-1}(\{\Sigma_k\} \cap \pi^{-1}(U))$$

(or, equivalently, $\overline{H}_{c-i-1}(\Sigma_k) \to \overline{H}_{c-i-1}(\Sigma_k \cap U))$ is an isomorphism for $i \leq ([d/k] + 1)(k - 2) - 1$.

By Alexander duality for P_c and U this isomorphism commutes with the restriction homomorphism $H^i(P_c - \Sigma_k) \to H^i(U - \Sigma_k)$, which, therefore, is also an isomorphism for $i \leq ([d/k]+1)(k-2)-1$. This proves the homology part of Theorem 3.

Proof of the homology part of Theorem 3′ will be given in 4.8 and differs from the proof just given only in that it uses finite-dimensional approximations of the space $\mathscr{F}$. The homotopy parts of these theorems follow from the homology ones by the Whitehead theorem.

THEOREM 4. (A) *For all $c > d \geq 3$ the homomorphisms*

$$(I_{d,c})_* : \pi_1(P_d - \Sigma_3) \to \pi_1(P_c - \Sigma_3)$$

and

$$(I_d)_* : \pi_1(P_d - \Sigma_3) \to \pi_1(\mathscr{F} - \Sigma_3)$$

are isomorphisms.

(B) *For all $d \geq 6$ the similar homomorphisms $\pi_2(P_d - \Sigma_3) \to \pi_2(\mathscr{F} - \Sigma_3)$ are epimorphisms.*

Statement (A) is proved in [Ar$_{18}$], and for the proof of (B) it is enough to prove that the unique homomorphism

$$(I_6)_*: \pi_2(P_6 - \Sigma_3) \to \pi_2(\mathscr{F} - \Sigma_3) \tag{11}$$

that factors through all others is an epimorphism. Recall (see §2) that $\pi_2(\mathscr{F} - \Sigma_k) \cong \mathbb{Z}$, and a generator of this group is given by the family of functions u from 2.4 shown in Figure 17. Let us construct a family of polynomials φ_λ, $\lambda \in \mathbb{R}^2$, in $P_6 - \Sigma_3$ that are equal to $x^6 + 1$ for sufficiently large $|\lambda|$ and are such that for any λ the polynomial φ_λ has roots at the same points as the function f_λ from the family in Figure 17: such a family exists since each function f_λ has no more than six zeros (counted with multiplicities). The resulting family φ_λ defines a spheroid in $P_6 - \Sigma_3$ (with base point at $\varphi \equiv x^6 + 1$) which is sent by the homomorphism (11) to the generator of $\pi_2(\mathscr{F} - \Sigma_k)$. □

§4. Proof of the first main theorem

This proof is based on the use of a spectral sequence similar to the one constructed in 3.3. The cohomology of both spaces $A(M, \mathfrak{A}, \varphi)$ and $B(M, \mathfrak{A}, \varphi)$ is computed using spectral sequences that are isomorphic starting from the term E_1. For the justification of these spectral sequences it is necessary to overcome new (compared to §3) difficulties related to the absence of Alexander duality in infinite-dimensional spaces; so we have to consider finite-dimensional approximations of the spaces A, B and stabilize the corresponding spectral sequences.

4.1. Statement of the main homological result. Let $M, \mathfrak{A}, \varphi, A(M, \mathfrak{A}, \varphi)$, and $B(M, \mathfrak{A}, \varphi)$ be the same as in §1, and let $M_1 = M - \partial M$. For any natural number t denote by M_t the set of unordered collections of t pairwise distinct points of the space M_1.

For any manifolds M, N denote by $J_t^k(M, N)$ the space of t-multijets of maps $M \to N$, i.e. the set of pairs of the form (t pairwise distinct points of M, collection of k-jets of maps $M \to N$ at these points).

Denote by $\Lambda_t(\mathfrak{A}, M)$ the subset in $J_t^k(M_1, \mathbb{R}^n)$ consisting of multijets such that all t k-jets in them belong to the set $\mathfrak{A}(M)$. Set

$$\Delta = \dim J^k(M, \mathbb{R}^n) - m \equiv \dim J^k|_z(M, \mathbb{R}^n)$$

for any point $z \in M$.

THEOREM 1. *Let the codimension of the set $\mathfrak{A}$ in $J^k(M, \mathbb{R}^n)$ be at least $m+2$. Then there exists a spectral sequence $E_r^{p,q} = E_r^{p,q}(M, \mathfrak{A}, \varphi)$ in the second quadrant, $p \leq 0 \leq q$, whose term E_∞ is associated to the cohomology* group of the space $A(M, \mathfrak{A}, \varphi)$, and

$$E_1^{p,q} \cong \overline{H}_{-\Delta p - q}(\Lambda_{-p}(\mathfrak{A}, M), T), \qquad (12)$$

where T is the local system of groups defined below; in particular, $E_1^{p,q} = 0$ if $p(\operatorname{codim} \mathfrak{A} - m) + q < 0$.

Note that formula (12) involves no φ.

THEOREM 2. *In the statement of Theorem 1 we can replace the space $A(M, \mathfrak{A}, \varphi)$ by $B(M, \mathfrak{A}, \varphi)$. The imbedding j^k of the space $C^\infty(M, \mathbb{R}^n)$ into the space of C^∞-sections of the bundle $J^k(M, \mathbb{R}^n) \to M$ induces an isomorphism of the corresponding spectral sequences (and hence also of the cohomology of $A(M, \mathfrak{A}, \varphi)$ and $B(M, \mathfrak{A}, \varphi))$.*

Now we describe the local system T. Define the cocycle $\sigma \in H^1(M_t, \mathbb{Z}_2)$ which takes the values 0 or 1 on each loop in M_t depending on whether it defines an even or odd permutation of points $x_1, \ldots, x_t \in M_1$ forming the base point in M_t. Define the cocycle $\chi = \chi_{k,M,t}^n \in H^1(M_t, \mathbb{Z}_2)$ as the sum of the cocycle σ and the first Stiefel-Whitney class of the natural vector bundle $J_t^k(M_1, \mathbb{R}^n) \to M_t$.

Consider the local system of groups over M_t that is locally isomorphic to $\mathbb{Z}$ but is multiplied by -1 under traversal over the loops in M_t on which the cocycle χ takes the value 1.

The local system T from Theorems 1, 2 is induced from this system by the natural projection $\Lambda_t(\mathfrak{A}, M) \to M_t$.

We give an explicit description of the first Stiefel-Whitney class $w_1(\{J_t^k(M_1, \mathbb{R}^n) \to M_t\})$. Any loop l in M_t is the simultaneous movement of t points $x_1, \ldots, x_t \in M_t$. The union of the paths traversed by these points is a closed path in M_1 which we will denote by $[l]$.

PROPOSITION. *The first Stiefel-Whitney class of the bundle $\{J_t^k(M_1, \mathbb{R}^n) \to M_t\}$ takes on each loop $l \in H_1(M_t, \mathbb{Z}_2)$ the value*

$$n\left(\binom{m+k}{m}\sigma(l) + \binom{m+k}{k-1}\langle w_1(M), [l]\rangle\right) \pmod{2}, \qquad (13)$$

where $w_1(M)$ is the first Stiefel-Whitney class of the tangent bundle of M.

It suffices to prove this for an arbitrary collection of closed curves l generating the group $H_1(M_t, \mathbb{Z}_2)$. Such a collection can be chosen to consist of curves of the following two types:

1) curves along which all $x_3, \ldots, x_t$ are fixed points, whereas x_1 and x_2 are interchanged without going out of a small disc in M_1 (see Figure 21);

2) curves along which $x_2, \ldots, x_t$ are fixed points.

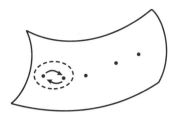

FIGURE 21

Here we can assume that $n = 1$: indeed, the bundle $J_t^k(M_1, \mathbb{R}^n)$ is isomorphic to the sum of n bundles $J_t^k(M_1, \mathbb{R}^1)$.

Over a curve of the first type the latter bundle splits into a sum of one-dimensional bundles, and the number of those that reverse orientation constitute $(1/t)$th part of the total number of one-dimensional bundles (so that this number is equal to the dimension of the space of k-jets of maps $M \to \mathbb{R}^1$ at one point, i.e. to $\binom{m+k}{m}$). Hence for curves of the first type formula (13) holds. Let l be a curve of the second type, and $[l]$ be the corresponding path traversed by the point x_1 in M_1. We can assume that $[l]$ is a smoothly imbedded circle. In a neighborhood of this path the manifold M_1 is diffeomorphic to the space of an $(m-1)$-dimensional vector bundle over the circle. Such a bundle splits into the sum of the $(m-2)$-dimensional trivial bundle and a one-dimensional bundle that is trivial or not, depending on the value of $\langle w_1(M), [l] \rangle$. In the first case the bundle $\{J(M_1, \mathbb{R}) \to M_1\}$ over $[l]$ can be trivialized, in the second case it splits into the sum of one-dimensional bundles indexed by monomials of degree $\leq k$ in m variables, and among them the nontrivial ones are precisely those that correspond to the monomials in which the first variable is raised to an odd power. It remains to prove that this number coincides with $\binom{m+k}{k-1}$ (mod 2). Denote by $\{m, k\}$ the cardinality of the complement, which consists of monomials of degree $\leq k$ containing an even power of the first variable. It is easy to see that the number $\binom{m+k}{k}$ of all monomials of degree $\leq k$ is equal to $2\{m, k\} - \{m-1, k\}$; in particular $\{m, k\} \equiv \binom{m+k+1}{k}$ (mod 2). This completes the proof of the proposition.

4.2. Finite-dimensional approximations in the space of functions on M. By the theorem on increasing smoothness (see, for example, [RF], §III.4) we can assume that M is an analytic manifold and define on M an analytic "collar" function $\omega: M \to \mathbb{R}_+^1$ such that $\omega^{-1}(0) = \partial M$, and the set $\omega^{-1}([0, 3])$ is diffeomorphic to $\partial M \times [0, 3]$ (where the projection onto the second factor is given by ω).

Let $\varkappa: \mathbb{R}^1 \to [0, 1]$ be a smooth piecewise-analytic cut-off C^∞-function such that $\varkappa(x) = 0$ for $x \leq 1$, and $\varkappa(x) = 1$ for $x \geq 2$; see, for example, [Hirsch$_3$].

Fix the space $\mathbb{R}^L$ with coordinates $y_1, \ldots, y_L$, and denote by Θ^d the space of polynomial maps $\mathbb{R}^L \to \mathbb{R}^n$ of degree $\leq d$. Let $i: M \to \mathbb{R}^L$ be a smooth imbedding.

For any two natural numbers d, s and function $\varphi: M \to \mathbb{R}^n$ with the same properties as in Theorem 1 define an affine space $\Theta_\varphi^{d,s}$ as the set of all maps $M \to \mathbb{R}^n$ representable in the form

$$(\vartheta_1 \circ i) \cdot (\varkappa \circ \omega) + (\vartheta_2 \circ i) \cdot (\varkappa \circ 2\omega) + \cdots + (\vartheta_s \circ i) \cdot (\varkappa \circ 2^s \omega) + \varphi, \qquad (14)$$

where all maps ϑ_j belong to the space Θ^d. In particular, for $d \leq d'$, $s \leq s'$ the space $\Theta_\varphi^{d,s}$ is imbedded in $\Theta_\varphi^{d',s'}$.

Let us show that computing the homology of the space $A(M, \mathfrak{A}, \varphi)$ reduces to the study of the spaces $A(M, \mathfrak{A}, \varphi) \cap \Theta_\varphi^{d,s}$ for sufficiently large (but finite) d, s.

DEFINITION. A pair of maps $i: M \to \mathbb{R}^L$, $\varphi: M \to \mathbb{R}^n$ is called *admissible* if i is an imbedding, and φ has no singular points of the class $\mathfrak{A}$ near ∂M. A pair (i, φ) is called (d, s)-*perfect* if it is admissible and for any natural t the set of maps of the form (14) in the space $\Theta_\varphi^{d,s}$ with at least t singular points of type $\mathfrak{A}$ is either empty or has codimension at least $t(\operatorname{codim} \mathfrak{A} - m)$ (in particular, it is empty if $t(\operatorname{codim} \mathfrak{A} - m) > \dim \Theta_\varphi^{d,s}$).

4.2.1. THEOREM. *For any natural numbers d and s, the set of (d, s)-perfect pairs of maps (i, φ) contains a residual subset of the space of all admissible pairs (see Appendix 4 for the definition of the term residual). In particular, there exist pairs of maps (i, φ) that are (d, s)-perfect for all d and s.*

4.2.2. THEOREM. *For any map $\varphi: M \to \mathbb{R}^n$ without singularities of type $\mathfrak{A}$ near ∂M and for any cycle $\gamma \subset A(M, \mathfrak{A}, \varphi)$ there is a pair of numbers d_0 and s_0 such that for all $d \geq d_0$, $s \geq s_0$ the space $A(M, \mathfrak{A}, \varphi) \cap \Theta_\varphi^{d,s}$ contains the cycle γ' homologous to the cycle γ in $A(M, \mathfrak{A}, \varphi)$. If also $\dim \gamma \leq ([d/(k+1)]+1)(\operatorname{codim} \mathfrak{A} - m - 1) - 2$ and the pair of maps (i, φ) is (d', s')-perfect for all natural d' and s' then the cycle γ is homologous to zero in $A(M, \mathfrak{A}, \varphi)$ if and only if γ' is homologous to zero in $A(M, \mathfrak{A}, \varphi) \cap \Theta_\varphi^{d,s}$.*

4.2.3. THEOREM. *Suppose the pair of maps (i, φ) is (d, s)-perfect and $\operatorname{codim} \mathfrak{A} \geq m + 2$. Then there exists a spectral sequence*

$$\mathscr{E}_r^{p,q} = \mathscr{E}_r^{p,q}(\mathfrak{A}, M, \varphi, i, d, s)$$

in the second quadrant converging to the cohomology of the space $A(M, \mathfrak{A}, \varphi) \cap \Theta_\varphi^{d,s}$ and such that for $p \geq -d/(k+1)$ the term $\mathscr{E}_r^{p,q}$ is given by (12), and $\mathscr{E}_1^{p,q} = 0$ for $q + p(\operatorname{codim} \mathfrak{A} - m) < 0$ and any p.

(In particular, the support of $\mathscr{E}_1^{p,q}$ belongs to the region shown in Figure 22 with $\tan \alpha = \operatorname{codim} \mathfrak{A} - m \geq 2$. This guarantees convergence.)

§4. PROOF OF THE FIRST MAIN THEOREM

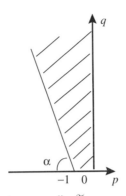

$\tan \alpha = \operatorname{codim} \mathfrak{A} - m$

FIGURE 22

4.2.4. THEOREM. *If $d' \geq d$, $s' \geq s$, and the pair (i, φ) is both (d, s)-perfect and (d', s')-perfect, then the corresponding spectral sequences from Theorem 4.2.3 computing the cohomology of the spaces*

$$A(M, \mathfrak{A}, \varphi) \cap \Theta_\varphi^{d,s} \quad \text{and} \quad A(M, \mathfrak{A}, \varphi) \cap \Theta_\varphi^{d',s'} \qquad (15)$$

are isomorphic at all terms $E_r^{p,q}$, $r \geq 1$, of total dimension $p + q$ at most $([d/(k+1)]+1)(\operatorname{codim} \mathfrak{A} - m - 1) - 1$. In particular, the cohomology of these spaces is isomorphic in dimensions that do not exceed the same number. The last isomorphism is given by the obvious imbedding of these spaces.

Theorem 1 of 4.1 is an immediate consequence of Theorems 4.2.1–4.2.4.

4.3. Proof of Theorem 4.2.1. Let an admissible pair of maps (i, φ) be given. For any natural t and any point $u = (u_1, \ldots, u_t) \in M_t$ (where all $u_j \in M_1$) denote by $\mathfrak{A}(u) = \mathfrak{A}(u, d, s, \varphi, i)$ the set of maps of the form (14) with singularities of type $\mathfrak{A}$ at all points u_j. For any natural number l denote by $V_{l,t}$ the set of points $u \in M_t$ such that in the space of maps (14) the codimension of the set $\mathfrak{A}(u)$ is at most $t \operatorname{codim} \mathfrak{A} - l$.

For a pair of maps (i, φ) to be (d, s)-perfect it suffices that for any t, l the codimension of the set $V_{l,t}$ in M_t is at least l (and all these codimensions are well defined). We prove that this is satisfied for a residual subset in the space of admissible pairs (i, φ).

For any point $u = (u_1, \ldots, u_t) \in M_t$ consider the set $J_{(u)}^k(M_1, \mathbb{R}^L) \times J_{(u)}^k(M_1, \mathbb{R}^n)$ consisting of collections of pairs of k-jets of maps $M_1 \to \mathbb{R}^L$ and $M_1 \to \mathbb{R}^n$ at all points $u_1, \ldots, u_t \in u$. The space $J_{(u)}^k(M_1, \mathbb{R}^n)$ is obviously isomorphic to $J_{(u_1)}^k(M_1, \mathbb{R}^n) \times \cdots \times J_{(u_t)}^k(M_1, \mathbb{R}^n)$ and belongs to $J_t^k(M_1, \mathbb{R}^n)$. Let $\Lambda_{(u)}(\mathfrak{A}, M)$ be its intersection with the set $\Lambda_t(\mathfrak{A}, M)$, i.e. $\Lambda_{(u)}(\mathfrak{A}, M)$ is the set of collections of k-jets of functions with singularities of type $\mathfrak{A}$. This is a semialgebraic subset in $J_{(u)}^k(M_1, \mathbb{R}^n)$ of codimension $t \cdot \operatorname{codim} \mathfrak{A}$.

Define a subset $W_{l,u}$ of the space $J_{(u)}^k(M_1, \mathbb{R}^L) \times J_{(u)}^k(M_1, \mathbb{R}^n)$ as follows. Let (I, Φ) be a point of this space. Consider an arbitrary imbedding $i: M \to \mathbb{R}^L$ whose t-fold k-jet at u is equal to I. Let $\Theta_0^{d,s}$ be the space of maps of the form (14) with $\varphi \equiv 0$ defined by i. Consider an affine map

$$v_{u,i,\Phi}: \Theta_0^{d,s} \to J_{(u)}^k(M_1, \mathbb{R}^n) \qquad (16)$$

that relates to any map of the class $\Theta_0^{d,s}$ the sum of the collection of its k-jets at the points $u_1, \ldots, u_t$ and the collection Φ. By definition, the pair (I, Φ) belongs to the set $W_{l,u}$ if the codimension of the set $v_{u,i,\Phi}^{-1}(\Lambda_{(u)}(\mathfrak{A}, M))$ in $\Theta_0^{d,s}$ is at most $t \cdot \operatorname{codim} \mathfrak{A} - l$: it is easy to see that this definition depends only on I and Φ but not i. Denote by $W_{l,t}$ the union of the sets $W_{l,u}$ over all $u \in M_t$.

LEMMA 1. *For any natural d, s the set $W_{l,t}$ is a semianalytic subset in $J_t^k(M_1, \mathbb{R}^L \times \mathbb{R}^n)$.*

This follows from the piecewise analyticity of the cut-off function $\varkappa$.

LEMMA 2. *For any t and any collection $u = (u_1, \ldots, u_t) \in M_t$ the codimension of $W_{l,u}$ in $J_{(u)}^k(M_1, \mathbb{R}^L) \times J_{(u)}^k(M_1, \mathbb{R}^n)$ is at least $t \cdot \operatorname{codim} \mathfrak{A} + l$ (and hence the codimension of $W_{l,t}$ in $J_t^k(M_1, \mathbb{R}^L \times \mathbb{R}^n)$ is at least $t \cdot \operatorname{codim} \mathfrak{A} + l$).*

In order to prove the lemma it suffices to show that for any element $I \in J_{(u)}^k(M_1, \mathbb{R}^L)$ the set of Φ such that $(I, \Phi) \in W_{l,u}$ has codimension at least $t \cdot \operatorname{codim} \mathfrak{A} + l$ in $J_{(u)}^k(M_1, \mathbb{R}^n)$. But this follows from the general fact: if a, b, c are natural numbers and $a < b + c$, then an a-dimensional semialgebraic set does not admit a b-parametric family of pairwise parallel (and disjoint) planes intersecting it in the sets of dimension $\geq c$.

Now consider the set of admissible pairs of maps $(i, \varphi): M \to \mathbb{R}^L \times \mathbb{R}^n$ such that for any t their multijet extensions $M_t \to J_t^k(M_1, \mathbb{R}^L \times \mathbb{R}^n)$ are transversal to all sets $W_{l,t}$, $l \geq 1$. By the multijet Thom transversality theorem this set is residual in the space of all admissible pairs (i, φ) (see, for example, [GG]). By Lemmas 1 and 2 (and definition of the sets $W_{l,u}$) each such pair (i, φ) is (d, s)-perfect, and Theorem 4.2.1 is proved.

4.4. Proof of Theorem 4.2.3. The required spectral sequence is constructed along the same lines as the sequence from §3 that computes the cohomology of the spaces $P_d - \Sigma_k$.

Let a (d, s)-perfect pair of maps i, φ and the corresponding space $\Theta_\varphi^{d,s}$ of maps $M \to \mathbb{R}^n$ be fixed. Denote by $[\mathfrak{A}]$ the set $\Theta_\varphi^{d,s} \setminus A(M, \mathfrak{A}, \varphi)$, i.e.

§4. PROOF OF THE FIRST MAIN THEOREM

the set of maps of the form (14) with singularities of type $\mathfrak{A}$. Set $\Gamma = \dim \Theta_\varphi^{d,s}$; then by the Alexander duality theorem

$$H^l(A(M, \mathfrak{A}, \varphi) \cap \Theta_\varphi^{d,s}) \cong \overline{H}_{\Gamma-l-1}([\mathfrak{A}]). \tag{17}$$

Construct a filtered space whose closed homology coincide with the closed homology of $[\mathfrak{A}]$. Let $[\mathfrak{A}_t]$ be the subset of maps of class $\Theta_\varphi^{d,s}$ with at least t critical points of type $\mathfrak{A}$ in M_1. Construct a partial resolution of singularities of the sets $[\mathfrak{A}_t]$. Namely, let $[\widetilde{\mathfrak{A}}_t]$ be the set of pairs of the form (map $\vartheta \in [\mathfrak{A}_t]$; unordered collection of t distinct points in M_1 at which ϑ has a singularity of type $\mathfrak{A}$). Forgetting the first element of such a pair we obtain the projection $\rho_t: [\widetilde{\mathfrak{A}}_t] \to M_t$, forgetting the second we obtain a map $\pi_t: [\widetilde{\mathfrak{A}}_t] \to [\mathfrak{A}]$.

Suppose the number $\nu = \nu(d, s)$ is so large that no map of class $\Theta_\varphi^{d,s}$ has ν distinct singular points of type $\mathfrak{A}$ on M (for example, by Theorem 4.2.1 we can take ν to be $[\Gamma/(\operatorname{codim} \mathfrak{A} - m)] + 1$).

LEMMA 1. *There exists an imbedding I_ν of the manifold M into the space $\mathbb{R}^N$ of sufficiently large dimension such that the images of no ν points of M lie in one $(\nu - 2)$-dimensional affine plane in $\mathbb{R}^N$.* □

Let us fix such imbedding $I_\nu: M \to \mathbb{R}^N$. The desired filtered space is a topological subspace of the product $\mathbb{R}^N \times \Theta_\varphi^{d,s}$. Let us construct it. To each point ϑ' of the set $[\widetilde{\mathfrak{A}}_t]$ associate the $(t-1)$-dimensional open simplex in $\mathbb{R}^N \times \Theta_\varphi^{d,s}$ spanned by t points whose projections to the factor $\Theta_\varphi^{d,s}$ coincide and are equal to the point $\pi_t(\vartheta')$, and the projections to the factor $\mathbb{R}^N$ are images of the t points of M that form the collection $\rho_t(\vartheta')$. The union of such simplices over all points of the set $[\widetilde{\mathfrak{A}}_t]$ is denoted by $\{\mathfrak{A}, t\}$; it is the space of the bundle over $[\widetilde{\mathfrak{A}}_t]$ with simplex as the fiber.

Define the set $\{\mathfrak{A}\} \subset \mathbb{R}^N \times \Theta_\varphi^{d,s}$ as the union of the sets $\{\mathfrak{A}, t\}$ over all $t = 1, 2, \ldots, \nu - 1$ and induce the subspace topology on it.

This set has a natural increasing filtration: for any positive integer τ, the term $F_\tau\{\mathfrak{A}\}$ is the union of the sets $\{\mathfrak{A}, t\}$ for all $t \leq \tau$.

LEMMA 2. *The map $\pi: \{\mathfrak{A}\} \to [\mathfrak{A}]$ given by the projection $\mathbb{R}^N \times \Theta_\varphi^{d,s} \to \Theta_\varphi^{d,s}$ is proper and continuously extends to a map of one-point compactifications of these spaces sending the added point to the added point. The homomorphism on homology with closed support*

$$\overline{H}_*(\{\mathfrak{A}\}) \to \overline{H}_*([\mathfrak{A}]) \tag{18}$$

defined by the last map is an isomorphism.

This is proved similarly to Lemma 1 in §3.3. □

Consider the homology spectral sequence $\mathscr{E}_{p,q}^{r}$ converging to the group $\overline{H}_*(\{\mathfrak{A}\})$ and generated by the filtration $\{F_\tau\{\mathfrak{A}\}\}$. Transform it into a cohomology spectral sequence by renaming the term $\mathscr{E}_{p,q}^{r}$ to be $\mathscr{E}_r^{-p,\Gamma-q-1}$. By Alexander duality (17), if this sequence converges (and it does), then it converges to precisely the cohomology group of the space $A(M, \mathfrak{A}, \varphi)$. We show that the sequence satisfies the Theorem 4.2.3.

It follows immediately from definitions that the term $\mathscr{E}_1^{p,q}$ of this sequence is isomorphic to the group $\overline{H}_{\Gamma-p-q-1}(\{\mathfrak{A}, -p\})$, and it remains to prove the following statement.

LEMMA 3. *Let p, q be integers, $p \leq 0 \leq q$. Then*
(a) *if $p \geq -d/(k+1)$ then the group $\overline{H}_{\Gamma-p-q-1}(\{\mathfrak{A}, -p\})$ is given by* (12);
(b) *if $q + p(\operatorname{codim} \mathfrak{A} - m) < 0$ then $\overline{H}_{\Gamma-p-q-1}(\{\mathfrak{A}, -p\}) = 0$.*

Part (b) follows directly from dimensional considerations and the definition of a (d, s)-perfect pair of maps (i, φ). Now let $p \geq -d/(k+1)$. Set $t = -p$ and let $M_{t,s}$ be the subset of M_t consisting of collections $(u_1, \ldots, u_t)$ such that $\omega(u_j) > 2^{-s}$ for all $j = 1, \ldots, t$. By formula (14), the maps of class $\Theta_\varphi^{d,s}$ can have singularities of class $\mathfrak{A}$ only at these points u_j.

For any collection $u = (u_1, \ldots, u_t) \in M_{t,s}$ consider the affine restriction map $\Theta_\varphi^{d,s} \to J_{(u)}^k(M_1, \mathbb{R}^n)$; the interpolation theorem implies that for $t \leq d/(k+1)$ this map is surjective, and the bundle over $M_{t,s}$ with the kernel of this map as fiber over u is a smooth affine bundle whose first Stiefel-Whitney class coincides with that for the bundle $\{J_t^k(M, \mathbb{R}^n) \to M_{t,s}\}$ and therefore is given by the formula (13). Hence the set $\{\mathfrak{A}, t\}$ is diffeomorphic to the space of the bundle over $M_{t,s}$ which is a product of three bundles: the restriction to $M_{t,s}$ of the natural bundle $\rho_t \colon \Lambda_t(\mathfrak{A}, M) \to M_t$, the affine bundle just defined, and the bundle whose fiber is the $(t-1)$-dimensional open simplex with the orientation reversed after traversing paths in M_t defining odd permutations of points $u_1, \ldots, u_t$. Thus by the Thom isomorphism

$$\overline{H}_{i+\beta}(\{\mathfrak{A}, t\}) \simeq \overline{H}_i(\Lambda_t(\mathfrak{A}, M) \cap \rho^{-1}(M_{t,s}); T), \tag{19}$$

where $\beta = t - 1 + \Gamma - \dim J_{(u)}^k(M_1, \mathbb{R}^n)$, and T is the restriction of the local system with the same name defined on $\Lambda_t(\mathfrak{A}, M)$ and used in (12). But $\Lambda_t(\mathfrak{A}, M) \cap \rho^{-1}(M_{t,s})$ is homeomorphic (and even ambient diffeomorphic) to $\Lambda_t(\mathfrak{A}, M)$, and this diffeomorphism is given by a deformation retraction, so that the right-hand side of (19) coincides with the right-hand side of (12). □

Theorem 4.2.3 is proved.

4.5. Proof of Theorem 4.2.4. In the construction of the complex $\{\mathfrak{A}\}$ we used an imbedding of M in the space $\mathbb{R}^N$ satisfying the conditions of

Lemma 1 of subsection 4.4. It is obvious, however, that complexes, constructed using different imbeddings, are homeomorphic as filtered complexes. Therefore, it suffices to consider an imbedding $M \to \mathbb{R}^N$ satisfying the conditions of Lemma 1 for both numbers $\nu(d, s)$ and $\nu(d', s')$.

Let $[\mathfrak{A}]$, $[\mathfrak{A}']$ be the sets $\Theta_\varphi^{d,s} \setminus A(M, \mathfrak{A}, \varphi)$, $\Theta_\varphi^{d',s'} \setminus A(M, \mathfrak{A}, \varphi)$, and let $\{\mathfrak{A}\}$, $\{\mathfrak{A}'\}$ be the corresponding filtered complexes constructed in the proof of Theorem 4.2.3 (whose closed homology coincides with that of $[\mathfrak{A}]$ and $[\mathfrak{A}']$ respectively). We set $\tau = [d/(k+1)]$ and $D = \dim \Theta_\varphi^{d',s'} - \dim \Theta_\varphi^{d,s}$.

PROPOSITION 1. *The one-point compactification of the term $F_\tau\{\mathfrak{A}\}'$ of the natural filtration of the set $\{\mathfrak{A}\}'$ is homotopy equivalent to the D-fold suspension of the one-point compactification of the similar term $F_\tau\{\mathfrak{A}\}$.*

First, we suppose that $s' = s$.

LEMMA 1. *For any $i = 1, \ldots, \tau$, the space $F_i\{\mathfrak{A}\}' - F_{i-1}\{\mathfrak{A}\}'$ is homeomorphic to the direct product $(F_i\{\mathfrak{A}\} - F_{i-1}\{\mathfrak{A}\}) \times \mathbb{R}^D$.*

The proof is based on the fact that the space $\Theta_\varphi^{d,s}$ is an affine subspace in $\Theta_\varphi^{d',s'}$ and, by construction, $F_i\{\mathfrak{A}\} = F_i\{\mathfrak{A}\}' \cap (\mathbb{R}^N \times \Theta_\varphi^{d,s})$.

Both of the spaces $F_i\{\mathfrak{A}\}' - F_{i-1}\{\mathfrak{A}\}'$, $F_i\{\mathfrak{A}\} - F_{i-1}\{\mathfrak{A}\}$ can be considered as spaces of affine bundles over the same base, which consists of all triples of the form {a subset $\{u_1, \ldots, u_i\} \subset M_{1,s}$; a point λ of the open simplex in $\mathbb{R}^N$ spanned by the points $I_\nu(u_1), \ldots, I_\nu(u_i)$; some k-jets of maps $M \to \mathbb{R}^n$ at the points $u_1, \ldots, u_i$ belonging to the class $\mathfrak{A}$}}. The fiber of the first bundle over such a point consists of all pairs $(\lambda, f) \in \mathbb{R}^N \times \Theta_\varphi^{d',s}$, where f is a function $M \to \mathbb{R}^n$ having these k-jets at the points $u_1, \ldots, u_i$. This fiber is an affine subspace in $\mathbb{R}^N \times \Theta_\varphi^{d',s}$ and intersects the subspace $\mathbb{R}^N \times \Theta_\varphi^{d,s}$ transversally; the intersection of these two subspaces is exactly the fiber of the second bundle over the same point of the base. Thus, $F_i\{\mathfrak{A}\} - F_{i-1}\{\mathfrak{A}\}$ is the space of an affine subbundle in the first bundle. The corresponding quotient bundle is trivial, because so is the normal bundle to the subspace $\Theta_\varphi^{d,s} \subset \Theta_\varphi^{d',s}$. This implies Lemma 1.

COROLLARY. *For any $p \leq \tau$, the group $E_{p,q}^1(d, s)$ of the homological spectral sequence calculating the closed homology of the space $F_\tau\{\mathfrak{A}\}$ is isomorphic to a similar group $E_{p,q-D}^1(d', s)$ calculating closed homology of $F_\tau\{\mathfrak{A}\}'$.* □

Let us fix such a homeomorphism for any $i = 1, \ldots, \tau$, let $p_i : (F_i\{\mathfrak{A}\}' - F_{i-1}\{\mathfrak{A}\}') \to (F_i\{\mathfrak{A}\} - F_{i-1}\{\mathfrak{A}\})$ be its composition with the projection onto the first factor in this direct product.

Denote by $\overline{F}_i\{\mathfrak{A}\}$ and $\overline{F}_i\{\mathfrak{A}\}'$ the one-point compactifications of $F_i\{\mathfrak{A}\}$ and $F_i\{\mathfrak{A}\}'$; let $\overline{F}_0\{\mathfrak{A}\}$ and $\overline{F}_0\{\mathfrak{A}\}'$ be the added (compactifying) points.

DEFINITION. A *collar* of the space $F_i\{\mathfrak{A}\} - F_{i-1}\{\mathfrak{A}\}$ is its intersection with a small closed neighborhood V_i of the space $\overline{F}_{i-1}\{\mathfrak{A}\}$ in $\overline{F}_i\{\mathfrak{A}\}$, such that the pair $(V_i \setminus \overline{F}_{i-1}\{\mathfrak{A}\}, \partial V_i)$ (where ∂V_i is the boundary of V_i) is homeomorphic to the pair $\{\partial V_i \times [0, 1); \partial V_i \times \{0\}\}$, and the obvious factorization map $(\overline{F}_i\{\mathfrak{A}\}/\overline{F}_{i-1}\{\mathfrak{A}\}) \to (\overline{F}_i\{\mathfrak{A}\}/V_i)$ is a homotopy equivalence.

LEMMA 2. *For any* $i = 1, \ldots, \tau$, *the space* $F_i\{\mathfrak{A}\} - F_{i-1}\{\mathfrak{A}\}$ *admits a collar.*

Indeed, we can suppose that M is an algebraic manifold, and the imbedding $I_\nu : M \to \mathbb{R}^N$ is algebraic. Then all spaces $F_i\{\mathfrak{A}\}$ are semialgebraic: this follows from the Tarski-Saidenberg lemma. Let $\widetilde{F}_i\{\mathfrak{A}\}$ be the closure of $F_i\{\mathfrak{A}\}$ in the projective compactification $\overline{\mathbb{R}^N \times \Theta_\varphi^{d,s}}$ of the space $\mathbb{R}^N \times \Theta_\varphi^{d,s}$, and let $\pi : \widetilde{F}_i\{\mathfrak{A}\} \to \overline{F}_i\{\mathfrak{A}\}$ be the obvious factorization map. Then the pair $(\widetilde{F}_i\{\mathfrak{A}\}, \pi^{-1}(\overline{F}_{i-1}\{\mathfrak{A}\}))$ is a compact semialgebraic pair; in particular, it has a subdivision which is a Whitney stratification of $\widetilde{F}_i\{\mathfrak{A}\}$. It is easy to construct a smooth real-valued function χ defined in $\overline{\mathbb{R}^N \times \Theta_\varphi^{d,s}}$ in a neighborhood of the boundary of the space $\widetilde{F}_i\{\mathfrak{A}\} \setminus \pi^{-1}(\overline{F}_{i-1}\{\mathfrak{A}\})$, which takes the value zero on this boundary and takes on positive values in other points of this neighborhood. Then the levels $\chi^{-1}(\varepsilon)$ of this function for sufficiently small positive ε are transversal to our stratification of $\widetilde{F}_i\{\mathfrak{A}\}$, and Lemma 2 follows from Thom's isotopy lemma (see for example [AVGL$_1$]). □

Let us fix such a collar for any i and denote the complement to it in $F_i\{\mathfrak{A}\}$ by L_i. Given a closed tubular neighborhood T of the subspace $\mathbb{R}^N \times \Theta_\varphi^{d,s} \subset \mathbb{R}^N \times \Theta_\varphi^{d',s}$, denote the one-point compactification of $F_i\{\mathfrak{A}\}' \cap T$ by $\overline{F}_i(\{\mathfrak{A}\}' \cap T)$.

LEMMA 3. *There exist a small closed tubular neighborhood* T *of the subspace* $\mathbb{R}^N \times \Theta_\varphi^{d,s}$ *in* $\mathbb{R}^N \times \Theta_\varphi^{d',s}$ *and a continuous retraction* $\mu : \overline{F}_\tau(\{\mathfrak{A}\}' \cap T) \to \overline{F}_\tau\{\mathfrak{A}\}$ *such that for any* $i = 1, \ldots, \tau$,

a) $\mu(\overline{F}_i(\{\mathfrak{A}\}' \cap T) \subset \overline{F}_i\{\mathfrak{A}\}$,

b) *the restriction of* μ *to the space* $(F_i\{\mathfrak{A}\}' - F_{i-1}\{\mathfrak{A}\}') \cap \mu^{-1}(L_i)$ *is a fiber bundle over* L_i *which coincides on this space with the projection* p_i *(defined after the proof of Lemma 1) and also with the restriction on* L_i *of the fibration described in the definition of the tubular neighborhood* T *(see* [Hirsch$_3$]*)*.

PROOF. First, we construct the structure of the tubular neighborhood T. Over the points of all spaces $L_1, \ldots, L_\tau$ it is uniquely determined by condition b) of the lemma, and there is no obstacle to extending it to all of T. The map μ will be constructed by induction on the filtration $\{F_i\}$. At the ith step we define its restriction to the space $\overline{F}_i(\{\mathfrak{A}\}' \cap T)$. Before this step, this map is already defined on the following three subspaces: $\overline{F}_{i-1}(\{\mathfrak{A}\}' \cap T)$ (by the inductive hypothesis); $p_i^{-1}(L_i) \cap T$ (by condition b) of our lemma);

and $\overline{F}_i\{\mathfrak{A}\}$ (by the condition that μ is a retraction). The quotient space of $\overline{F}_i(\{\mathfrak{A}\}' \cap T)$ by the union of these three subspaces is homotopy equivalent to the quotient space of the suspension of the space $p_i^{-1}(\partial V_i) \cap T \cong \partial V_i \times B^D$ (where B^D is a closed D-dimensional disk) by the suspension of its subspace ∂V_i; in particular, it is contractible. Thus, there exists a retraction of $\overline{F}_i(\{\mathfrak{A}\}' \cap T)$ onto this union. The desired extension of μ on $\overline{F}_i(\{\mathfrak{A}\}' \cap T)$ is defined as the composition of this retraction and the map defined earlier. □

The normal bundle of an affine subspace is trivial; therefore, there exists a diffeomorphism $T \to (\mathbb{R}^N \times \Theta_\varphi^{d,s}) \times B^D$ whose composition with the projection onto the first factor coincides with the canonical projection of the tubular neighborhood. Let h be the composition of this diffeomorphism and the projection onto the second factor B^D. Consider the map $(\mu, h) : (F_\tau\{\mathfrak{A}\}' \cap T) \to \overline{F}_\tau\{\mathfrak{A}\} \times B^D$ and the map $F_\tau\{\mathfrak{A}\}' \cap T \to \Sigma^D \overline{F}_\tau\{D\}$ defined as the composition of this map (μ, h) with the factorization of the product $\overline{F}_\tau\{\mathfrak{A}\} \times B^D$ by the union of spaces (the compactifying point in $\overline{F}_\tau\{\mathfrak{A}\}) \times B^D$ and $\overline{F}_\tau\{\mathfrak{A}\} \times \partial B^D$. This composition map can be extended by continuity to a map $\overline{(\mu, h)} : \overline{F}_\tau(\{\mathfrak{A}\}' \cap T) \to \Sigma^D \overline{F}_\tau\{\mathfrak{A}\}$.

Consider also the map $\overline{F}_\tau\{\mathfrak{A}\}' \to \Sigma^D \overline{F}_\tau\{\mathfrak{A}\}$ defined as the composition of the obvious factorization $\overline{F}_\tau\{\mathfrak{A}\}' \to \overline{F}_\tau(\{\mathfrak{A}\}' \cap T)$ and this map $\overline{(\mu, h)}$.

LEMMA 4. *This composition map is a homotopy equivalence.*

PROOF. The spaces $\overline{F}_\tau\{\mathfrak{A}\}'$, $\Sigma^D \overline{F}_\tau\{\mathfrak{A}\}$ are filtered, respectively, by the spaces $\overline{F}_i\{\mathfrak{A}\}'$ and $\Sigma^D \overline{F}_i\{\mathfrak{A}\}$. The terms $E^1_{p,q}$ of the corresponding spectral sequences are naturally isomorphic to the group $H_{p+q-D}(\overline{F}_p\{\mathfrak{A}\}, \overline{F}_{p-1}\{\mathfrak{A}\})$, in particular they are isomorphic to each other. Our composition map respects these filtrations. It follows from the choice of the spaces L_i and the map μ that the corresponding morphism of our sequences is isomorphic in the term E^1; hence, our map is a homology equivalence. But both spaces are simply-connected (and even D-connected); thus, the lemma follows from the Whitehead theorem.

This lemma finishes the proof of Proposition 1 in the case when $s' = s$.

Finally, suppose that $s' > s$. Consider in the space $F_\tau\{\mathfrak{A}\}' \subset \mathbb{R}^N \times \Theta_\varphi^{d',s'}$ the subspace $F_\tau^{(s)}\{\mathfrak{A}\}$ which consists only of the pairs of the form (a function of the class $\Theta_\varphi^{d',s'}$, an interior point in a simplex in $\mathbb{R}^N$) that all vertices of this simplex in $\mathbb{R}^N$ are the images under the map I_ν of some points from the manifold $M_{1,s}$. The previous considerations prove that the one-point compactification of $F_\tau^{(s)}\{\mathfrak{A}\}'$ is homotopy equivalent to the D-fold suspension of the one-point compactification of $F_\tau\{\mathfrak{A}\}$. But this space $F_\tau^{(s)}\{\mathfrak{A}\}'$ is obviously homeomorphic to $F_\tau\{\mathfrak{A}\}'$: this homeomorphism is lifted from any standard retraction of the space $M_{1,s'}$ onto $M_{1,s}$. Proposition 1 is completely proved.

REMARK. In fact, a stronger assertion holds: given the hypotheses of Proposition 1, the space $F_\tau\{\mathfrak{A}\}'$ is homeomorphic to the direct product $F_\tau\{\mathfrak{A}\} \times \mathbb{R}^D$. We present here only the statement of Proposition 1 because it can (and will) be generalized immediately (together with the proof) to more complicated situations in which such homeomorphisms do not exist.

This proposition implies that the terms $\mathscr{E}_1^{p,q}$ of the considered spectral sequences are isomorphic for $p \geq -d/(k+1)$, and these isomorphisms commute with all subsequent (for $r > 1$) differentials of the terms $\mathscr{E}_r^{p,q}$ with these p.

By the definition of a (d,s)-perfect pair (see Figures 20 and 22), the minimal total dimension $p+q$ of those terms $\mathscr{E}_1^{p,q}$ where these spectral sequences can differ is $([d/(k+1)]+1)(\operatorname{codim}\mathfrak{A} - m - 1)$, and the only suspect term $\mathscr{E}_1^{p,q}$ of this dimension corresponds to $p = -[d/(k+1)] - 1$, $q = (\operatorname{codim}\mathfrak{A} - m)p$, and no nontrivial maps of the terms $\mathscr{E}_r^{p',q'}$, $p'+q' = p+q-1$, have this term as a target. Therefore, for

$$p + q \leq ([d/(k+1)] + 1)(\operatorname{codim}\mathfrak{A} - m - 1) - 1 \qquad (20)$$

these spectral sequences are isomorphic. In particular, the $(p+q)$-dimensional cohomology groups of the spaces (15) are isomorphic. It remains to prove that this isomorphism is induced by the imbedding of the first of the spaces (15) into the second.

The Alexander isomorphism (17) is given by linking indices. Let γ be an arbitrary cycle in $A(M, \mathfrak{A}, \varphi) \cap \Theta_\varphi^{d,s}$ whose dimension is smaller than the right-hand side of (20). Consider a chain in general position with respect to the subset $[\mathfrak{A}]$ in $\Theta_\varphi^{d,s}$ with boundary γ; in particular, such a chain does not intersect the set $[\mathfrak{A}_{\tau+1}]$, where $\tau = [d/(k+1)]$, whose codimension equals $([d/(k+1)] + 1)(\operatorname{codim}\mathfrak{A} - m)$.

This chain can be considered as a chain in $\Theta_\varphi^{d,s}$ as well as in $\Theta_\varphi^{d',s'}$, and we can consider its intersection index with cycles in $\overline{H}_*([\mathfrak{A}])$ or $\overline{H}_*([\mathfrak{A}'])$, respectively. The intersection indices, with cycles corresponding to each other by the isomorphism of spectral sequences, coincide. This proves that the obvious imbedding of spaces (15) defines an isomorphism of the groups $\operatorname{Hom}(H_*(\cdot, \mathbb{Z}), \mathbb{Z})$ of these spaces. But all our constructions are preserved under the coefficient homomorphism $\mathbb{Z} \to \mathbb{Z}/q\mathbb{Z}$ for any q. Thus, by the universal coefficient formula the map in cohomology is also an isomorphism; Theorem 4.2.4 is proved.

4.6. Proof of Theorem 4.2.2. An arbitrary singular chain γ in $A(M, \mathfrak{A}, \varphi)$ can be viewed as a continuous map into $A(M, \mathfrak{A}, \varphi)$ of a compact polyhedron $\tilde{\gamma}$ regularly (but, in general, nonlinearly) imbedded in the Euclidean space $\mathbb{R}^T$ of suitable dimension; if γ is a cycle, then the polyhedron $\tilde{\gamma}$ can be chosen to be without boundary. The continuous map $\chi: \tilde{\gamma} \to A(M, \mathfrak{A}, \varphi)$

§4. PROOF OF THE FIRST MAIN THEOREM

defining the cycle γ is a map $X \colon \tilde{\gamma} \times M \to \mathbb{R}^n$, and there is a natural number s such that for all $z \in \tilde{\gamma}$ the corresponding map $X(z, \cdot) \colon M \to \mathbb{R}^n$ coincides with φ in the domain where $\omega < 2^{-s+1}$ (see 4.2).

Using the Weierstrass theorem on approximation of functions by polynomials in the C^k-topology (see [RF]), we can construct a polynomial $\Pi \colon \mathbb{R}^T \times \mathbb{R}^L \to \mathbb{R}^n$ such that for any $z \in \tilde{\gamma} \subset \mathbb{R}^T$ the map $X(z, \cdot) \colon M \to \mathbb{R}^n$ is approximated (uniformly with respect to z) by a map

$$(\Pi(z, \cdot) \circ i) \cdot \varkappa(2^s \omega) + \varphi \qquad (21)$$

with any prescribed accuracy. This map belongs to the space $\Theta_\varphi^{d,s}$, where $d = \deg \Pi$. Since the prohibited set $\mathfrak{A}(M) \subset J^k(M, \mathbb{R}^n)$ is closed, for a sufficiently close approximation Π the cycle $\tilde{\gamma} \subset \Theta_\varphi^{d,s}$ formed by maps (21) is homologous in $A(M, \mathfrak{A}, \varphi)$ to the initial cycle γ.

Let us now assume that the pair of maps $(i, \varphi) \colon M \to \mathbb{R}^L \times \mathbb{R}^n$ is (d, s)-perfect for all d, s, and the cycle γ is homologous to zero in $A(M, \mathfrak{A}, \varphi)$. Applying the arguments of the previous paragraph to chains spanned by γ, we deduce that there are numbers $d' \geq d$ and $s' \geq s$ such that there exists a chain in $A(M, \mathfrak{A}, \varphi) \cap \Theta_\varphi^{d', s'}$ spanned by γ'. On the other hand, if $\dim \gamma \leq ([d/(k+1)] + 1)(\operatorname{codim} \mathfrak{A} - m - 1) - 2$ then by Theorem 2.4.4 the $(\dim \gamma)$-dimensional homology groups of the spaces $A(M, \mathfrak{A}, \varphi) \cap \Theta_\varphi^{d,s}$ and $A(M, \mathfrak{A}, \varphi) \cap \Theta_\varphi^{d', s'}$ are isomorphic, and γ' spans a chain in $A(M, \mathfrak{A}, \varphi) \cap \Theta_\varphi^{d,s}$. $\square$

4.7. Proof of Theorem 2 of 4.1. The cohomology of the space $B(M, \mathfrak{A}, \varphi)$ is computed using a spectral sequence similar to the previous one. To construct it we fix linear coordinate systems in $\mathbb{R}^L$ and $\mathbb{R}^n$ and consider the trivial bundle $J^k(\mathbb{R}^L, \mathbb{R}^n) \to \mathbb{R}^L$. Denote by $\Theta^d(k)$ the space of all polynomial sections of this bundle of degree $\leq d$. (For example, the space Θ^d used in 4.2 is naturally imbedded in $\Theta^d(k)$: this imbedding is given by k-jet extensions.)

For any smooth section ξ of this bundle, $\xi \colon \mathbb{R}^L \to J^k(\mathbb{R}^L, \mathbb{R}^n)$, and any imbedding $i \colon M \to \mathbb{R}^L$ we define the restriction $i[\xi]$ of the section ξ to M which is a section of the bundle $J^k(M, \mathbb{R}^n) \to M$. Namely, we choose arbitrary local coordinates in M near $a \in M$. Then for any map $F \colon \mathbb{R}^L \to \mathbb{R}^n$ the coordinate representation of the k-jet of the map $F \circ i$ at the point a can be expressed in terms of partial derivatives of order $\leq k$ of the map i at a and the map F at $i(a)$. The value of $i[\xi]$ at the point a is obtained by replacing partial derivatives of F in this expression by the corresponding components of the section ξ.

LEMMA. *The operation $i[\cdot]$ is invariant with respect to the choice of local coordinates in M.*

This is obvious in the case when this operation is applied to an integrable section ξ (i.e. to a k-jet extension of $\mathbb{R}^L \to \mathbb{R}^n$). But for any section ξ and any point $a \in M$ we can choose a map $\mathbb{R}^L \to \mathbb{R}^n$ whose k-jet at $i(a)$ is equal to $\xi(i(a))$, and it remains to remark that the changes of local coordinates act locally on the coordinate representations of sections $M \to J^k(M, \mathbb{R}^n)$ (i.e. for any points $a \in M$, $\beta \in J^k_{(a)}(M, \mathbb{R}^n)$ the expressions for β in different coordinate systems in M are related to each other by formulas that depend explicitly only on the partial derivatives at a of expressions relating these coordinate systems to each other). □

Similarly we define the formal "multiplication" of sections of the bundle $J^k(M, \mathbb{R}^n)$ by smooth functions. Namely, for any map $f: M \to \mathbb{R}^n$ and any function $\lambda: M \to \mathbb{R}^1$, the k-jet of the map $f \cdot \lambda$ in local coordinates is expressed in terms of k-jets of f and λ. Replacing the partial derivatives of f by the corresponding components of the section $\eta: M \to J^k(M, \mathbb{R}^n)$ in this expression we obtain a new section denoted by $\eta \boxtimes \lambda$.

Fix an imbedding $i: M \to \mathbb{R}^L$. Define the space $\Theta^{d,s}_\varphi(k) \subset B(M, \mathfrak{A}, \varphi)$ as the set of sections $M \to J^k(M, \mathbb{R}^n)$ of the form

$$i[\Phi_1] \boxtimes (\varkappa \circ \omega) + i[\Phi_2] \boxtimes (\varkappa \circ 2\omega) + \cdots + i[\Phi_k] \boxtimes (\varkappa \circ 2^s \omega) + j^k \varphi,$$

where $\Phi_j \in \Theta^d(k)$ for all j, and ω is the same as in 4.2.

The operation of k-jet extension defines the imbedding

$$\Theta^{d,s}_\varphi \to \Theta^{d,s}_\varphi(k). \tag{22}$$

The imbedding $i: M \to \mathbb{R}^L$ is called (d, s, k)-perfect if for any $t \geq 1$ the subset in $\Theta^{d,s}_\varphi(k)$ consisting of the sections intersecting the set $\mathfrak{A}(M) \subset J^k(M, \mathbb{R}^n)$ at least t times has codimension at least $t(\operatorname{codim} \mathfrak{A} - m)$.

The construction given below of the spectral sequence promised in Theorem 2 of 4.1 repeats the steps used in 4.2–4.6: we construct the sequences computing the cohomology of the spaces $B(M, \mathfrak{A}, \varphi) \cap \Theta^{d,s}_\varphi(k)$ and prove their stabilization, as $d, s \to \infty$, to the cohomology of the space $B(M, \mathfrak{A}, \varphi)$ in the case of generic imbedding $i: M \to \mathbb{R}^L$. Here the stabilization of the spectral sequence with respect to d is much faster than before: in the analogues of Theorems 4.2.2–4.2.4 we can take $p \geq -d$ instead of $p \geq -d/(k+1)$.

The imbeddings of the form (22) together with the obvious imbeddings of spaces (15) and the analogous imbeddings

$$B(M, \mathfrak{A}, \varphi) \cap \Theta^{d,s}_\varphi(k) \to B(M, \mathfrak{A}, \varphi) \cap \Theta^{d',s'}_\varphi(k)$$

for any $d' \geq d$, $s' \geq s$, and j define a commutative diagram

$$\begin{array}{ccc} H^j(B(M,\mathfrak{A},\varphi) \cap \Theta_\varphi^{d',s'}(k)) & \longrightarrow & H^j(B(M,\mathfrak{A},\varphi) \cap \Theta_\varphi^{d,s}(k)) \\ \downarrow & & \downarrow \\ H^j(A(M,\mathfrak{A},\varphi) \cap \Theta_\varphi^{d',s'}) & \longrightarrow & H^j(A(M,\mathfrak{A},\varphi) \cap \Theta_\varphi^{d,s}). \end{array} \quad (23)$$

THEOREM. *Let a pair of maps $i: M \to \mathbb{R}^L$, $\varphi: M \to \mathbb{R}^n$ defining the spaces $\Theta_\varphi^{d,s}$, $\Theta_\varphi^{d,s}(k)$ be (d,s)-perfect and (d,s,k)-perfect for all d,s. Then for any $j \leq ([d/(k+1)]+1) \cdot (\operatorname{codim}\mathfrak{A} - m - 1) - 2$ all arrows of diagram (23) are isomorphisms.*

THEOREM. *If the pair (i,φ) is (d,s,k)-perfect for all d,s then the natural map $H^j(B(M,\mathfrak{A},\varphi)) \to H^j(B(M,\mathfrak{A},\varphi) \cap \Theta_\varphi^{d,s}(k))$ is an isomorphism for $j \leq (d+1)(\operatorname{codim}\mathfrak{A} - m - 1) - 2$.*

These two theorems are proved in exactly the same way as Theorems 4.2.4 and 4.2.2. Theorem 2 from 4.1 is their direct consequence; thus we have completed the proof of the first and second main theorems of §1.

4.8. On the proof of the homological part of Theorem 3' of §3. By Theorem 3 of §3 it suffices to consider the case when d is even and $I_d: P_d \to \mathscr{F}$ is the imbedding described after the statement of Theorem 3'. By Proposition 2 of 3.2 we can replace the space P_d by a compact ball V centered at $\{x^d\} \subset P_d$. Then there is a closed interval $[-T,T]$ such that for all polynomials f from the ball V the functions $I_d(f)$ are identically 1 outside this interval, hence the ball V is imbedded in the space of smooth functions on $[-T-1, T+1]$ equal to 1 outside $[-T,T]$. It is easy to see that in the class of maps that respect all sets Σ_k this imbedding is homotopic to an imbedding whose image belongs to a suitable subspace $\Theta_1^{d,s}$ of this space. By construction, the spectral sequences in 3.3 and 4.4 for computing the cohomology of the spaces $V - \Sigma_k \sim P_d - \Sigma_k$ and $\Theta_1^{d,s} - \Sigma_k$, respectively, are isomorphic in necessary dimensions. The fact that the corresponding isomorphism in cohomology is induced by the imbedding is obvious (and is proved similarly to the last assertion of Theorem 4.2.4; see the end of 4.5). □

§5. Cohomology of spaces of maps from m-dimensional spaces to m-connected spaces

Let X and Y be finite simplicial polyhedra, with X m-dimensional and Y m-connected. Denote by Y^X the space of continuous maps $X \to Y$. In this section we construct a spectral sequence converging to the cohomology of the space Y^X whose first term E_1 can be expressed in terms of the cohomology of Y and the configuration spaces $X(t)$. Its components $E_1^{p,q}$ are finitely generated and vanish unless $p \leq 0$, $q \geq -2p$ (and unless

$q \geq -p(2+k-m)$ if Y is not only m-connected, but also k-connected for $k \geq m$).

To illustrate our method we will prove several known results about the homology of loop spaces of spheres. Let $X = S^m$, $Y = S^n$, $m < n$. Then the term $E_1^{p,q}$ of our spectral sequence $E_r^{p,q} \to H^*(\Omega^m S^n)$ is equal to

$$H^{q+p(n+1-m)}(\mathbb{R}^m(-p), (\pm\mathbb{Z})^{\otimes(n-m)}),$$

where the sheaf $\pm\mathbb{Z}$ was defined in Chapter I. This spectral sequence degenerates at the term E_1 for odd n (and also for even n in the case of cohomology with coefficients in $\mathbb{Z}_2$). Hence

$$H^i(\Omega^m S^n) \cong \bigoplus_{t=0}^{\infty} H^{i-t(n-m)}(\mathbb{R}^m(t), (\pm\mathbb{Z})^{\otimes(n-m)})$$

for odd n, and, in particular,

$$H^i(\Omega^\infty S^{\infty+j}) = \bigoplus_{t=0}^{\infty} H^{i-tj}(S(t), (\pm\mathbb{Z})^{\otimes j}), \qquad (24)$$

cf. [Mil$_1$], [May$_2$], [CMT$_1$], [Snaith].

It seems that in our constructions we can replace polyhedra by CW-complexes and weaken the finiteness conditions.

5.1. Description of the term $E_1^{p,q}$. Let X, Y be simplicial polyhedra as above, and $X(t)$ be the space of unordered collections of t pairwise distinct points in X.

LEMMA 1. *There exist a regular C^∞-imbedding of Y into a sphere S^N of sufficiently high dimension and a Whitney stratified subset Λ of codimension $m+2$ in $S^N - Y$, such that Y is a deformation retract of $S^N - \Lambda$.*

Proof will be given in 5.3 below.

The homology of the spaces Λ and Y are related by the Alexander duality. Let $C\Lambda$ be the open cone on Λ, i.e. the quotient space $(\Lambda \times [0, \infty))/(\Lambda \times 0)$. Consider the trivial bundle $C\Lambda \times X \to X$ over X and a locally trivial bundle $\Pi_t \to X(t)$ over $X(t)$ whose fiber over the collection $(x_1, \ldots, x_t) \in X(t)$, $x_j \in X$, is the product of fibers of this trivial bundle $C\Lambda \times X \to X$ over the points $x_1, \ldots, x_t$. Let $\pm\mathbb{Z}$ be the system of groups on Π_t locally isomorphic to $\mathbb{Z}$ and changing orientation after traversing paths whose projection to $X(t)$ defines an odd permutation of the points $x_1, \ldots, x_t \in X$.

THEOREM 1. *There exists a spectral sequence $E_r^{p,q}$, $p \leq 0 \leq q$, such that for any i the group $\bigoplus_{p+q=i} E_\infty^{p,q}$ is associated to the group $H_i(Y^X)$, and*

$$E_1^{p,q} = \overline{H}_{-p(N+1)-q}(\Pi_{-p}, (\pm\mathbb{Z})^{\otimes N}); \qquad (25)$$

in particular, $E_r^{p,q} = 0$ for $q < -2p$.

Now we construct this sequence.

5.2. Consider S^N as the standard sphere in $\mathbb{R}^{N+1}$, and the cone $C\Lambda$ as the union of rays intersecting this sphere at points of the set Λ. Then the space Y^X is homotopy equivalent to the space $(\mathbb{R}^{N+1} - C\Lambda)^X$ that is obtained from the contractible space $(\mathbb{R}^{N+1})^X$ by deleting the set of codimension ≥ 2 consisting of the maps $X \to \mathbb{R}^{N+1}$ intersecting $C\Lambda$. This set is naturally stratified according to the number of intersections of each map with $C\Lambda$. The following construction of the spectral sequence in fact repeats the construction used to compute the cohomology of the spaces $P_d - \Sigma_k$, $A(M, \mathfrak{A}, \varphi)$, $B(M, \mathfrak{A}, \varphi)$; nevertheless we recall its main steps.

a) Finite-dimensional approximations of the space $(\mathbb{R}^{N+1})^X$: the polyhedron X (or its appropriate subdivision) is C^∞-imbedded into the space $\mathbb{R}^L$, a map $\varphi: X \to \mathbb{R}^{N+1} - C\Lambda$ is fixed, and we consider the spaces Θ_φ^d of maps $X \to \mathbb{R}^{N+1}$ of the form $\varphi + p$, where p is the restriction to X of a polynomial map $\mathbb{R}^L \to \mathbb{R}^{N+1}$ of degree $\leq d$. If the imbedding $X \to \mathbb{R}^L$ and the map φ are generic then for any d and any t the set of maps $f \in \Theta_\varphi^d$ intersecting $C\Lambda$ at least t times has codimension at least $t(N + \dim \Lambda - m)$ in Θ_φ^d.

b) We construct resolutions of these sets as in 4.4 and use Alexander duality in Θ_φ^d and the natural filtration of the resolution space to construct a spectral sequence converging to the cohomology of the set of those maps of class Θ_φ^d that do not intersect $C\Lambda$.

c) As d increases these sequences stabilize to a spectral sequence $E_r^{p,q} \equiv E_r^{p,q}(X, Y) \to H^*(Y^X)$; by the interpolation theorem, its terms $E_1^{p,q}$ are given by (25). In particular, the sheaf of coefficients $(\pm \mathbb{Z})^{\otimes N}$ (or better the isomorphic sheaf $(\pm \mathbb{Z})^{(N+1)} \otimes (\pm \mathbb{Z})$) appears for the following reason. Consider a bundle on $X(t)$ whose fiber over the collection $x_1, \ldots, x_t \in X$ is the space of maps $X \to \mathbb{R}^{N+1}$ sending these points to 0. This bundle changes the orientation exactly when the sheaf $(\pm \mathbb{Z})^{\otimes(N+1)}$ does. The $(N+2)$th factor $\pm \mathbb{Z}$ arises from the bundle of $(t-1)$-dimensional simplices used to construct the resolution.

5.3. Proof of Lemma 1. Let $\dim Y = n$ and $N > 2n$. Replacing if necessary the simplicial polyhedron Y by its barycentric subdivision, imbed it linearly into $\mathbb{R}^N$ (and hence into S^N). For each of its maximal faces (i.e. faces that do not lie on the boundary of other faces) it is easy to construct a C^2 function on S^N equal to 0 on this face, positive outside, and bounded from below (in some neighborhood) by the third power of the distance to this face. Let χ be the product of such functions over all maximal faces. For sufficiently small ε the set $\chi^{-1}([\varepsilon, +\infty])$ is a C^2 manifold (with boundary) homeomorphic to $S^N - Y$. By the Alexander duality theorem its cohomology groups H^i are trivial for $i > N - m - 2$. Moreover, increasing N if necessary, we can assume that it is simply connected. By Smale's theorem

[Smale$_1$], [Milnor$_2$], we can choose the function χ away from the boundary of this manifold in such a way that it has only Morse singular points of index $\leq N-m-2$. Choose a gradientlike field in general position for this function; the attractor of this field satisfies the requirements of the lemma. □

5.4. Spaces of maps with restrictions on subsets. Let X, Y be as before, K be a subpolyhedron in X, and we fix a map $\psi: X \to Y$. Let (Y^X, ψ) be the set of continuous maps coinciding with ψ on K. (The most popular example: X and Y are pointed spaces, and maps send base point to base point.)

Denote by (Π_t, K) the subset of the space of the bundle $\Pi_t \to X(t)$ (see 5.1) defined as the preimage of the set $(X - K)(t)$ (consisting of collections of t points of the set $X - K$).

THEOREM 2. *There exists a spectral sequence $E_r^{p,q}$ converging to $H^*(Y^X, \psi)$ such that*

$$E_1^{p,q} = \overline{H}_{-p(N+1)-q}((\Pi_{-p}, K), (\pm \mathbb{Z})^{\otimes N}). \tag{26}$$

Proof repeats 5.2 with only minor variations as follows. The imbedding $i: X \to \mathbb{R}^L$ must send the subcomplex K (and only this subcomplex) to a plane in $\mathbb{R}^L$, and all closed cells of the polyhedron X not in K must be transversal to that plane. The approximating spaces $\widetilde{\Theta}_\varphi^d$ consist of maps $p_\varphi: X \to \mathbb{R}^{N+1}$ of the form $\varphi + p \circ i$, where p is a polynomial of degree $\leq d$ equal to 0 on the distinguished plane in $\mathbb{R}^L$, and φ is a fixed map $X \to \mathbb{R}^{N+1}$ coinciding with ψ on K. Similarly to 4.3 we prove that, for any ψ and generic ("perfect") i and φ, the set of maps $p_\varphi \in \widetilde{\Theta}_\varphi^d$ for which $p_\varphi(X)$ intersects $C\Lambda$ at least t times has codimension at least $t(N - \dim \Lambda - m)$ in $\widetilde{\Theta}_\varphi^d$. The rest of the proof repeats the arguments from the corresponding part of 5.2.

5.5. Example: loop spaces of spheres. Let $X = S^m$, $Y = S^n$, $m < n$, and $\Omega^m S^n$ be the space $\Omega^m S^n$ of maps $(S^m, *) \to (S^n, *)$. In this case $(X - K)(t)$ is the configuration space $\mathbb{R}^m(t)$ (see Chapter I), and formula (26) leads to the following result.

THEOREM. *For any $n > m \geq 1$ there is a spectral sequence $E_r^{p,q} \to H^*(\Omega^m S^n)$ such that $E_1^{p,q} = 0$ for $p \geq 0$ and*

$$E_1^{p,q} = H^{q+p(n+1-m)}(\mathbb{R}^m(-p), (\pm \mathbb{Z})^{\otimes(n-m)}) \tag{27}$$

for $p < 0$.

Indeed, in this case we can take $N = n$ in formula (26). Then Λ is the empty subset of S^N, $C\Lambda$ is the origin in $\mathbb{R}^{N+1}$, and $(\Pi_t, K) \sim \mathbb{R}^m(t)$. Applying formula (26) and Poincaré duality in $\mathbb{R}^m(t)$ we get

$$\begin{aligned} E_1^{-t,q} &= \overline{H}_{t(n+1)-q}(\mathbb{R}^m(t), (\pm \mathbb{Z})^{\otimes n}) \\ &= H^{q-t(n+1-m)}(\mathbb{R}^m(t), (\pm \mathbb{Z})^{\otimes(n-m)}) \quad \square \end{aligned} \tag{28}$$

In Chapter IV we will meet this formula again: if $m = 2k$, $n = 2k+1$, then this spectral sequence converges to the stable cohomology of complements of discriminants of singularities of holomorphic functions $(\mathbb{C}^k, 0) \to (\mathbb{C}, 0)$.

5.6. Degeneration of the spectral sequence for $\Omega^m S^n$.

5.6.1. THEOREM. *Let $m < n$. Then the formula*

$$H^i(\Omega^m S^n) \cong \bigoplus_{t=0}^{\infty} H^{i-t(n-m)}(\mathbb{R}^m(t), (\pm\mathbb{Z})^{\otimes(n-m)}) \tag{29}$$

holds in each of the following three cases:
(1) for odd n and any i;
(2) for any n and $i < 3n - 2m - 2$;
(3) for any n, i if on both sides of this formula we consider only cohomology with coefficients in $\mathbb{Z}_2$.

In all these cases the spectral sequence $E_1^{p,q} \to H^{p+q}(\Omega^m S^n)$ of 5.5 degenerates at the term E_1 so that $E_1^{p,q} = E_\infty^{p,q}$ for any p, q in cases (1) and (3) and for $p+q < 3n-2m-2$ in case (2).

5.6.2. COROLLARY. *Let $i < 2n - m - 1$, $m < n$; then*

$$H^i(\Omega^m S^n) = \bigoplus_{t=0}^{\infty} H^{i-t(n-m)}(S(t), (\pm\mathbb{Z})^{\otimes(n-m)}), \tag{30}$$
$$H^i(\Omega^m S^n) = H^i(\Omega^{m+1} S^{n+1}).$$

The second formula of this corollary follows directly from the first, and the first follows from part (2) of Theorem 5.6.1 and the fact that the map $H^l(S(t)) \to H^l(\mathbb{R}^m(t))$ of the cohomology with arbitrary coefficients given by the inclusion $\mathbb{R}^m \to \mathbb{R}^\infty$ is an isomorphism for $l \leq m-2$ and any t, and for $t = 0, 1$ and any l.

5.6.3. *Sketch of proof of Theorem 5.6.1.* First we prove the theorem in case (2), which implies that the stable spectral sequence converging to $H^*(\Omega^\infty S^{\infty+j})$ degenerates at the term E_1. Then we show that for odd n and any N (or for any n in the case of $\mathbb{Z}_2$-coefficients) the Freudenthal inclusion $\Omega^m S^n \to \Omega^{m+N} S^{n+N}$ defines an epimorphism of the corresponding cohomology spectral sequences. But by the assertion in case (2), for any d there is an N such that this sequence for the last space degenerates at all terms $E_1^{p,q}$ with $p+q \leq d$. Hence the sequence for $\Omega^m S^n$ also degenerates.

It will be convenient to prove part (2) of Theorem 5.6.1 in the following dual formulation.

LEMMA 1. *For $i < 3n - 2m - 2$, $m < n$,*

$$H_i(\Omega^m S^n) = \bigoplus_{t=0}^{\infty} H_{i-t(n-m)}(\mathbb{R}^m(t), (\pm\mathbb{Z})^{\otimes(n-m)}). \tag{31}$$

Part (2) of Theorem 5.6.1 follows from this lemma because of the relation between homology and cohomology of the same space.

PROOF OF LEMMA 1. Consider the homology spectral sequence $E^r_{p,q} \to H_*(\Omega^m S^n)$ dual to the sequence (27); by construction, $E^r_{p,q} = 0$ for $p \geq 0$ and $q < 0$, and

$$E^1_{p,q} = H_{q+p(n+1-m)}(\mathbb{R}^m(t), (\pm\mathbb{Z})^{\otimes(n-m)}). \tag{32}$$

Let $\mathfrak{A}$ be the set of maps $(S^m, *) \to (\mathbb{R}^{n+1}, *)$ passing through $0 \neq * \in \mathbb{R}^{n+1}$ that is used in the construction ([5]) of this sequence. Let $\{\mathfrak{A}\}$ be a resolution of this set similar to the one described in 4.4, and $F_1 \subset F_2 \subset \cdots$ be its natural filtration.

The set $\{\mathfrak{A}, t\} = F_t - F_{t-1}$ is a locally trivial bundle whose fiber is a $(t-1)$-dimensional simplex and whose base is the set of pairs (map $(S^m, *) \to (\mathbb{R}^{n+1}, *)$ passing through 0 at least t times; some t points in S^m sent to 0 under this map).

Forgetting the first element of such a pair defines a map

$$\{\mathfrak{A}, t\} \to \mathbb{R}^m(t), \tag{33}$$

which is a locally trivial bundle with contractible $(\omega - tn - 1)$-dimensional fiber. The orientation of the manifold $\{\mathfrak{A}, t\}$ is reversed over the paths whose projections to $\mathbb{R}^m(t)$ flip the system $(\pm\mathbb{Z})^{\otimes(n-m)}$; this degree $n - m$ is the sum of the numbers $-m$ (orientation of $\mathbb{R}^m(t)$), -1 (orientation of the bundle of $(t-1)$-dimensional simplices), and $n+1$ (orientation of the bundle of pairs of the form (map, collection of points in $\mathbb{R}^m$)).

To each element of each of the groups (32), $p = -t$, corresponds an element of the isomorphic group

$$H_{q-t(n+1-m)}(\{\mathfrak{A}, t\}, (\pm\mathbb{Z})^{\otimes(n-m)}) \cong \overline{H}^\sigma(\{\mathfrak{A}, t\}), \tag{34}$$

where $\sigma = \dim\{\mathfrak{A}, t\} - (q - t(n+1-m))$, and $\overline{H}^*(\cdot)$ is the cohomology group of the one-point compactification modulo the compactifying point. We must prove that for $q - t < 3n - 2m - 2$ all maps of such groups to $\overline{H}^*(F_t\{\mathfrak{A}\})$ given by the factorization {compactification of F_t}/(point) → {compactification of F_t}/{compactification of F_{t-1}} ≡ {compactification of $\{\mathfrak{A}, t\}$}/(point) are monomorphisms. This means that no element γ of the groups (34) with $q - t \leq 3n - 2m - 2$ is killed at any stage of the spectral sequence by an element that comes from a similar group $E_{-t-r, q+r+1}$. By the construction of the

---

([5]) I will take the liberty not to present precise constructions of finite-dimensional approximations, etc.: we work with the space of maps $S^m \to \mathbb{R}^n$ as a linear space of very high but finite dimension ω. For example, each of the sets $F_t - F_{t-1}$ introduced below will be considered as a smooth manifold of dimension $\omega - t(n-m) - 1$. To make the arguments precise one has to apply them to sufficiently good finite-dimensional (ω-dimensional) approximations of the space of maps and then justify the passage to the limit; the latter step is performed as in §4 of this chapter.

filtered space $\{\mathfrak{A}\}$ this is guaranteed if $\gamma \in H_{q-t(n+1-m)}(\{\mathfrak{A}, t\}, (\pm \mathbb{Z})^{\otimes(n-m)})$ can be realized by a cycle consisting of maps $S^m \to \mathbb{R}^{n+1}$ passing through 0 exactly t times. Such a realization for $q - t < 3n - 2m - 2$ will be constructed in Lemmas 2, 3 below, thus completing the proof of Lemma 1 and statement (2) of Theorem 5.6.1.

For any $t \leq m+1$ denote by $\nu(m, t)$ the subset in $\mathbb{R}^m(t)$ consisting of collections of t points in $\mathbb{R}^m$ that do not lie in the same $(t-2)$-dimensional affine plane.

LEMMA 2. *The homomorphism*

$$H_i(\nu(m, t), (\pm \mathbb{Z})^{\otimes(n-m)}) \to H_i(\mathbb{R}^m(t), (\pm \mathbb{Z})^{\otimes(n-m)}),$$

given by inclusion is an isomorphism for $t = 1, 2$ and any i, and for any t and $i \leq m - t$.

Indeed, the set $\nu(m, t)$ is obtained from the manifold $\mathbb{R}^m(t)$ by removing a set of codimension $m + 2 - t$ which is empty if $t = 1$ or 2.

Lemma 2 implies that any element of the group

$$E^1_{-t, q} = H_{q-t(n+1-m)}(\mathbb{R}^m(t), (\pm \mathbb{Z})^{\otimes(n-m)})$$

with $-t + q < 3n - 2m - 2$ can be realized by a cycle lying in the set $\nu(m, t)$. Indeed, the pair (t, q) for which this is not true satisfies the conditions $-t \leq 3$, $q - t(n - m) > m$ by Lemma 2; moreover, the number $q - t(n+1-m)$ must be nonnegative. The minimal value of the function $(-t + q)$ over all pairs t, q satisfying these conditions is $3n - 2m - 2$.

LEMMA 3. *For any t the restriction of the bundle (33) to the preimage of the set $\nu(m, t) \subset \mathbb{R}^m(t)$ has a section whose image consists of maps sending exactly t points to 0.*

SKETCH OF PROOF. First we associate to each collection $z = (z_1, \ldots, z_t) \in \nu(m, t)$ a smooth function $\varphi_z \colon \mathbb{R}^m \to \mathbb{R}$ such that
  (a) for any z, $\varphi_z \equiv 1$ outside a sufficiently large ball in $\mathbb{R}^m$;
  (b) $\varphi_z \geq 0$ everywhere in $\mathbb{R}^m$;
  (c) $\varphi_z = 0$ only at points of the collection z.

Fix a coordinate system $x_1, \ldots, x_m$ in $\mathbb{R}^m$; then any such function φ_z defines a map $\lambda_z \colon \mathbb{R}^m \to \mathbb{R}^{n+1}$ with value

$$(\varphi_z(x), \partial \varphi_z / \partial x_1, \ldots, \partial \varphi_z / \partial x_m, 0, \ldots, 0)$$

at $x \in \mathbb{R}^m$. Outside some ball in $\mathbb{R}^m$ this map is identically equal to $(1, 0, \ldots, 0)$, and hence extends to a map $\lambda_z \colon (S^m, *) \to (\mathbb{R}^{n+1}, *)$. The rule that associates to the point $z \in \nu(m, t)$ the map λ_z is the desired section of the bundle $\{\mathfrak{A}, t\} \to \mathbb{R}^m(t)$.

It remains to construct the function φ_z. Choose any smooth function $\zeta \colon \mathbb{R}^1 \to \mathbb{R}^1$ equal to $y^2(y-1)^2$ for $y \in [-1/2, 3/2]$, positive outside this

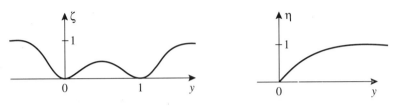

FIGURE 23

interval, and equal to 1 outside the interval $[-2, 3]$; see Figure 23. Let η be a smooth function $\mathbb{R}^1_+ \to \mathbb{R}^1$ equal to y for $y \in [0, 1/2]$, positive outside this interval, and equal to 1 for $y > 2$.

Introduce the orthogonal metric in $\mathbb{R}^m$. Let $z = (z_1, \ldots, z_t) \in \nu(m, t)$. Let $\mathbb{R}^{t-1}_z$ be the affine subspace in $\mathbb{R}^m$ spanned by the points $z_1, \ldots, z_t$. Imbed $\mathbb{R}^{t-1}_z$ into the space $\mathbb{R}^t$ as a plane $\{y_1 + \cdots + y_t = 1\}$ so that the points $z_1, \ldots, z_t$ are sent to the ends of the coordinate basis elements, and restrict the function $\zeta(y_1) + \cdots + \zeta(y_t) \colon \mathbb{R}^t \to \mathbb{R}$ to this plane $\mathbb{R}^{t-1}_z$. Let $\tilde{\zeta}_z$ be the function on $\mathbb{R}^m$ induced from this restriction by the orthogonal projection $\mathbb{R}^m \to \mathbb{R}$. Next, define a function $\tilde{\eta}_z \colon \mathbb{R}^m \to \mathbb{R}^1$ whose value at $x \in \mathbb{R}^m$ equals the value of the function η on the distance from x to $\mathbb{R}^{t-1}_z$. The desired function φ is defined as $\frac{1}{t+1}(\tilde{\zeta}_z + \tilde{\eta}^2_z)$; it obviously satisfies the conditions (a)-(c).

Case (2) of Theorem 5.6.1 is proved completely; now we prove cases (1) and (3).

5.7. The action of the Freudenthal homomorphism on spectral sequences. Again let $m < n$, and let $E^{p,q}_r(m, n)$ be the spectral sequence constructed in 5.4–5.5 and converging to the cohomology of the space $\Omega^m S^n$; in particular, the term $E^{p,q}_1$ is equal to

$$H^{q+p(n+1-m)}(\mathbb{R}^m(-p), (\pm \mathbb{Z})^{\otimes(n-m)}). \tag{34}$$

5.7.1. THEOREM. *For any natural number N there exists a homomorphism of spectral sequences*

$$E^{p,q}_r(m+N, n+N) \to E^{p,q}_r(m, n) \tag{35}$$

such that the action of this homomorphism on the group $E^{p,q}_1$ coincides with the homomorphism of the corresponding groups (34) induced by the obvious imbedding $\mathbb{R}^m(-p) \to \mathbb{R}^{m+N}(-p)$; see 3.1 of Chapter I.

The assertion of Theorem 5.6.1 in cases (1) and (3) follows immediately from this theorem and case (2) just considered. Indeed, fix some p and q. In order to prove that all differentials d_r acts trivially on the groups $E^{p,q}_r(m, n)$ choose a number N so large that $p + q < 3(n + N) - 2(m + N) - 2$. Then by part (2) any differential d_r acts trivially on the cell $E^{p,q}_r(m+N, n+N)$.

§5. MAPS FROM m-DIMENSIONAL TO m-CONNECTED SPACES

But by Theorem 4.3 and Lemma 3.3.3 of Chapter I, in cases (1) and (3) the homomorphism (35) of groups $E_1^{p,q}$ is an epimorphism; hence d_r acts trivially on the cell $E_r^{p,q}(m,n) \equiv E_1^{p,q}(m,n)$. □

5.7.2. *Proof of Theorem* 5.7.1. The homomorphism (35) is constructed similarly to the homomorphism from Theorem 4.2.4 (see 4.5) and is induced by the standard inclusion $\Omega^m(\mathbb{R}^{n+1}-0) \to \Omega^{m+N}(\mathbb{R}^{n+N+1}-0)$; to give its precise construction we will use imbeddings of sufficiently large subspaces approximating these spaces.

It suffices to prove that for any natural T the desired homomorphism of spectral sequences is defined in the domain $\{(p,q)|p+q<T\}$. We do this now.

Again let $\widetilde{\Theta}_\varphi^d$ be the affine subspace of the space of maps $(S^m,*) \to (\mathbb{R}^{n+1},(1,0,\ldots,0))$ given by generic ("perfect") imbedding $i: S^m \to \mathbb{R}^L$ and map $\varphi: (S^m,*) \to (\mathbb{R}^{n+1},(1,0,\ldots,0))$ defined in 5.1–5.4; $\widetilde{\Theta}_\varphi^d$ consists of all maps of the form $\varphi + p \circ i$, where p is an arbitrary polynomial map $\mathbb{R}^L \to \mathbb{R}^{n+1}$ of degree $\leq d$ sending the point $i(*)$ to 0. Let $\widetilde{\Theta}_{\varphi'}^{\prime d}$ be the similar subspace approximating the space of maps $(S^{m+N},*) \to (\mathbb{R}^{n+N+1},(1,0,\ldots,0))$. Let $\mathfrak{A}, \mathfrak{A}'$ be subsets of the spaces $\widetilde{\Theta}_\varphi^d, \widetilde{\Theta}_{\varphi'}^{\prime d}$ consisting of maps whose images pass through the origin; let $\{\mathfrak{A}\}$ and $\{\mathfrak{A}'\}$ be the resolutions of these sets constructed as in 3.3 and 4.4 so that $\overline{H}_*(\{\mathfrak{A}\}) = \overline{H}_*(\mathfrak{A})$, $\overline{H}_*(\{\mathfrak{A}'\}) = \overline{H}_*(\mathfrak{A}')$. Let $\mathscr{E}_r^{p,q}$ and $\mathscr{E}_r^{\prime p,q}$ be the corresponding spectral sequences converging to the cohomology of the spaces $\widetilde{\Theta}_\varphi^d - \mathfrak{A}$, $\widetilde{\Theta}_{\varphi'}^{\prime d} - \mathfrak{A}'$ constructed according to our usual scheme.

Suppose that the number d is so large that all nonzero terms $\mathscr{E}_r^{p,q}, \mathscr{E}_r^{\prime p,q}$ of these spectral sequences lying in the domain $\{(p,q)|p+q<T\}$ are also in the domain $\{(p,q)|p \geq -d\}$. Then everywhere in the domain $\{(p,q)|p+q<T\}$ our spectral sequences coincide with their "truncated" versions $\widehat{\mathscr{E}}_r^{p,q}, \widehat{\mathscr{E}}_r^{\prime p,q}$ obtained by making all the terms with $p<-d$, $r \geq 1$ zero. It remains to construct a homomorphism between these truncated sequences. But these sequences have the following geometrical description. Consider the dth term $F_d\{\mathfrak{A}\}$ ($F_d\{\mathfrak{A}'\}$, respectively) of the standard filtration of the space $\{\mathfrak{A}\}$ ($\{\mathfrak{A}'\}$, respectively) and the homology spectral sequence $\widehat{\mathscr{E}}_{p,q}^r$ ($\widehat{\mathscr{E}}_{p,q}^{\prime r}$, respectively) converging to the closed homology of these terms. Our truncated spectral sequences are obtained from these spectral sequences by a transformation of indices sending the pair (p,q) to $(-p, \dim \widetilde{\Theta}_\varphi^d - q - 1)$ (to $(-p, \dim \widetilde{\Theta}_{\varphi'}^{\prime d} - q - 1)$, respectively).

Now we construct the homomorphism of homology spectral sequences $\widehat{\mathscr{E}}_{p,q}^{\prime r} \to \widehat{\mathscr{E}}_{p,q-\Delta}^r$, $\Delta = \dim \widetilde{\Theta}_{\varphi'}^{\prime d} - \dim \widetilde{\Theta}_\varphi^d$. Note that these homology spectral sequences do not depend on the choice of the maps i, φ (or, i', φ') used in the construction of spaces $\mathfrak{A}$ ($\mathfrak{A}'$); moreover, we can drop the requirement that these maps are "perfect". Indeed, the interpolation theorem implies that a continuous homotopy of these maps gives a homotopy equivalence of the filtered spaces $F_d\{\mathfrak{A}\}$ ($F_d\{\mathfrak{A}'\}$, respectively) they define. (Earlier we needed the "perfectness" of φ and i only to insure that the spectral sequences have no nontrivial terms in the domain $\{p, q | p(n - m + 1) + q < 0\}$ for the nonstable values of p.)

Thus, we can choose φ, φ', i, and i' arbitrarily, for example taking φ and φ' to be the maps $1, 1'$ whose values are identically $(1, 0, \ldots, 0) \in \mathbb{R}^{n+1}$ and $(1, 0, \ldots, 0) \in \mathbb{R}^{n+N+1}$, and i (i') to be the standard imbedding of S^m (S^{m+N}) as the unit sphere in the subspace $\mathbb{R}^{m+1} \subset \mathbb{R}^L$ (in $\mathbb{R}^{m+1+N} \subset \mathbb{R}^{L+N}$, respectively).

Let $\{x_i\}$ and $\{y_j\}$ be linear coordinate systems in $\mathbb{R}^L$ and $\mathbb{R}^N$; then the product structure in $\mathbb{R}^{L+N} = \mathbb{R}^L \oplus \mathbb{R}^N$ defines a coordinate system $\{x_i, y_j\}$ on $\mathbb{R}^{L+N}$. Similarly, let X_i and Y_j be the linear coordinates in the direct summands $\mathbb{R}^{n+1}$ and $\mathbb{R}^N$ of the space $\mathbb{R}^{n+1+N}$. Imbed the space $\widetilde{\Theta}_1^d$ in $\widetilde{\Theta}_{1'}^{'d}$ by sending the polynomial map $\{x \to X = p(x)\}$ to the polynomial map $\{(x, y) \to (X = p(x), Y = y)\}$. This imbedding lifts to the imbedding of the resolutions $\{\mathfrak{A}\} \to \{\mathfrak{A}'\}$ and, in particular, to the imbedding $I: F_d\{\mathfrak{A}\} \to F_d\{\mathfrak{A}'\}$. It follows from the interpolation theorem that the imbedded subspace $I(F_d\{\mathfrak{A}\})$ has a neighborhood in $F_d\{\mathfrak{A}'\}$ homeomorphic to its product by R^Δ, $\Delta = \dim \widetilde{\Theta}_{1'}^{'d} - \dim \widetilde{\Theta}_1^d$. The desired homomorphism of spectral sequences is the composition of the restriction homomorphism of the closed chains in $F_d\{\mathfrak{A}'\}$ to this neighborhood and the Thom isomorphism in this neighborhood. Its agreement with the standard map of groups (34) follows immediately from the construction. □

§6. On the cohomology and stable homotopy of complements of arrangements of planes in $\mathbb{R}^n$

6.1. Main theorem.

DEFINITION. An *affine plane arrangement* is an arbitrary finite collection of affine planes of arbitrary (possibly different) dimensions in affine space.

Let $\{V_1, \ldots, V_s\}$ be such an arrangement in $\mathbb{R}^n$; denote by V the union $V_1 \cup \cdots \cup V_s$. The group

$$H^*(\mathbb{R}^n - V) \tag{36}$$

has been considered in [Ar$_3$], [Br$_4$], [D$_2$], [OS], [GM] and elsewhere. In [GM] an effective method is given to compute such groups using only data about the dimensions of all (multiple) intersections of the spaces V_i. Here we prove that the stable homotopy type of the space $\mathbb{R}^n - V$ also depends only on these data.

Since Spanier-Whitehead duality preserves stable homotopy equivalence classes (see [W]), it is sufficient to express the homotopy type of the one-point compactification of V (which is Spanier-Whitehead dual to the space

$\mathbb{R}^n - V$) through these dimensional data. This expression is as follows.

Let Λ be any affine plane which is the intersection of several planes V_i. Consider the simplex $\Delta(\Lambda)$ whose vertices correspond formally to all the planes V_i containing Λ. A face of this simplex, i.e., a collection of such planes, is called *marginal*, if the intersection of these planes is strictly greater than Λ. The quotient space of the simplex $\Delta(\Lambda)$ by the union of all marginal faces will be denoted by $K(\Lambda)$. Denote by $\overline{V}$ the one-point compactification of V.

THEOREM 1 (see [V_{11}]). *The space $\overline{V}$ is homotopy equivalent to the wedge of $\dim(\Lambda)$-fold suspensions of the spaces $K(\Lambda)$ taken over all planes Λ which are the intersections of some planes of the set $(V_1, \ldots, V_s)$.*

REMARK 1. An equivalent description of the homotopy type of the space $\overline{V}$ was independently and simultaneously (in late 1991) obtained by Günter Ziegler and Rade Živaljević [ZZ].

COROLLARY 1. *The stable homotopy type of the complement $\mathbb{R}^n - V$ of the arrangement V is completely determined by its combinatorial (dimensional) characteristics. In particular, all extraordinary cohomology groups of these spaces depend only on these data.*

COROLLARY 2. *The homology group of the space $\overline{V}$ is expressed by the formula*

$$H_j(\overline{V}) \simeq \bigoplus H_{j-\dim(\Lambda)}(K(\Lambda)) \tag{37}$$

(the summation over all planes Λ which are the intersections of several planes V_i), and the reduced cohomology group of the space $\mathbb{R}^n - V$ is given by the formula

$$\widetilde{H}^i(\mathbb{R}^n - V) \simeq \bigoplus \widetilde{H}_{n-i-1-\dim(\Lambda)}(K(\Lambda)). \tag{38}$$

Indeed, the formula (37) is a direct corollary of Theorem 1, and formula (38) follows from formula (37) by Alexander duality.

Formula (38) was previously obtained by Goresky and MacPherson, see [GM].

REMARK 2. The (nonstable) homotopy type of the space $\mathbb{R}^n - V$ does not, in general, depend only on the dimensions of the intersections of spaces V_i, see [Z].

The remainder of this section is devoted to the proof of Theorem 1.

NOTATIONS. Let $V_1, \ldots, V_s$ be a finite set of affine planes in $\mathbb{R}^n$, and let V be the union of all the V_i. Denote by S the set of natural numbers $1, \ldots, s$. For any subset $J \subset S$, denote by V_J the intersection of planes V_j, $j \in J$. The dimension of this intersection is denoted by $|J|$. For any J, J' is the maximal subset in S such that $V_J = V_{J'}$, $\overline{V}_J$ and $\overline{V}$ are the notations for the one-point compactifications of the complexes V_J and V, respectively.

A set J is called *geometrical* if $J = J'$.

For any geometrical set $J \subset S$, denote by $\Delta(J)$ (or by $\Delta(V_J)$) the simplex whose vertices are in one-to-one correspondence with the elements of J. Let $M(J)$ be the subcomplex in $\Delta(J)$ consisting of all marginal faces, and recall the notation $K(V_J)$ for the quotient complex $\Delta(J)/M(J)$.

6.2. Geometrical resolution of the complex V. Without loss of generality, we shall assume that the dimension of each of the planes V_j is positive.

Consider a space $\mathbb{R}^N$, where N is sufficiently large, and s affine embeddings $I_j: V_j \to \mathbb{R}^N$, $j \in S$. For any point $x \in V$, consider all its images in $\mathbb{R}^N$ under all maps I_j such that $x \in V_j$. Denote by $\|x\|$ the number of such j and by x' the convex hull of these images in $\mathbb{R}^N$.

LEMMA 1. *If N is sufficiently large and the system of embeddings I_j is generic, then for any point $x \in V$ the polyhedron x' is a simplex with $\|x\|$ vertices, and the intersection of simplices x', y' is empty if $x \neq y$.*

The proof is trivial.

We shall suppose that the maps I_j satisfy the conditions of this lemma. Denote by V' the union of all simplices x' over all $x \in V$, and by $\overline{V}'$ the one-point compactification of V'. These spaces V', $\overline{V}'$ will be called the geometrical resolutions of V and $\overline{V}$. The natural projection $\pi: V' \to V$ (which maps any simplex x' in the point x) is obviously proper and can be extended to a continuous map $\overline{V}' \to \overline{V}$ which will be denoted by the same letter π.

LEMMA 2. *This projection $\pi: \overline{V}' \to \overline{V}$ induces a homotopy equivalence of these spaces.*

This is (a special case of) the principal fact of the theory of simplicial resolutions, see [D$_1$].

Hence, we have only to prove the following theorem.

THEOREM 1'. *For any affine plane arrangement V, the one-point compactification of its resolution V' is homotopy equivalent to the wedge indicated in Theorem 1.*

6.3. Proof of Theorem 1'. This proof is based on a variant of stratified Morse theory, see [GM].

DEFINITION. A function $f: \mathbb{R}^n \to \mathbb{R}^1$ is called a *generic quadratic function* if it can be expressed in the form $x_1^2 + \cdots + x_n^2$ in some affine coordinate system in $\mathbb{R}^n$, whose origin does not belong to V, and any level set $f^{-1}(t)$ of this function is tangent to at most one plane V_j.

Let us fix such a function f. For any set J denote by t_J the unique number t such that $f^{-1}(t)$ is tangent to V_J; such numbers t_J are called *singular values*, and the other values are called regular.

§6. COMPLEMENTS OF ARRANGEMENTS OF PLANES IN $\mathbb{R}^n$

For any value $t \in \mathbb{R}^1$ denote by $V'(t)$ the space
$$\pi^{-1}(V \cap f^{-1}([t, \infty))).$$

LEMMA 3. a) *If t is greater than all singular values t_J, then the quotient space $V'/V'(t)$ is homotopy equivalent to $\overline{V}'$;*

b) *If t is less than all values t_J, then this quotient space is a point;*

c) *if the segment $[t, s]$ does not contain singular values, then the obvious factorization mapping*
$$(V'/V'(s)) \to (V'/V'(t))$$
is a homotopy equivalence;

d) *if the segment $[t, s]$ contains exactly one singular value t_J, $t < t_J < s$, and the set J is geometrical, then the space $V'/V'(s)$ is homotopy equivalent to the wedge of spaces $V'/V'(t)$ and $\Sigma^{|J|}(\Delta(J)/M(J))$, where Σ^i is the notation for the i-fold reduced suspension.*

Theorem $1'$ follows immediately from this lemma.

Items a), b), and c) of this lemma are obvious, let us prove d). For any geometrical set J, define the *proper inverse image* of the plane V_J to be the closure in $\mathbb{R}^N$ of the union of simplices x' over all points $x \in V_J$ which do not belong to subplanes V_I of lower dimensions in V_J. Denote this closure by V'_J, and denote by $V'_J(t)$ its intersection with $V'(t)$.

Any space V'_J is naturally homeomorphic to the complex $\Delta(J) \times V_J \simeq \Delta(J) \times \mathbb{R}^{|J|}$.

Now let J be the geometrical set considered in Lemma 3(d). Then the space $V'_J/V'_J(s)$ is naturally homeomorphic to the space $(\Delta(J) \times V_J)/(\Delta(J) \times (V_J \cap f^{-1}([s, \infty)))) \sim \Sigma^{|J|}\Delta(J)$.

Let W be the union of inverse images of all other planes $V_{I'}$, $I' \neq J$, having nonempty intersections with the disk $f^{-1}([0, s])$. The intersection of varieties V'_J and W can be naturally identified with the complex $M(J) \times V_J \simeq M(J) \times R^{|J|}$. Let φ be the identical imbedding of this intersection into W. Let ψ be the map of the quotient space $(M(J) \times V_J)/(M(J) \times V_J(s))$ into $W/(W \cap V'(s))$ induced by φ. The space $V'/V'(s)$ can be considered as the space
$$(W/(W \cap V'(s))) \cup_\psi (V'_J/V'_J(s)),$$
where $\cup_\psi$ is the topological operation "paste together by the map ψ", see [FV].

But the quotient space $W/(W \cap V'(s))$ is naturally homotopy equivalent to the space $V'/V'(t) = W/(W \cap V'(t))$: this homotopy equivalence is realized by the obvious factorization map which contracts all points z at which $f(\pi(z)) \in [t, s]$.

The composition of the map ψ with this factorization is a map into one point. Since the operation $\cup_\psi$ is homotopy invariant (see [FV], §I.2.11), the space $V'/V'(s)$ is homotopy equivalent to the composite space

$$(V'/V'(t)) \bigcup_{\psi'} (V'_J/V'_J(s)),$$

where the map ψ' is defined on the same subspace as ψ and takes this subspace into one point $[V'(t)] \in V'/V'(t)$.

Hence, the space $V'/V'(s)$ is homotopy equivalent to the wedge of spaces $V'/V'(t)$ and

$$\begin{aligned}
&V'_J/(V'(s) \cup (M(J) \times V_J)) \\
&\simeq (\Delta(J) \times V_J)/((\Delta(J) \times V_J(s)) \cup (V'_J \cap W)) \\
&\simeq (\Delta(J)/M(J)) \wedge (V_J/V_J(s)) \simeq \Sigma^{|J|}(\Delta(J)/M(J)). \quad \square
\end{aligned}$$

Theorems $1'$ and 1 are completely proved.

CHAPTER IV

Stable Cohomology of Complements of Discriminants and Caustics of Isolated Singularities of Holomorphic Functions

There is a ring associated with a singular point of finite multiplicity of a holomorphic function; namely, the cohomology ring of the complement of the discriminant set (= bifurcation diagram of zeros) of any of its versal deformation. (⁶) Adjacencies of singularities define ring homomorphisms: the complement of the discriminant for a simpler singularity is imbedded in the complement of the discriminant for a more complicated one (in Figure 1 we showed an imbedding corresponding to the adjacency of real singularities $A_2 \to A_3$).

Using these rings we construct limit objects: stable cohomology rings. An element of such a ring is a rule that associates to each singularity a cohomology class of the complement of its discriminant in such a way that, if two singularities are adjacent, the class corresponding to a simpler singularity is induced from the class corresponding to a more complicated singularity by any such imbedding of complements of the discriminants.

THEOREM 1. *The stable cohomology ring $_n\mathscr{H}^*$ of complements of discriminants of isolated singularities of holomorphic functions in n complex variables is naturally isomorphic to the cohomology ring of the $2n$th iterated loop space of the $(2n+1)$-dimensional sphere*:

$$_n\mathscr{H}^* \cong H^*(\Omega^{2n} S^{2n+1});$$

the same is true for cohomology with any coefficient group.

This theorem, together with Theorem 5.6.1 of Chapter III, gives the following representation.

COROLLARY 1.
$$_n\mathscr{H}^i = \bigoplus_{t=0}^{\infty} H^{i-t}(\mathbb{R}^{2n}(t), \pm \mathbb{Z}).$$

(⁶) Definitions of versal deformation, bifurcation diagram, and other notions of singularity theory used here are given in §1 below; see also [AVG₁], [AVGL₁].

COROLLARY 2. *All groups $_n\mathcal{H}^i$ for any n are finite, except the groups $_n\mathcal{H}^0 \cong {_n\mathcal{H}^1} \cong \mathbb{Z}$.*

COROLLARY 3. *Several low-dimensional stable cohomology groups $\mathcal{H}^i \equiv {_\infty\mathcal{H}^i}$ are given in the following table:*

dimension	0	1	2	3	4	5
group	$\mathbb{Z}$	$\mathbb{Z}$	0	$\mathbb{Z}_2$	$\mathbb{Z}_2$	$\mathbb{Z}_6 + \mathbb{Z}_2$

Several low-dimensional stable cohomology groups with coefficients in $\mathbb{Z}_2$ are given in the following table:

dimension	0	1	2	3	4	5	6	7	8
dimension of the group	1	1	1	2	3	4	6	9	12

These tables can easily be extended: the cohomology of the spaces $\Omega^m S^r$, $m < r$, were computed in [DL], [Mil$_{1,2}$], [CLM], and the groups

$$H^*(S(t), \pm \mathbb{Z} \otimes \mathbb{Z}_2) = H^*(S(t), \mathbb{Z}_2)$$

are described in Chapter I.

COROLLARY 4 (cf. [Segal$_1$]). *The cohomology ring of the braid group on infinite number of strings is isomorphic to the ring $H^*(\Omega^2 S^3)$: this is a special case of Theorem 1 corresponding to $n = 1$.*

The isomorphism from Theorem 1 is constructed using natural imbeddings of complements of discriminants in the space $\Omega^{2n} S^{2n+1}$. Namely, if λ is a nondiscriminant value of the deformation parameter, then the corresponding perturbation f_λ of the initial function $f: (\mathbb{C}^n, 0) \to (\mathbb{C}, 0)$ has, by definition, no critical points with critical value 0 in a given spherical neighborhood B_ε of the point 0. Associate to such a λ a map $X_\lambda: B_\varepsilon \to \mathbb{C}^{n+1} - \{0\} \sim S^{2n+1}$ that sends each point x to the $(n + 1)$-tuple of complex numbers $f_\lambda(x)$, $\partial f_\lambda / \partial x_1, \ldots, \partial f_\lambda / \partial x_n$. We can assume that the maps corresponding to various λ coincide on the boundary of the ball B_ε. Identifying B_ε with the lower half of S^{2n} and extending all these maps in a common way to the upper half we get an imbedding of the set of nondiscriminant λ in the space $(S^{2n+1}, *)^{(S^{2n}, *)} \simeq \Omega^{2n} S^{2n+1}$. It turns out that, for a sufficiently complicated singularity f, such an imbedding of the complement of the discriminant of the versal deformation defines an isomorphism of the first several groups H^*, and the number of these groups increases with the complexity of f.

THEOREM 2. *For any positive integers c and n there is a singularity f in n variables such that the obvious homomorphism from the stable ring $_n\mathcal{H}^*$ to the cohomology ring $J^*(f)$ of the complement of the discriminant of the* *versal deformation of f is an isomorphism in dimensions less than c. (An equivalent formulation: the homomorphism $H^i(\Omega^{2n}S^{2n+1}) \to J^i(f)$ given by the above imbedding is an isomorphism for $i < c$.) In fact, it is enough to set $f = x_1^N + \cdots + x_n^N$, where $N > n(2c - 1)$.*

To prove Theorems 1 and 2 we construct spectral sequences $E_r^{p,q}(n)$, $E_r^{p,q}(f)$ converging to the groups $_n\mathscr{H}^*$, $J^*(f)$ (they are again versions of the spectral sequence from §III.4) and prove that for all p, q (for all p, q with $p + q < c$, respectively) starting with the term E_1 these spectral sequences are naturally isomorphic to the sequence from §III.5, which computes the cohomology of $\Omega^{2n}S^{2n+1}$.

With increasing n the stable rings $_n\mathscr{H}^*$ naturally stabilize to a bistable ring $\mathscr{H}^*$. Stabilization of the groups $_n\mathscr{H}^i$ (with fixed i) happens no later than at $n = i/2$ (i.e., $_n\mathscr{H}^i \cong \mathscr{H}^i$ for $n \geq i/2$); this stabilization agrees with the stabilization of the groups $H^i(\Omega^{2n}S^{2n+1})$, $n \to \infty$, which follows from Freudenthal's theorem.

In §6 of this chapter we compute the stable cohomology rings $_n\widetilde{\mathscr{H}}^*$ of complements of caustics (see [AVG$_1$], [AVGL$_1$]).

THEOREM 3. $_n\widetilde{\mathscr{H}}^* \cong H^*(\Omega^{2n}\Sigma^{2n}\Lambda(n))$, where Σ^{2n} denotes the $2n$-fold suspension, and $\Lambda(n)$ is the nth Lagrange Grassmannian $\Lambda(n) = U(n)/O(n)$.

The homology isomorphisms of Theorems 1 and 3 have no analogues for homotopy groups: for example, for $n = 1$ both stable fundamental groups of complements of discriminants and of caustics are isomorphic to $\mathrm{Br}(\infty) \neq \pi_1(\Omega^2 S^3) = \mathbb{Z}$.

The two isomorphisms of Theorems 1 and 3 are special cases of the general "homological Smale-Hirsch principle" for stable complements of singular strata in deformation spaces of holomorphic functions; we state this principle in §6.

In §7 we generalize the theorem [CCMM] to the effect that the stable homotopy types of complements of the discriminant of polynomials of degree $2d$ in $\mathbb{C}^1$ and of the resultant of pairs of polynomials of degree d coincide. Namely, consider the space of polynomial systems

$$x^{d_1} + a_{1,1}x^{d_1-1} + \cdots + a_{1,d_1},$$
$$\cdots\cdots\cdots\cdots\cdots\cdots\cdots\cdots\cdots\cdots$$
$$x^{d_k} + a_{k,1}x^{d_k-1} + \cdots + a_{k,d_k},$$

(1)

$a_{i,j} \in \mathbb{C}$. Denote the space of all such systems by $\mathbb{C}^{d_1+\cdots+d_k}$.

DEFINITION. The *resultant* $\Sigma_{d_1,\ldots,d_k}$ is the set of systems (1) having a common root.

It is easy to check that the space $\mathbb{C}^{d_1+\cdots+d_k} - \Sigma_{d_1,\ldots,d_k}$ of nonresultant systems is homotopy equivalent to a similar space where each $d_1, \ldots, d_k$ is replaced by $\min(d_i)$.

THEOREM 4. *The space* $\mathbb{C}^{d+\cdots+d} - \sum_{d,\ldots,d}$ *(where d is iterated k times in both cases) is stable homotopy equivalent to the space* $\mathbb{C}^{k\cdot d} - \sum(k\cdot d, k)$ *of complex polynomials of the form* (I.1) *of degree $k\cdot d$ having no roots of multiplicity* $\geq k$; *in particular,* $H^*(\mathbb{C}^{d+d} - \sum_{d,d}) \simeq H^*(\mathrm{Br}(2d))$.

In the case $k = 2$, this result was previously obtained in [CCMM].

On the other hand, the topology of the space $\mathbb{C}^{d+\cdots+d} - \sum_{d,\ldots,d}$ is partially described by the following theorem by G. Segal.

THEOREM (see [Segal$_2$]). *There exists a map* $\mathbb{C}^{d+\cdots+d} - \sum_{d,\ldots,d} \to \Omega^2 S^{2k-1}$ *that induces an isomorphism of homotopy groups* π_i, $i < (d+1)(2k-3)$, *an epimorphism of the groups* $\pi_{(d+1)(2k-3)}$, *an isomorphism of the groups* H^i, $i \leq (d+1)(2k-3)$, *and an epimorphism of all other cohomology groups.*

Many stable cohomology classes of complements of discriminants can be realized as characteristic classes of cohomological Milnor bundles. Over the complement of the discriminant of any singularity f there is an infinite collection of vector bundles M_k, $k = n, n+1, \ldots$, whose fibers over nondiscriminant values of the parameter λ are middle $((k-1)$-dimensional) vanishing real cohomology groups of the manifolds

$$\{(x, y) \in \mathbb{C}^n \times \mathbb{C}^{k-n} | f_\lambda(x) + y_1^2 + \cdots + y_{k-n}^2 = 0\}.$$

The Picard-Lefschetz formulas imply that all bundles M_k corresponding to odd or to even numbers k are isomorphic to each other. So we will consider only two bundles: the even bundle M_e and the odd bundle M_o.

THEOREM 5. *The Stiefel-Whitney classes of both (even and odd) cohomological Milnor bundles are well-defined classes of the stable $\mathbb{Z}_2$-cohomology of complements of discriminants.*

THEOREM 6. *The even cohomological Milnor bundle of any isolated singularity is stably equivalent to the trivial bundle; in particular, all its characteristic classes are trivial.*

(This theorem was obtained by the author together with A. N. Varchenko and A. B. Givental.)

REMARK (A. B. Givental, D. B. Fuchs). All Pontryagin classes and the squares of all Stiefel-Whitney classes of the odd (and even) cohomological Milnor bundle are trivial.

This follows from the fact that the complexifications of these bundles are trivializable; see [Br$_1$], [Var$_2$].

On the contrary, all basic Stiefel-Whitney classes of the odd cohomological Milnor bundles are nontrivial stable classes; here is their realization.

Consider any nonsingular point of a multiplicity k self-intersection of a discriminant (such points correspond to functions with precisely k Morse critical points with zero critical values). Near such a point the pair (deformation base, discriminant) is diffeomorphic to a product of a linear space and the pair ($\mathbb{C}^k$, the union of coordinate planes in $\mathbb{C}^k$). Hence the k-dimensional homology group of the complement of the discriminant in this neighborhood is generated by the k-dimensional torus (the distinguished boundary of the polydisk in $\mathbb{C}^k$).

THEOREM 7. *The value of the kth Stiefel-Whitney class of the odd cohomological Milnor bundle on this torus is nontrivial.*

COROLLARY. *For $k \neq 0, 1$ the kth Stiefel-Whitney class $w_k(M_0)$ is not integral (does not lie in the image of the coefficient homomorphism of the ring $_n\mathscr{H}^k$ into the similar stable cohomology ring $\mathrm{mod}\, 2$). In particular, its image under the Bockstein homomorphism is an integral class (recall that this image is equal to $w_k \cdot w_1 + \varepsilon w_{k+1}$, where $\varepsilon = 1$ for even k and $\varepsilon = 0$ for odd k).*

This follows from the finiteness of all groups $_n\mathscr{H}^i$, $i > 1$; see Corollary 2 to Theorem 1.

In §1.5 we prove the conjecture [Ar$_{11, 16}$] about the stable irreducibility of strata of the discriminant. Let f, g be two isolated singularities of holomorphic functions in $\mathbb{C}^n$, and G be the versal deformation of g. In the discriminant of G consider the set $\{f\}$ consisting of functions g_λ with singular points equivalent to f. In general this set is reducible: for example, in the space of the miniversal deformation of the singularity D_4 the set $\{A_3\}$ consists of three lines. Upon passing to stable objects this effect disappears according to the following theorem on stable irreducibility: if a "more complicated" singularity g is complicated enough (with respect to the "simpler" singularity f), then there can be only one such component. The analogous theorem is proved for multisingularities of functions (see the end of §1).

§1. Singularities of holomorphic functions, their deformations and discriminants

1.1. Deformations and discriminants. Consider a holomorphic function f of n complex variables defined in a neighborhood of the origin $0 \in \mathbb{C}^n$; let $f(0) = 0$. The point 0 is a *singular point* of the function f if $df(0) = 0$. The singularity of f at 0 is *isolated* if $df \neq 0$ at all nearby points. In this case the manifold $V_0 = f^{-1}(0)$ is nonsingular in a punctured neighborhood of 0.

A *deformation* of the function f is a holomorphic function $F(x, \lambda)$ defined in a neighborhood of 0 in the space $\mathbb{C}^n \times \mathbb{C}^l$ such that $F(\cdot, 0) \equiv f$. For any $\lambda \in \mathbb{C}^l$ the corresponding function $F(\cdot, \lambda): \mathbb{C}^n \to \mathbb{C}$ is denoted by f_λ; in particular, $f_0 = f$. The space $\mathbb{C}^l$ of parameters of the deformation is called its *base*. The *discriminant* of the deformation F is the set of parameters $\lambda \in \mathbb{C}^l$ such that the function f_λ has a critical point with critical value 0 near 0; the discriminant is denoted by $\Sigma(F)$. The *caustic* of the deformation F is the set of $\lambda \in \mathbb{C}^l$ such that f_λ has a non-Morse critical point near 0.

EXAMPLE. Let $n = 1$, $f = x^m$. Then the set of polynomials

$$x^m + a_1 x^{m-1} + \cdots + a_{m-1} x + a_m \tag{2}$$

can be considered as a deformation $F(x, a)$ of the function f, $a \in \mathbb{C}^m$. The complement of the discriminant of this deformation is the set of polynomials without multiple roots, studied in Chapters I and II; the complement of the caustic is the set of polynomials without multiple roots of the first derivatives.

PROPOSITION. *The discriminant and caustic of every deformation of an isolated singularity are analytic subsets in $\mathbb{C}^l$ (in general, nonsmooth); if the deformation $F(x, \lambda)$ is polynomial, then they are algebraic subsets.*

1.2. The Milnor bundle. The importance of a discriminant is determined by the fact that, over its points, the local (near 0) topology of the sets $f_\lambda^{-1}(0)$ changes; over the complement of the discriminant, such sets define a locally trivial bundle, the *Milnor bundle* (see [Milnor$_3$], [AVG$_2$]). Here are precise definitions. Suppose an isolated singularity of $f: (\mathbb{C}^n, 0) \to (\mathbb{C}, 0)$ and its deformation $F: \mathbb{C}^n \times \mathbb{C}^l \to \mathbb{C}$ are given.

LEMMA 1 (see [Milnor$_3$]). *There exists a number $\varepsilon > 0$ such that the set $V_0 = f^{-1}(0) \subset \mathbb{C}^n$ is transversal to the sphere of radius ε centered at 0 in $\mathbb{C}^n$ and to all spheres with radii $\varepsilon' \in (0, \varepsilon]$.*

Fix such a sphere S_ε.

LEMMA 2 (see [Milnor$_3$], [AVG$_2$]). *There exists $\delta = \delta(\varepsilon) > 0$ such that, for all $\lambda \in \mathbb{C}^l$ in the δ-neighborhood of the origin, the set $f_\lambda^{-1}(0) \subset \mathbb{C}^n$ intersects the sphere S_ε transversally (and is regular at these intersection points).*

For any λ from this δ-neighborhood D_δ denote by V_λ the intersection of the set $f_\lambda^{-1}(0)$ with the ball B_ε bounded by the sphere S_ε. The Thom isotopy theorem (see [Ph], [AVGL$_1$]) implies that for all such λ the sets $\partial V_\lambda = V_\lambda \cap S_\varepsilon$ are diffeomorphic to each other, and for all nondiscriminant λ (i.e. for $\lambda \in D_\varepsilon - \Sigma(F)$) the sets V_λ are also diffeomorphic to each other.

THEOREM (see [Milnor$_3$]). *For any isolated singularity, any deformation of it, and any nondiscriminant value λ of the parameter of this deformation the set V_λ is homotopy equivalent to a bouquet of finitely many spheres of dimension $n - 1$.*

The number of the spheres does not depend on the choice of deformation (if this deformation has at least one nondiscriminant value); it is called the *Milnor number* (or the *multiplicity*) of the singularity f and is denoted by $\mu(f)$.

The sets V_λ define a smooth bundle over the complement to the discriminant, the *Milnor bundle*: the space of this bundle is the set of points

$(x, \lambda) \in B_\varepsilon \times (D_{\delta(\varepsilon)} - \Sigma(F))$ such that $F(x, \lambda) = 0$; the bundle structure is given by the projection onto the second factor.

To this locally trivial bundle we associate a locally trivialized bundle, the *cohomological Milnor bundle*: its fiber over $\lambda \notin \Sigma(F)$ is the group $\tilde{H}^{n-1}(V_\lambda, \mathbb{R})$, i.e., the $(n-1)$-dimensional cohomology group of V_λ reduced modulo the fundamental cocycle. Another name for this bundle is the *bundle of vanishing cohomology* (the cohomology of V_λ is called vanishing since for $\lambda = 0$ the set V_λ is contractible). The local trivialization of the latter bundle is given by any local trivialization of the Milnor bundle; it does not depend on the latter choice, and is called the *Gauss-Manin connection* of the deformation F (see [Br$_1$], [AVG$_2$]). This connection determines the action of the group $\pi_1(D_\varepsilon - \Sigma(F))$ on the fibers of the cohomological bundle; this representation of the fundamental group is called the *local monodromy group* of the singularity f (and deformation F).

EXAMPLE. Again let $n = 1$, $f = x^m$, and suppose $F(x, a)$ is given by the formula (2). Then $\mu(f) = m - 1$, the Milnor bundle is the covering φ_m considered in 2.3 of Chapter II, and the monodromy group coincides with the reduced Coxeter representation considered in §5 of Chapter I.

Together with the cohomological Milnor bundle one can consider the homological bundle and similar (co)homological bundles with coefficients in $\mathbb{C}$ (and in any other group).

THEOREM (see [Br$_1$]). *For any deformation of any isolated singularity the complex (with fibers $\tilde{H}^{n-1}(V_\lambda, \mathbb{C})$) cohomological Milnor bundle is isomorphic to the trivial bundle.*

1.3. The sufficient jet theorem. Denote by $\mathfrak{M}$ the algebra of germs of holomorphic functions $(\mathbb{C}^n, 0) \to (\mathbb{C}, 0)$. Then $\mathfrak{M}^k$ is the space of germs whose Taylor series contain only monomials of degree k and higher. The group R_0 of local diffeomorphisms $(\mathbb{C}^n, 0) \to (\mathbb{C}^n, 0)$ acts on the space $\mathfrak{M}^2$ of germs of functions with a singularity at 0.

THEOREM (see [AVG$_1$], [AVGL$_1$]). *Let $f \in \mathfrak{M}^2$ be an isolated singularity with Milnor number μ. Then*

(a) *The fact that a function $\varphi \in \mathfrak{M}^2$ belongs to the R_0-orbit of f can be determined from its $(\mu + 1)$-jet: if $\varphi - \varphi' \in \mathfrak{M}^{\mu+2}$, then the functions φ and φ' both do or do not belong to this orbit.*

(b) *The codimension of this orbit in $\mathfrak{M}^2$ is equal to $\mu - 1$.*

EXAMPLE. Let $n = 1$. Any two functions $\varphi, \varphi' : (\mathbb{C}^1, 0) \to (\mathbb{C}^1, 0)$ are locally biholomorphically equivalent if and only if their Taylor series start with nontrivial terms of the same degree.

1.4. Versal deformations. Among all the deformations of an isolated singularity there are the so-called versal deformations to which (to each of which) all the other deformations reduce; for a majority of statements about deformations of a singularity it is sufficient to prove the case of an arbitrary versal deformation (for example, this is true for the last theorem of subsection 1.2).

DEFINITION. A deformation $F'(x, \lambda)$ of the function f is *equivalent* to the deformation $F(x, \lambda)$ if it can be represented in the form $F'(x, \lambda) = F(U(x, \lambda), \lambda)$, where $U: (\mathbb{C}^n \times \mathbb{C}^l, 0) \to (\mathbb{C}^n, 0)$, $U(\cdot, 0) = \mathrm{id}$, is a family of diffeomorphisms $\mathbb{C}^n \to \mathbb{C}^n$ defined near 0 and depending smoothly on the parameter $\lambda \in \mathbb{C}^l$.

The deformation $\Phi(x, \varkappa)$, $\varkappa \in \mathbb{C}^k$, is *induced* from the deformation F by a smooth transformation of the parameters $\theta: (\mathbb{C}^k, 0) \to (\mathbb{C}^l, 0)$ if $\Phi(x, \varkappa) = F(x, \theta(\varkappa))$.

A deformation $F(x, \lambda)$ of the germ $f(x)$ is called *versal* if any other deformation $\Phi(x, \varkappa)$ is equivalent to a deformation induced from it, i.e. can be represented in the form

$$\Phi(x, \varkappa) \equiv F(U(x, \varkappa), \theta(\varkappa)) \tag{3}$$

for some U and θ.

THEOREM (see [T], [AVG$_1$]). *Every isolated singularity has a versal deformation depending on μ parameters, where μ is the Milnor number of the singularity, and does not have a versal deformation depending on a smaller number of parameters.*

A versal deformation depending on exactly μ parameters is called *miniversal*.

A versal deformation of the singularity f can be chosen in the form

$$F(x, \lambda) = f + \sum_\alpha \lambda_\alpha x^\alpha, \tag{4}$$

where α runs over a finite collection of multi-indices, $\alpha = (\alpha_1, \ldots, \alpha_n)$, λ_α are parameters, and x^α is the monomial $x_1^{\alpha_1} \cdots x_n^{\alpha_n}$.

EXAMPLE. Deformation (2) of the singularity x^m is versal but not miniversal: it depends on m parameters while the Milnor number is $m - 1$. For a miniversal deformation we can choose a subdeformation in (2) depending only on $a_2, \ldots, a_m$.

In general, for a miniversal deformation of the function $f = x_1^{m_1} + \cdots + x_n^{m_n}$ we can take the deformation (4) in which the multi-indices α range over the parallelepiped $\{(\alpha_1, \ldots, \alpha_n) \in \mathbb{Z}^n | 0 \leq \alpha_i \leq m_i - 2 \ \forall i\}$.

A versal deformation of an arbitrary singularity f can be constructed as follows. Consider the ideal in the algebra of formal series $\mathbb{C}[[x_1, \ldots, x_n]]$ generated by all partial derivatives of the function f. If the singularity f at 0 is isolated then this ideal has finite codimension. A remarkable fact is that this codimension equals the Milnor number of the singularity f; see [P], [K], [AVG$_1$]. Choose arbitrary $\mu(f)$ functions $\varphi_1, \ldots, \varphi_\mu$ generating the quotient space of the space of formal series modulo this ideal. Then the deformation

$$F(x, \lambda) = f + \sum_{i=1}^{\mu} \lambda_i \varphi_i(x) \qquad (4')$$

is versal. For these $\mu(f)$ basis functions φ_i we can always choose a family of monomials x^α; in this case the deformation (4') has the form (4).

1.5. Adjacency of singularities. Stable irreducibility of singularities and multisingularities.

DEFINITIONS. Two germs $f: (\mathbb{C}^n, a) \to (\mathbb{C}, 0)$, $f': (\mathbb{C}^n, b) \to (\mathbb{C}, 0)$ with singularities at points a, b respectively are called *equivalent* if they are mapped to each other by some local diffeomorphism $u: (\mathbb{C}^n, a) \to (\mathbb{C}^n, b)$, i.e. $f = f' \circ u$. The germs f, f' are called *weakly equivalent* if there is a complex number $c \neq 0$ such that f is equivalent to $c \cdot f'$.

Suppose there are given arbitrary singularities $f, g: (\mathbb{C}^n, 0) \to (\mathbb{C}, 0)$ and a deformation of the singularity g. Denote by $\{f\}$ the set of parameters λ of this deformation such that the corresponding perturbation g_λ of the singularity g has a critical point near 0 equivalent to the singularity f. By definition, the singularity g is *adjacent* to f (or is "more complicated than f") if in the base of an arbitrary (and so any other) versal deformation of g the set $\{f\}$ is nonempty in any neighborhood of the origin. This is denoted by $f \succcurlyeq g$. (For example, any singularity is adjacent to itself.)

We say that an isolated singularity g is *covered by* a singularity f if for any isolated singularity h adjacent to g the set $\{f\}$ in the base of the versal deformation of h is irreducible and nonempty.

THEOREM (on stable irreducibility). *For any isolated singularity f in n variables there is a singularity g covered by f. Namely, it suffices to take $g = x_1^N + \cdots + x_n^N$, where $N \geq \mu(f) + 2$ (and $\mu(f)$ is the Milnor number).*

(This theorem was formulated as a conjecture in [Ar$_{11, 16}$].)

PROOF. Let $h \in \mathfrak{M}^2$ be an isolated singularity adjacent to g, and $\mu(h)$ be the Milnor number. Let $D(h)$ be an arbitrary $(\mu(h) - 1)$-dimensional complex manifold that passes through the point $h \in \mathfrak{M}^2$ and is transversal to the orbit of this point under the action of the group R_0 of local diffeomorphisms; see 1.3. This transversal submanifold is a deformation of the function h and hence is equivalent to the deformation induced from the miniversal deformation of h by some map of the parameter spaces of these deformations $\gamma: (D(h), h) \to (\mathbb{C}^{\mu(h)}, 0)$. Since $D(h)$ belongs to $\mathfrak{M}^2$, the image of this map belongs to the discriminant $\Sigma \subset \mathbb{C}^{\mu(h)}$. It was proved in [G] that this image $\gamma(D(h))$ coincides with Σ (and the pair $(D(h), \gamma)$ is a normalization of Σ). Here the set $\{f\} \subset \Sigma$ is the image of the intersection of $D(h)$ with the orbit of f under the action of R_0. Hence it remains to prove the irreducibility of this intersection; of course, this property does not depend on the choice of $D(h)$. We can choose $D(h)$ in the form

$$h + \sum \lambda_\alpha x^\alpha, \qquad (5)$$

where the summation is taken over some collection of multi-indices α such that the monomials x^α form a basis in the quotient space of the algebra $\mathfrak{M}^2$ by the ideal spanned by various functions of the form $x_i \partial h/\partial x_j$; see [Ar$_7$], [AVG$_1$]. (This ideal coincides with the tangent space at h to the orbit of the group D_0; its codimension is $\mu(h) - 1$.)

Since h is adjacent to g, the $(\mu(f)+1)$th jet of the function h is 0, and hence all monomials x^α with $\sum \alpha_i \leq \mu(f)+1$ occur in the deformation (5). The irreducibility of the orbit of f in the set of functions of the form (5) now follows from the connectivity of the group R_0 and the sufficient jet theorem (see 1.3).

THEOREM (on stable irreducibility of multisingularities). *For any finite family of isolated singularities $f_1, \ldots, f_k$ and g in n variables there is an isolated singularity h such that h is adjacent to g, and in the versal deformation of h the set of points corresponding to functions with k distinct critical points with critical value 0 that are weakly equivalent to the singularities $f_1, \ldots, f_k$ is irreducible and nonempty. Namely, it suffices to take $h = x_1^N + \cdots + x_n^N$, where $N > n(k(2+\max \mu(f_i))-1)$ and N is large enough so that h is adjacent to g.*

Proof is given in §5.

§2. Definition and elementary properties of the stable cohomology of complements of discriminants

2.1. Local cohomology of complements of discriminants. Let

$$f: (\mathbb{C}^n, 0) \to (\mathbb{C}, 0)$$

be a holomorphic function with a singularity at 0, let

$$F(x, \lambda): \mathbb{C}^n \times \mathbb{C}^l \to \mathbb{C}$$

be its holomorphic deformation, let $\Sigma = \Sigma(F)$ be its discriminant, and let D_δ be the ball of radius δ in $\mathbb{C}^l$.

PROPOSITION. 1. *For all sufficiently small $\delta > 0$ the pairs $(D_\delta, D_\delta \cap \Sigma)$ are topologically the same; in particular, the corresponding rings $H^*(D_\delta - \Sigma)$ are canonically isomorphic (for $\delta' < \delta$ this isomorphism is given by the inclusion $D_{\delta'} - \Sigma \to D_\delta - \Sigma$).*

2. *For any sufficiently small δ the set $\Sigma \cap D_\delta$ is homeomorphic to the cone over its boundary:*

$$\Sigma \cap D_\delta \cong C(\Sigma \cap S_\delta).$$

All these statements follow from the existence of an analytic Whitney stratification of the set Σ; see [Loj$_2$], [Wall].

DEFINITION. The ring of the *local cohomology of the complement of the discriminant* of the deformation F is the ring $H^*(D_\delta - \Sigma(F))$ for any sufficiently small δ.

This ring is denoted by $J^*(F)$.

PROPOSITION. *All rings $J^*(\cdot)$ for all versal deformations of the given singularity f are isomorphic to each other.*

(One can claim even more: the discriminant of any versal deformation is diffeomorphic to the product of the discriminant of any miniversal deformation and a linear space; see [AVG$_2$].)

This isomorphism is given by maps of the bases of deformations used in the definition of the induced deformation (and hence in the definition of the versal deformation); see 1.4 above.

2.2. Maps of local cohomology groups given by adjacency of singularities. To give a precise meaning to the operation of imbedding the complement of a discriminant of a "simpler" singularity in that of a "more complicated" singularity, we must generalize the notions of induced and of equivalent deformations to the case when the corresponding germs do not coincide, or are even applied at different points.

Let $f\colon (\mathbb{C}^n, a) \to (\mathbb{C}, 0)$, $\varphi\colon (\mathbb{C}^n, b) \to (\mathbb{C}, 0)$ be two holomorphic functions defined near (in general distinct) points $a, b \in \mathbb{C}^n$. Let their germs at these points be equivalent, i.e. in suitable local coordinate systems x_i (y_i, respectively) centered at a and b they can be written as equal power series. Suppose we are given a deformation $F(x, \lambda)$ of the function f, $F\colon (\mathbb{C}^n \times \mathbb{C}^l, a \times 0) \to (\mathbb{C}, 0)$, and a deformation $\Phi(y, \varkappa)$ of the function φ, $\Phi\colon (\mathbb{C}^n \times \mathbb{C}^k, b \times 0) \to (\mathbb{C}, 0)$.

DEFINITION. The deformation Φ is *generated* by the deformation F if there exist a holomorphic transformation of parameters $\theta\colon (\mathbb{C}^k, 0) \to (\mathbb{C}^l, 0)$ and a map $U\colon (\mathbb{C}^n \times \mathbb{C}^k, a \times 0) \to (\mathbb{C}^n, b)$ defined near points 0 and $a \times 0$ such that for $\varkappa = 0$ (and hence also for all nearby $\varkappa \in \mathbb{C}^k$) the corresponding map $U(\cdot, \varkappa)\colon \mathbb{C}^n \to \mathbb{C}^n$ is a diffeomorphism near $0 \in \mathbb{C}^n$, and

$$\Phi(x, \varkappa) \equiv F(U(x, \varkappa), \theta(\varkappa)).$$

This formula coincides with formula (3) from the definition of versal deformation. However, even in the case when $a = b = 0$ and $f = \varphi$, this definition of generated deformation does not coincide with the definition of deformation "equivalent to the deformation induced from ... ," since the local diffeomorphism $U(\cdot, \varkappa)\colon \mathbb{C}^n \to \mathbb{C}^n$ used in the construction of the generated deformation does not have to be the identity for $\varkappa = 0$.

Assume now that f, g are two holomorphic functions $(\mathbb{C}^n, 0) \to (\mathbb{C}, 0)$ defined near 0, and g is adjacent to f. Let $G\colon \mathbb{C}^n \times \mathbb{C}^r \to \mathbb{C}$ be a deformation of g, and $\{f\} \subset \Sigma(G) \subset \mathbb{C}^r$ be the set of points of its discriminant corresponding to functions with a singularity near 0 (holomorphically) equivalent to f. Let f' be an arbitrary point of the set $\{f\}$ lying in a sufficiently small neighborhood D_δ of $0 \in \mathbb{C}^r$ (so that $H^*(D_\delta - \Sigma(G)) \equiv J^*(G)$; see 2.1 above) and the corresponding function does not have other critical points with critical value 0 near 0. Then the deformation G can be considered as a deformation of the corresponding singular point of the function f'. Suppose the deformation $F(x, \lambda)$ of the singularity f is generated by this deformation for a suitable map of parameters $\theta \colon (\mathbb{C}^l, 0) \to (\mathbb{C}^r, f')$ (and some family of diffeomorphisms $U(x, \lambda)$). Then θ maps the discriminant of the deformation F to $\Sigma(G)$ and hence defines a homomorphism

$$J^*(F) \leftarrow H^*(D_\delta - \Sigma(G)) \equiv J^*(G). \tag{6}$$

This homomorphism depends on the choice of the point f' (more precisely, of the component of the set $\{f\}$ containing this point) and the generating map θ.

2.3. Definition of stable cohomology for singularities in a fixed number of variables. The *stable cohomology class of complements of discriminants* for singularities of functions in n variables is a rule that assigns to each deformation F of any isolated singularity of any function $(\mathbb{C}^n, 0) \to (\mathbb{C}, 0)$ an element of the ring $J^*(F)$ so that for each pair of adjacent singularities $f \succeq g$ any homomorphism $J^*(G) \to J^*(F)$ constructed as in 2.2 sends the class corresponding (by this rule) to the deformation G to the class corresponding to F.

The set of such rules forms a ring with an obvious grading; this ring is denoted by $_n\mathscr{H}^*$.

EXAMPLE. Let $n = 1$, $f = x^m$. Then for any versal deformation F of the singularity f, $J^*(F)$ is the cohomology ring of the braid group on m strings, the homomorphism (6) corresponding to the adjacency $x^m \to x^{m+1}$ coincides with the homomorphism corresponding to the imbedding $\mathrm{Br}(m) \to \mathrm{Br}(m+1)$ (see §1 of Chapter I), and the ring $_1\mathscr{H}^*$ is the stable cohomology ring of braid groups (i.e. the cohomology ring of the braid group with infinite number of strings).

2.4. Stabilization with respect to the number of variables. Let $f \colon (\mathbb{C}^n, 0) \to (\mathbb{C}, 0)$ be an isolated singularity, and $F(x_1, \ldots, x_n, \lambda)$ be its deformation. Then the function $f(x_1, \ldots, x_n) + x_{n+1}^2 \colon (\mathbb{C}^{n+1}, 0) \to (\mathbb{C}, 0)$ is again isolated (and has the same Milnor number $\mu(f)$). Consider its deformation F_1 given by the formula

$$F_1(x_1, \ldots, x_{n+1}, \lambda) \equiv F(x_1, \ldots, x_n, \lambda) + x_{n+1}^2. \tag{7}$$

This deformation has the same discriminant as F (and is versal together with F). In particular, the cohomology of the complements of the discriminants of these deformations is the same. Thus, the complement of the discriminant of any deformation of a function f in n variables can be considered as the complement of the discriminant of the corresponding deformation of the function $f + x_{n+1}^2$. This gives a homomorphism

$$_{n+1}\mathscr{H}^* \to {}_n\mathscr{H}^*. \tag{8}$$

DEFINITION. The inverse limit ring $\mathscr{H}^* = \varprojlim {}_n\mathscr{H}^*$ with respect to maps (8) is called the *bistable cohomology ring of complements of discriminants*.

Equivalent definition. A bistable cohomology class of complements of discriminants is a rule that associates to each deformation F of any isolated singularity (in any number of variables) an element of the ring $J^*(F)$ in such a way that for adjacent singularities the condition from Definition 2.3 is satisfied, and the class associated to each deformation F coincides with the class associated to its stabilization F_1.

THEOREM. *The homomorphism* (8) *is compatible with the Freudenthal homomorphism: for any n the diagram*

$$\begin{array}{ccc} {}_{n+1}\mathscr{H}^* & \xrightarrow{\sim} & H^*(\Omega^{2n+2}S^{2n+3}) \\ \downarrow & & \downarrow \\ {}_n\mathscr{H}^* & \xrightarrow{\sim} & H^*(\Omega^{2n}S^{2n+1}) \end{array} \tag{9}$$

is commutative; here the horizontal arrows are taken from Theorem 1 of this chapter, and the right vertical arrow is given by the double Freudenthal (*suspension*) *imbedding*.

This will be proved in §3.

2.5. Consequences of the definitions. Suppose the deformation F generates itself by some local transformation of parameters $\theta: (\mathbb{C}^l, 0) \to (\mathbb{C}^l, 0)$. This transformation maps the discriminant into itself and hence acts on $J^*(F)$.

Denote by $I^*(F)$ the subring of $J^*(F)$ formed by the elements invariant under all such transformations used in self-generations of the deformation F. For example, for any versal deformation of the singularity $x^m: \mathbb{C}^1 \to \mathbb{C}^1$, $I^* = J^*$.

PROPOSITION 1. *For any deformation F of a singularity in n variables the image of the tautological homomorphism ${}_n\mathscr{H}^* \to J^*(F)$ belongs to $I^*(F)$.* □

If f, φ are biholomorphically equivalent singularities, and F, Φ are their versal deformations then the rings $J^*(F)$, $J^*(\Phi)$ are isomorphic. In general, this isomorphism is not canonical, but its restriction to the subring $I^*(F)$ is uniquely determined, and the image of this subring is $I^*(\Phi)$.

In Definition 2.3 we use various deformations of functions. We can obtain an equivalent definition by considering only versal deformations: they form a representative subset among all deformations. Here any deformation of a simpler singularity is generated from any versal deformation of a more complicated one by the scheme from 2.2. Indeed, again let $G: \mathbb{C}^n \times \mathbb{C}^r \to \mathbb{C}$

be a deformation of the singularity g, $\{f\} \subset \Sigma(G)$ be the set of functions with a singularity equivalent to the fixed singularity f, and f' be a function from the set $\{f\}$ without any other discriminant critical points near 0. Let $\theta\colon (\mathbb{C}^l, 0) \to (\mathbb{C}^r, f')$ be an arbitrary holomorphic map. Then there exists a deformation $F\colon \mathbb{C}^n \times \mathbb{C}^l \to \mathbb{C}$ of f generated from the family of functions $G(\cdot, \rho)$ (considered as a deformation of a singular point of the function f') using the map θ. (For example, it can be defined by the formula $F(x, \lambda) = G(D(x), \theta(\lambda))$, where $D\colon (\mathbb{C}^n, 0) \to (\mathbb{C}^n, b)$ is an arbitrary local diffeomorphism such that $f = f' \circ D$.) For such θ and f' the following statements are true.

PROPOSITION 2. *Let G be a versal deformation of g. Then*
(1) *the set $\{f\}$ is nonsingular near f';*
(2) *if the map $\theta\colon (\mathbb{C}^l, 0) \to (\mathbb{C}^r, f')$ is transversal to the stratum $\{f\}$ then any deformation of the singularity f generated using the map θ is again versal.*

This follows from the theorem claiming that the notions of a versal and of an infinitesimally versal deformation coincide; see [AVG$_1$].

2.6. PROBLEM. Suppose the miniversal deformation $F(x, \lambda)$ of a singularity of finite multiplicity is equivalent to the one induced from itself: $F(x, \lambda) \equiv F(U(x, \lambda), \theta(\lambda))$, $U(\cdot, 0) = \mathrm{id}$. Is it then true that $\theta = \mathrm{id}$ or at least that the map $\theta\colon (D_\delta, \Sigma(F)) \to (D_\delta, \Sigma(F))$ is homotopic to the identity as a map of pairs?

§3. Stable cohomology of complements of discriminants, and the loop spaces

3.1. In this section we prove Theorems 1 and 2 from the introduction to this chapter. Here is the plan of the proof:

(A) Let f be an isolated singularity, and $F(x, \lambda)$ be its deformation. Using any value of the parameter $\lambda \in D_\delta$ of this deformation, we construct a map $X_\lambda\colon S^{2n} \to \mathbb{R}^{2n+2}$ depending continuously on λ and sending the base point $* \in S^{2n}$ to the point $(1, 0, \ldots, 0) \in \mathbb{R}^{2n+2}$; here the image of X_λ contains the point $0 \in \mathbb{R}^{2n+2}$ if and only if λ is a discriminant value of the parameter. In particular, we get the map

$$\omega_F \colon (D_\delta - \Sigma(F)) \to \Omega^{2n}(\mathbb{R}^{2n+2} - 0) \xrightarrow{\sim} \Omega^{2n} S^{2n+1} \tag{10}$$

and the maps on the cohomology level

$$\omega_F^* \colon H^i(\Omega^{2n}(\mathbb{R}^{2n+2} - 0)) \to H^i(D_\delta - \Sigma(F)). \tag{11}$$

(B) The maps ω_F corresponding to different deformations homotopy commute with the maps of the bases of the deformations which generate these deformations from each other (see 2.2). More precisely, if $f \succcurlyeq g$ is a pair of adjacent singularities, F and G are their deformations, and F is generated from G by some map of the bases $\theta: D_{\delta(F)} \to D_{\delta(G)}$ then the corresponding maps ω_F and $\omega_G \circ \theta: (D_{\delta(F)} - \Sigma(F)) \to \Omega^{2n}(\mathbb{R}^{2n+2} - 0) \sim \Omega^{2n} S^{2n+1}$ are homotopic.

In particular, this allows us to define a homomorphism $H^*(\Omega^{2n} S^{2n+1}) \to {}_n\mathscr{H}^*$ properly: for any deformation F the corresponding composition $H^*(\Omega^{2n} S^{2n+1}) \to {}_n\mathscr{H}^* \to J^*(F)$ coincides with ω_F^*. At the next stages (C) and (D) we will show that this homomorphism is an isomorphism.

(C) For any versal deformation F we construct a spectral sequence that computes the cohomology of $D_{\delta(F)} - \Sigma(F)$ and is similar to the spectral sequences from §III.5 converging to the cohomology of the loop spaces of spheres. In particular, all its nonzero terms $E_r^{p,q}$ for $r \geq 1$ lie in the second quadrant $p \leq 0 \leq q$; moreover, $E_1^{p,q} = 0$ if $q + 2p < 0$. All such spectral sequences corresponding to versal deformations of one singularity f are isomorphic to each other starting with the term E_1.

(D) Suppose the function f has the form

$$f = x_1^N + \cdots + x_n^N, \tag{12}$$

where $N > n(2c - 1)$ as in Theorem 2; let F be its versal deformation. Let $E_r^{p,q}(F)$ be the corresponding spectral sequence constructed at stage (C), and $E_r^{p,q}(\Omega^{2n} S^{2n+1})$ be the spectral sequence constructed in §III.5 and converging to the cohomology of $\Omega^{2n} S^{2n+1}$; let $E_r^{p,q}(F, c)$ and $E_r^{p,q}(\Omega^{2n} S^{2n+1}, c)$ be the truncated versions of these spectral sequences obtained by setting $E_1^{p,q} = 0$ for all (p, q) with $p < -c$. Then these truncated spectral sequences are canonically isomorphic.

In particular, this implies that both the groups in the homomorphism (11) are isomorphic for $i \leq c$: indeed, all terms $E_1^{p,q}$ of both original spectral sequences with $p + q \leq c$ lie in the strip $\{p \geq -c\}$ not affected by the truncation, and all differentials $d_r: E_r^{p,q} \to E_r^{p+r, q-r+1}$ of these sequences do not exit this strip. The resulting isomorphism between the groups $H^i(D_\delta - \Sigma(F))$, $H^i(\Omega^{2n} S^{2n+1})$ with $i \leq c$ is induced by the imbedding ω_F, i.e. for $i \leq c$ the map (11) is an isomorphism.

This statement, together with (B), implies that, for the singularity (12) with $N > n(2c - 1)$, any versal deformation of it, and any $i < c$, the local cohomology group $J^i(F)$ of the complement of $\Sigma(F)$ coincides with its invariant subgroup $I^*(F)$ (see 2.5), i.e. the group of all self-generations of the deformation F acts trivially on $J^i(F)$.

These facts imply the equality ${}_n\mathscr{H}^* \cong H^*(\Omega^{2n} S^{2n+1})$. Such equalities for various n are compatible with the Freudenthal homomorphisms generated by the natural imbeddings of suspensions $\Omega^{2n} S^{2n+1} \to \Omega^{2n+2} S^{2n+3}$ (see [Ad], [Fuchs$_1$]); in particular, $\mathscr{H}^* \cong H^*(\Omega^\infty S^{\infty+1})$. More formally the following statement is valid.

(E) Let $f: (\mathbb{C}^n, 0) \to (\mathbb{C}, 0)$ be an isolated singularity, $F(x_1, \ldots, x_n, \lambda)$

be its deformation, and $F_1 \equiv F(x_1, \ldots, x_n, \lambda) + x_{n+1}^2$ be a deformation stably equivalent to F of the function $f_1 \equiv f(x_1, \ldots, x_n, \lambda) + x_{n+1}^2$ (see 2.4). Then the corresponding maps ω_F, ω_{F_1} are compatible with the Freudenthal map of (double) suspensions: the diagram

$$\begin{array}{ccc} D_{\delta(F)} - \Sigma(F) & \xrightarrow{\omega_F} & \Omega^{2n} S^{2n+1} \\ \downarrow \sim & & \downarrow \\ D_{\delta(F_1)} - \Sigma(F_1) & \xrightarrow{\omega_{F_1}} & \Omega^{2n+2} S^{2n+3} \end{array}$$

is homotopy commutative, where the right vertical arrow is the double-Freudenthal suspension map. This proves the theorem of 2.4.

The realization of this program occupies subsections 3.2–3.4.

3.2. Stages (A), (B), and (E) of the program from 3.1. Again let B_ε be a small ball in $\mathbb{C}^n$ with center at 0, D_δ be a very small (even when compared to B_ε) ball in the base of the deformation F so that the requirements of Lemmas 1 and 2 of 1.2 are satisfied. To each value $\lambda \in D_\varepsilon$ we associate a map $\chi_\lambda \colon B_\varepsilon \to \mathbb{C}^{n+1}$ given by the formula

$$\chi_\lambda(x) = \left(f_\lambda(x), \left. \frac{\partial f_\lambda}{\partial x_1} \right|_x, \ldots, \left. \frac{\partial f_\lambda}{\partial x_n} \right|_x \right). \tag{13}$$

Now we will extend these maps to a suitable family of maps of the sphere S^{2n} (that is, the one-point compactification of the space $\mathbb{C}^n \supset B_\varepsilon$).

If δ is sufficiently small with respect to ε, then for all x from $B_\varepsilon - B_{\varepsilon/2}$ we have the inequality

$$|f(x) - f_\lambda(x)| + |\operatorname{grad}|_x (f - f_\lambda)| \leq \frac{1}{2} |\operatorname{grad}|_x f|. \tag{14}$$

Let s be any smooth map $S^{2n} - B_{\varepsilon/2} \to \mathbb{C}^{n+1} - 0$ coinciding on $B_\varepsilon - B_{\varepsilon/2}$ with the map χ_0 (given by formula (13) with $\lambda = 0$) and sending the added point $*$ to the base point $(1, 0, \ldots, 0)$ of the space $\mathbb{C}^{n+1}$. Fix any smooth function $\psi \colon S^{2n} \to [0, 1]$ such that $\psi \equiv 1$ on $B_{\varepsilon/2}$ and $\psi \equiv 0$ outside B_ε. Finally, for any $\lambda \in D_\varepsilon$ define a map $X_\lambda \colon S^{2n} \to \mathbb{C}^{n+1}$ by the formula

$$X_\lambda(x) = \psi(x)\chi_\lambda(x) + (1 - \psi(x))s(x). \tag{15}$$

This map is an element of the space $\Omega^{2n}\mathbb{C}^{n+1}$. Therefore, we have constructed an imbedding $D_\delta \to \Omega^{2n}\mathbb{C}^{n+1}$: to each $\lambda \in D_\delta$ it associates the map X_λ. By construction (and by inequality (14)), the image of the map X_λ contains the point 0 if and only if $\lambda \in \Sigma(F)$. In particular, we get an imbedding

$$\omega_F \colon (D_\delta - \Sigma(F)) \to \Omega^{2n}(\mathbb{C}^{n+1} - 0);$$

this is exactly the imbedding used in formula (10).

The fact that such maps ω_F satisfy (B) and (E) of 3.1 follows directly from the construction and from the connectivity of the group of germs of diffeomorphisms $(\mathbb{C}^n, 0) \to (\mathbb{C}^n, 0)$.

3.3. The spectral sequence for the group $H^*(D_\delta - \Sigma(F))$ is constructed using the same scheme as for the similar sequences from §§3, 4, and 5 of Chapter III; nevertheless we must give a formal description.

Again let $f: (\mathbb{C}^n, 0) \to (\mathbb{C}, 0)$ be an isolated singularity, F its versal deformation, and $\Sigma = \Sigma(F)$ its discriminant. For all versal deformations of the singularity f (and any singularity equivalent to f) the rings $J^*(F)$ are isomorphic to each other (and, moreover, the corresponding spaces $D_\delta - \Sigma(F)$ are homotopy equivalent, see [AVG$_1$]). Thus we can assume that f is a polynomial (any isolated singularity is equivalent to a polynomial by the sufficient jet theorem, see 1.3), and the deformation F has the form $(4')$.

In this section we denote by Σ the part of the discriminant $\Sigma(F)$ in the (open) ball D_δ used in the definition of the Milnor bundle (see 1.2). Let $\overline{\Sigma}$ be the closure of Σ, $\partial\Sigma = \overline{\Sigma} - \Sigma \equiv \overline{\Sigma} \cap \partial D_\delta$. As before, $\overline{H}_*$ denotes the closed homology, i.e. the homology of the one-point compactification modulo the added point.

By the Alexander duality theorem

$$H^i(D_\delta - \Sigma) = H_{2l-i-1}(\overline{\Sigma}, \partial\Sigma) = \overline{H}_{2l-i-1}(\Sigma); \qquad (16)$$

we will compute the latter group.

NOTATION. Σ^t is the set of values of the parameter $\lambda \in D_\delta$ such that the function $F(\cdot, \lambda)$ has at least t distinct critical points near 0 with critical value 0.

$\widetilde{\Sigma}^t$ is the set of pairs of the form (point $\lambda \in \Sigma^t$; t points at which $F(\cdot, \lambda)$ has critical points with value 0). This is a complex $(l-t)$-dimensional analytical closed subset in the product $D_\delta \times \mathbb{C}^n(t)$ (recall that $\mathbb{C}^n(t)$ is the set of subsets of cardinality t in $\mathbb{C}^n$). The projections of $\widetilde{\Sigma}^t$ to the factors of this product define the maps

$$\begin{aligned}\gamma_t &: \widetilde{\Sigma}^t \to \Sigma^t, \\ \pi_t &: \widetilde{\Sigma}^t \to \mathbb{C}^n(t).\end{aligned} \qquad (17)$$

All fibers of the map π_t are contractible and canonically oriented: they are defined in the ball D_δ by a system of linear nonhomogeneous complex equations for λ.

THEOREM. *There exists a spectral sequence* $E_r^{p,q} \to H^{p+q}(D_\delta - \Sigma(F))$, $p \leq 0 \leq q$, *such that* $E_1^{p,q} = \overline{H}_{2l-q}(\widetilde{\Sigma}^{-p}, \pm\mathbb{Z})$, *where* $\pm\mathbb{Z}$ *is the local system on* $\widetilde{\Sigma}^{-p}$ *induced from the system* $\pm\mathbb{Z}$ *on* $\mathbb{C}^n(-p)$ *by the projection* π_{-p}; *see* §I.4. *In particular,* $E_1^{p,q} = 0$ *for* $q + 2p < 0$.

To prove this theorem we construct a filtered space whose closed homology coincides with that of Σ. Let the number M exceed the maximal multiplicity of self-intersections of Σ (it suffices to take $M \geq \mu(f)$). Let τ be an arbitrary smooth imbedding of the space $\mathbb{C}^n$ into the space $\mathbb{R}^L$ of sufficiently large dimension such that the images of no M points lie on the same $(M-2)$-dimensional affine plane in $\mathbb{R}^L$. To any point of the set $\widetilde{\Sigma}^t$, i.e. a certain pair of the form $(\lambda \in \Sigma^t, t \text{ points } a_1, \ldots, a_t \text{ in } \mathbb{C}^n)$, we associate a simplex in the set $D_\delta \times \mathbb{R}^L$ whose vertices are the points $(\lambda, \tau(a_1)), \ldots, (\lambda, \tau(a_t))$. The union of such simplices over all sets $\widetilde{\Sigma}^t$ with $t \leq m$ is denoted by σ_m. The set $\sigma_M \equiv \bigcup \sigma_m$ is denoted simply by σ.

PROPOSITION. (A) *The homomorphism $\overline{H}_*(\sigma) \to \overline{H}_*(\Sigma)$ given by the projection $D_\delta \times \mathbb{R}^L \to D_\delta$ is an isomorphism.*
(B) *For any $t = 1, 2, \ldots, M$ we have*

$$\overline{H}_{i+t-1}(\sigma_t - \sigma_{t-1}) \cong \overline{H}_i(\widetilde{\Sigma}^t, \pm \mathbb{Z}).$$

Part (A) is proved in exactly the same way as Lemma 1 from 3.3 of Chapter III; part (B) is the Thom isomorphism for the bundle $\sigma_t - \sigma_{t-1} \to \widetilde{\Sigma}^t$.

This proposition immediately implies the previous theorem: consider the homology spectral sequence computing the group $\overline{H}_*(\sigma) \cong \overline{H}_*(\Sigma)$ and generated by the filtration of the space σ by the sets σ_m. Rename its term $E^r_{t,q}$ by $E^{-t, 2l-q-1}_r$ and set $E^{0,0}_r = \mathbb{Z}$, $E^{0,q}_r = 0$ for $q > 0$ and all r. We have obtained the required spectral sequence.

3.4. The isomorphism of the truncated spectral sequences for the cohomology of $D_\delta - \Sigma(F)$ and $\Omega^{2n} S^{2n+1}$; stage (D) of the program from 3.1.

3.4.1. Proof of the isomorphism from 3.1(D) essentially repeats the proof of Theorems 4.2.1–4.2.4 of Chapter III; we mention the crucial points of this proof.

To construct the spectral sequence $E^{p,q}_r \to H^{p+q}(\Omega^{2n} S^{2n+1})$ in §III.5 we used finite-dimensional approximations of the space of maps $(S^{2n}, *) \to (\mathbb{R}^{2n+2}, *)$. Here we consider another system of approximating subspaces W^d, $d \to \infty$, that satisfies all necessary properties of the old system and contains as a subspace the set of maps of the form (15) for an appropriate versal deformation of f. The imbedding of this subspace into W^d induces an imbedding of the standard resolution σ of the set $\Sigma(F) \subset D_\delta$ into the resolution $\{\Sigma(d)\}$ of the discriminant of the space W^d (this discriminant is the set of maps $S^{2n} \to \mathbb{R}^{2n+2}$ of class W^d passing through 0). This imbedding respects the natural filtrations of the resolutions. For any d, the one-point compactification of the cth term $\{\Sigma(d)\}_c$ of the resolution $\{\Sigma(d)\}$ is homotopy equivalent (as a filtered space) to a multiple suspension of the one-point compactification of the corresponding term σ_c of the resolution σ. This implies the desired isomorphism of truncated spectral sequences considered in part (D) of our program.

Now we give a detailed proof.

3.4.2. *Nondegenerate subspaces in* $\Omega^{2n}\mathbb{R}^{2n+2}$. The space $\Omega^{2n}\mathbb{R}^{2n+2}$ is considered as the space of continuous maps $(S^{2n}, *) \to (\mathbb{R}^{2n+2}, (1, 0, \ldots, 0))$ and is endowed with the corresponding affine structure.

DEFINITION. A finite-dimensional affine subspace $\Xi \subset \Omega^{2n}\mathbb{R}^{2n+2}$ is called *nondegenerate* if for any natural number k the maps $\vartheta \in \Xi$ with card $\vartheta^{-1}(0) \geq k$ form a subset that has codimension $\geq 2k$ in Ξ or is empty.

PROPOSITION. *For any natural* Δ, *the* Δ-*dimensional nondegenerate subspaces are dense in the set of all* Δ-*dimensional affine subspaces of the space* $\Omega^{2n}\mathbb{R}^{2n+2}$. *Moreover, for any nondegenerate subspace* $\Xi \subset \Omega^{2n}\mathbb{R}^{2n+2}$ *and any* $\Delta \geq \dim \Xi$ *the nondegenerate subspaces are dense in the set of all* Δ-*dimensional subspaces containing* Ξ.

This follows from the Thom multijet transversality theorem; see the proof of Theorem 4.2.1 of Chapter III.

3.4.3. *Nondegenerate versal deformations.* Now let f be of the form (12) and F be an arbitrary versal deformation of f of type $(4')$. Then the set of all functions of the form $(4')$ (without the condition that λ be small) is an affine space $\{\lambda\} \sim \mathbb{C}^\mu$, $\mu = (N-1)^n$. The map of this space to the space $\Omega^{2n}\mathbb{R}^{2n+2}$ given by formula (15) preserves the affine structure. However, for a poor choice of the deformation $(4')$ and function ψ used in the definition of this map, its image may happen to be degenerate: the k-fold self-intersection of the discriminant may have too large a dimension away from D_δ. This difficulty can be overcome by the following result.

PROPOSITION. 1. *All spectral sequences* $E_r^{p,q} \to H^{p+q}(D_\delta - \Sigma(F))$ *constructed in 3.3 for arbitrary versal deformations of the singularity f are isomorphic starting with the term* E_1.

2. *For any singularity f of finite multiplicity there exist its miniversal deformation of the form* $(4')$ *and a smooth function* $\psi: S^{2n} \to [0, 1]$ *such that the space of all maps* X_λ, $\lambda \in \mathbb{C}$, *of the form* (15) *is nondegenerate in* $\Omega^{2n}\mathbb{R}^{2n+2}$.

Part 1 follows directly from the definition of versal deformation: the desired isomorphism for various versal deformations is induced by the map of their discriminants (and the resolutions of these discriminants) given by a map of the bases of these deformations inducing them from each other. Part 2 reduces to the Thom multijet transversality theorem, just as Theorem 4.2.1 of Chapter III does.

Fix such a versal deformation $(4')$ and function ψ.

3.4.4. *The approximating affine spaces* $W^d \subset \Omega^{2n}\mathbb{R}^{2n+2}$ will be defined inductively. Imbed S^{2n} as the unit sphere into $\mathbb{R}^{2n+1}$. Let Θ_*^d be the space of maps $S^{2n} \to \mathbb{R}^{2n+2}$ given by restricting to S^{2n} the polynomial maps

$\mathbb{R}^{2n+1} \to \mathbb{R}^{2n+2}$ of degree $\leq d$ sending the base point to the base point (equal to $(1, 0, \ldots, 0)$). Define W^{-1} as the space of maps X_λ of the form (15) constructed using the above versal deformation and ψ. Let $\widetilde{W}^d$ be an affine space consisting of maps $S^{2n} \to \mathbb{R}^{2n+2}$ of the form $\tau\theta + (1-\tau)w$, $\tau \in [0, 1]$, where $\theta \in \Theta_*^d$, $w \in W^{d-1}$. By the induction hypothesis the subspace W^{d-1} is nondegenerate in $\Omega^{2n}\mathbb{R}^{2n+2}$. Hence by the proposition of 3.4.2 we can perturb the space $\widetilde{W}^d$ so slightly that it still contains W^{d-1} and is nondegenerate; the resulting perturbed space is denoted by W^d.

3.4.5. For an arbitrary d consider the set $\Sigma(d)$ consisting of all maps of class W^d passing through the point $0 \in \mathbb{R}^{2n+2}$. Similarly to the previous construction we construct a resolution $\{\Sigma(d)\}$ of the set $\Sigma(d)$, and a spectral sequence $E_r^{p,q}[d] \to H^{p+q}(W^d - \Sigma(d))$; in particular

$$E_1^{p,q}[d] = \overline{H}_{\dim W^d - p - q - 1}(\{\Sigma(d)\}_{-p} - \{\Sigma(d)\}_{-p-1}),$$

where $\{\Sigma(d)\}_i$ is the ith term of the standard filtration of this resolution.

As d increases, these spectral sequences stabilize to the same spectral sequence $E_r^{p,q} = E_r^{p,q}(\Omega^{2n}\mathbb{R}^{2n+1})$ that was constructed in §III.5 to compute the cohomology of the space $\Omega^{2n}\mathbb{R}^{2n+1}$; indeed, the approximating spaces W^d are just as good as the spaces $\widetilde{\Theta}^d$ used there.

PROPOSITION. *If the perturbed space W^d is sufficiently close to $\widetilde{W}^d$, then the spectral sequence $E_r^{p,q}[d]$ is isomorphic to the stable sequence $E_r^{p,q}(\Omega^{2n}\mathbb{R}^{2n+1})$ everywhere in the domain $\{p + q \leq d\}$. In particular,*

$$H^i(W^d - \Sigma(d)) \cong H^i(\Omega^{2n}\mathbb{R}^{2n+1}) \quad \text{for } i \leq d.$$

The last isomorphism is given by the identity imbedding.

The proof repeats that of Theorem 4.2.4 from Chapter III. □

3.4.6. Let us consider in more detail the spectral sequence $E_r^{p,q}(F) \to H^{p+q}(D_\delta - \Sigma(F))$ from §3.3 in the case when the function f has the special form (12).

THEOREM. *Let $f = x_1^N + \cdots + x_n^N$, $N > n(2c - 1)$, and let F be an arbitrary versal deformation of f of the form $(4')$. Then for any $t \leq c$ the set $\widetilde{\Sigma}^t \subset D_\delta \times \mathbb{C}^n(t)$ is a smooth complex manifold (with boundary), and the obvious projection $\pi_t: \widetilde{\Sigma}^t \to \mathbb{C}^n(t)$ (see (17)) gives a homotopy equivalence between these spaces.*

Proof is given in §5.

The imbedding $D_\delta \to W^d$ that assigns to each $\lambda \in D_\delta$ a map X_λ by formula (15) sends $\Sigma(F)$ to $\Sigma(d)$. This imbedding can be lifted to an imbedding $I: \sigma \to \{\Sigma(F)\}$ of the standard resolutions of these sets that respects the filtrations $I(\sigma_i) \subset \{\Sigma(F)\}_i$. The last theorem immediately gives the

COROLLARY. *Under the conditions of the theorem, for any $t \leq c$ the imbedding $I: (\sigma_t - \sigma_{t-1}) \to (\{\Sigma(d)\}_t - \{\Sigma(d)\}_{t-1})$ is a homotopy equivalence.*

PROPOSITION. *Under the conditions of the last theorem, for any $d \geq -1$ the spectral sequences*

$$E_r^{p,q}(d) \to H^{p+q}(W^d - \Sigma(d)) \quad \text{and} \quad E_r^{p,q}(F) \to H^{p+q}(D_\delta - \Sigma(F))$$

are isomorphic in the domain $\{p + q \leq c\}$. In particular, $H^i(D_\delta - \Sigma(F)) \cong H^i(W^d - \Sigma(d))$ for $i \leq c$; the last isomorphism is given by the imbedding (15) of these spaces.

This proposition reduces to the previous one in exactly the same way as Theorem 4.2.4 in III.4.5 reduces to Theorem 4.2.3: the one-point compactification of the cth term $\{\Sigma(d)\}_c$ of the resolution $\{\Sigma(d)\}$ is homotopy equivalent to the $(\dim W^d - \dim D_\delta)$ fold suspension of the one-point compactification of the cth term of the natural filtration of the resolution σ. This homotopy equivalence respects natural filtrations of these spaces and induces an isomorphism of our spectral sequences.

This completes stage (D) of the program from 3.1.

§4. Cohomological Milnor bundles

Here we prove Theorems 5–7 from the introduction to this chapter, where we considered characteristic classes of cohomological real Milnor bundles over complements of discriminants.

4.1. Proof of Theorem 5. Let F, G be deformations of singularities f, g, with g adjacent to f, and let F be generated from the deformation G by some map of parameters θ and a family of local diffeomorphisms U (see 2.2); here we can assume that the ball $D_{\delta(F)} \subset \mathbb{C}^l$ used in the definition of Milnor bundle for F is mapped by the map θ into a similar ball $D_{\delta(G)}$, and for any $\lambda \in D_{\delta(F)}$ the corresponding local diffeomorphism $U(\cdot, \lambda)$ maps the ball $B_{\varepsilon(F)}$ into the ball $B_{\varepsilon(G)}$; in particular, if λ is a nondiscriminant value of the parameter, then the level manifold $V = f_\lambda^{-1}(0) \cap B_{\varepsilon(F)}$ is imbedded in the manifold $V_{\theta(\lambda)} = g_{\theta(\lambda)}^{-1}(0) \cap B_{\varepsilon(G)}$. This imbedding determines a homomorphism of the real cohomology of these manifolds. These homomorphisms taken for all $\lambda \in D_{\delta(F)} - \Sigma(F)$ define a morphism of two vector bundles on $D_{\delta(F)} - \Sigma(F)$: the real Milnor cohomological bundle $M(F)$ and the bundle induced from the similar bundle $M(G)$ by the map θ. We must prove that this morphism is an epimorphism everywhere (i.e. the bundle $\theta^* M(G)$ is isomorphic to the direct sum of the bundle $M(F)$ and the bundle of kernels of this morphism) and that the bundle of kernels is isomorphic to the trivial one. We can assume that both deformations F and G are versal: otherwise we could complete them to versal. In this case our morphism $\theta^* M(G) \to M(F)$ is an epimorphism because the basis in the (co)homology of nonsingular fibers is generated by (co)cycles vanishing along suitable paths in the base of the versal deformation; see, for example, [AVG$_2$], [AVGL$_1$]. Triviality of the bundle of kernels follows from the Picard-Lefschetz formula (ibid.): the cohomology classes of the manifold $V_{\theta(\lambda)} \equiv g_{\theta(\lambda)}^{-1}(0) \cap B_{\varepsilon(G)}$ that vanish on the vanishing cycles obtained by the decay of the singular point $f' \in \{f\} \subset \Sigma(G)$ do not change in traversing the loop in the space $D_{\theta(G)} - \Sigma(G)$ lying in a neighborhood of this point. □

4.2. Proof of Theorem 7. Let $\lambda \in \Sigma$ be a point of a k-fold transversal intersection of nonsingular branches of Σ; i.e. the function $f_\lambda = F(\cdot, \lambda)$ has k Morse critical points with zero value. Fix a disk $\nabla \subset \mathbb{C}^1$ centered at 0 that does not contain nonzero critical values of the function f_λ; let $\varkappa \approx \lambda$ so that k critical values of $f_\varkappa$ are again in ∇ but are distinct and nonzero, and the other critical values are still outside ∇. Fix a basis in the homology of the set $f_\varkappa^{-1}(0)$ (see [AVGL$_1$], II.1.5) given by a system of paths such that the path connecting any of our k points with zero lies entirely in ∇. Then the corresponding k vanishing cycles are pairwise orthogonal with respect to the intersection index. If W is a small spherical neighborhood of the point λ in $D_{\delta(F)}$ then $W - \Sigma$ is homotopy equivalent to the torus considered in Theorem 7, and k independent generators of the group $H_1(W - \Sigma)$ are given by k loops such that when the function $f_\varkappa$ moves along any of them then one of critical values of $f_\varkappa$ turns around 0 along a loop not leaving ∇, while the other $k-1$ critical values stay unmoved (see [AVGL$_1$], §II.1). As an easy consequence of the Picard-Lefschetz formula we see that in the restriction to $W - \Sigma$ the bundle M_0 splits into the sum of the trivial $(\mu(f) - k)$-dimensional bundle and k one-dimensional bundles, each flipped in traversal around one of the critical values and not flipped in traversal around the others. Therefore, denoting by α_i the dual generators in the one-dimensional cohomology of $W - \Sigma$ we see that $w_k = \alpha_1 \cdot \ldots \cdot \alpha_k \neq 0$ in the restriction to $W - \Sigma$. q.e.d.

4.3. Proof of Theorem 6. First we prove the theorem for certain special deformations and functions. Namely, let φ have the form

$$x_1^{N_1} + \cdots + x_n^{N_n} \tag{18}$$

(n even). Consider a deformation Φ of the function φ consisting of functions of the form $\varphi_\lambda = f + \sum \lambda_\alpha x^\alpha$, where the λ_α are complex parameters, and the summation is carried out only over the multi-indices $\alpha = (\alpha_1, \ldots, \alpha_n)$ such that $\alpha_1/N_1 + \cdots + \alpha_n/N_n < 1$ (i.e. the vector α lies below the hyperplane spanned by the vector of exponents of the monomials $x_i^{N_i}$).

LEMMA. *If all numbers $N_1, \ldots, N_n$ are relatively prime, then the real Milnor cohomological bundle $M_e(F)$ over the complement of the discriminant of the deformation F is trivializable.*

PROOF OF THE LEMMA. For any value of the parameter $\lambda = \{\lambda_\alpha\}$ outside the discriminant $\Sigma(F)$, the affine algebraic variety $W_\lambda = \varphi_\lambda^{-1}(0)$ is nonsingular everywhere in $\mathbb{C}^n$ (and not only near 0), and the space $H(\lambda) = H^{n-1}(\varphi_\lambda^{-1}(0), \mathbb{C})$ is isomorphic to $\mathbb{C}^\mu$, where μ is the Milnor number of φ. The space $H(\lambda)$ carries a mixed Hodge structure (see [D], [St], [Var$_2$], [AVG$_2$]). If all numbers N_i are relatively prime, then this structure is pure, i.e. the space $H(\lambda)$ splits canonically into the sum of complex subspaces $H^{(i)}(\lambda)$, where $i = 0, 1, \ldots, n-1$, and the space $H^{(i)}(\lambda)$ is complex conjugate to $H^{(n-1-i)}(\lambda)$. These subspaces split the complex Milnor cohomological bundle $M_e \otimes \mathbb{C}$ into the direct sum of n complex subbundles. Denote by $H^<(\lambda)$ ($H^>(\lambda)$) the sum of the subspaces $H^{(i)}(\lambda)$ over all $i < (n-1)/2$ ($i > (n-1)/2$, respectively); in particular, $H^<(\lambda) = \overline{H^>(\lambda)}$. The real subspace $H^{n-1}(\varphi_\lambda^{-1}(0), \mathbb{R}) \subset H(\lambda)$ projects isomorphically onto any of these two subspaces along the other. Now the lemma follows from the next statement.

THEOREM (A. B. Givental, A. N. Varchenko). *The complex bundle over the set of nondiscriminant values of the parameter λ whose fiber over the point λ is the space $H^<(\lambda)$ is trivializable. Its trivialization is given by classes of various differential Gelfand-Leray $(n-1)$-forms $x^\alpha (dx_1 \wedge \cdots \wedge dx_n)/d\varphi_\lambda$ (see* [AVG$_2$]), *where α ranges over the set $\{0 \leq \alpha_i \leq N_i - 2 \;\; \forall i, \; \Sigma(\alpha_i + 1)/N_i < n/2\}$.* □

Finally, for the real cohomological bundle M_e over the space of an arbitrary deformation F of any singularity, Theorem 6 follows from the fact that this deformation is generated (in the sense of 2.2) by a deformation of the special type considered in the lemma for singularity (18) with arbitrary sufficiently large numbers N_i, and the induced bundle on the complement of $\Sigma(F)$ splits into the sum of the bundle M_e and the trivial bundle similarly to 4.1.

§5. Proofs of technical results

In this section we prove the theorem on stable irreducibility of multisingularities (see end of §1) and the theorem from 3.4.6. Both follow from Theorem 5.4 below. To formulate this theorem we need several definitions.

5.1. Let q be a natural number, and SG be the set of germs of holomorphic functions $f: (\mathbb{C}^n, 0) \to \mathbb{C}$ with a singularity at 0. Obviously, $SG \equiv \mathbb{C}^1 \oplus \mathfrak{M}^2$; see 1.3.

DEFINITION. A finite family of subsets $L_1, \ldots, L_t$ of the space SG is called a q-family if for every $i = 1, \ldots, t$:

(1) L_i together with any singularity f contains all singularities weakly equivalent to it (that is, singularities R_0-equivalent to $a \cdot f$ for some complex constant $a \neq 0$; see 1.5);

(2) Any set L_j either coincides with L_i or is disjoint from it;

(3) Whether or not a germ f belongs to L_i is determined by its q-jet;

(4) The image of L_i under the projection $SG \to SG/\mathfrak{M}^{q+1}$ is an ana-

lytic (not necessarily smooth) set (and hence by (3) the same is true of the projection to any $SG/\mathfrak{M}^r$, $r > q$).

(Keeping in mind the proof of Theorem 3.4.6 it suffices to assume that $q = 1$ and all L_i coincide with $\mathfrak{M}^2$.)

5.2. In the parameter space of a versal deformation of type (4) we introduce the norm $\|\lambda\| = \sum |\lambda_\alpha|$. The ball of radius δ in this norm will be denoted by D_δ. (This disagrees with §§2 and 3, where this notation was used for the ball in the Hermitian metric, but such a change is inessential, since for sufficiently small δ the sets $D_\delta - \Sigma$ defined for the balls D_δ in both metrics are homotopy equivalent.) Fix such a ball D_δ.

5.3. Given a q-family $L = (L_1, \ldots, L_t)$. Denote by $[L]$ the set of unordered collections of t pairs of the form (point $a \in \mathbb{C}^n$, q-jet of a function $(\mathbb{C}^n, a) \to \mathbb{C}$) such that these t points are distinct, and suitably ordered q-jets in them are jets of functions $(\mathbb{C}^n, a) \to \mathbb{C}$ of the types $L_1, \ldots, L_t$ (that is, functions that become functions of classes $L_1, \ldots, L_t$ after the translations $(\mathbb{C}^n, a) \to (\mathbb{C}^n, 0)$).

If f is an isolated singularity and F is its deformation, then by $\{L\}(F)$ we denote the set of all points of the discriminant $\Sigma(F) \subset D_\delta$ corresponding to functions with t singular points of the types $L_1, \ldots, L_t$ near 0. (For example, if all the L_i are equal to $\mathfrak{M}^2$, then $\{L\}(F) = \Sigma^t$.) We denote by $L(F)$ the set of pairs of the form (point λ of the set $\{L\}(F)$, collection of t points such that the corresponding function f_λ has singularities of the types $L_1, \ldots, L_t$ at these points). (For example, $(\mathfrak{M}^2, \ldots, \mathfrak{M}^2)(F) = \tilde{\Sigma}^t$.)

Associating to each such pair the collection of q-jets at these t points we get a map $L(F) \to [L]$.

The function $\|\cdot\|$ is defined on the set $L(F)$: it is induced from the usual norm in the parameter space of the deformation F (see 5.2) by the obvious projection $L(F) \to \{L\}(F)$.

5.4. THEOREM. *Let $f = x_1^N + \cdots + x_n^N$, with $N > n(t(q+1) - 1)$, and let the versal deformation F of the function f be of the form* (4), *where α ranges over any finite set of multi-indices $(\alpha_1, \ldots, \alpha_n) \in Z_+^n$ containing the cube $\{\alpha | \max \alpha_i \leq N - 2\}$, and the number δ used in the definition of the ball D_δ (and hence the sets $\{L\}(F)$ and $L(F)$) is sufficiently small. Then for any q-family $L = (L_1, \ldots, L_t)$ the obvious map $L(F) \to [L]$ is a homotopy equivalence.*

Proof of this theorem is given in §§5.5–5.7. The theorem on the stable irreducibility of multisingularities is its immediate consequence, and the theorem from 3.4.6 is deduced from it in §5.8.

5.5. Recall the notation for the norm in the space of multi-indices $\alpha = (\alpha_1, \ldots, \alpha_n)$: $|\alpha| = \max \alpha_i$. Decompose the parameter space Λ of the deformation F into the sum of subspaces Λ^- and Λ^+ consisting of polynomials $\sum \lambda_\alpha x^\alpha$ such that $|\alpha| < (q+1)t - 1$ and $|\alpha| \geq (q+1)t$, respectively.

For any q-family L define the set $\{L\}^-(F)$ as the intersection of $\{L\}(F)$ with Λ^-, and define the set $L^-(F)$ as a subset in $L(F)$ consisting of the pairs ($\lambda \in \{L\}(F)$, t points) such that $\lambda \in \{L\}^-(F)$. Now Theorem 5.4 is a consequence of the following two theorems.

5.5.1. THEOREM. *If δ is sufficiently small, then the set $L^-(F)$ is a deformation retract of the set $L(F)$, and the corresponding deformation $I\colon L(F) \times [0, 1] \to L(F)$, $I(\cdot \times 1) = \mathrm{id}$, $I(L(F) \times 0) \subset L^-(F)$, preserves the map $L(F) \to [L]$ at each moment τ (i.e., the image of any point $I(u, \tau)$ under this map coincides with that of $I(u, 1)$ for all $u \in L(F)$, $\tau \in [0, 1]$) and does not increase the function $\|\cdot\|\colon L(F) \to \mathbb{R}_+$ ($\|I(u, \tau)\|$ is monotonically nonincreasing as τ decreases).*

5.5.2. THEOREM. *Under the assumptions of Theorem 5.4 the restriction of the natural map $L(F) \to [L]$ to the subset $L^-(F) \subset L(F)$ is a homotopy equivalence $L^-(F) \xrightarrow{\sim} [L]$.*

5.6. Proof of Theorem 5.5.1.

5.6.1. LEMMA. *For every monomial $x^\beta = x_1^{\beta_1} \cdot \dots \cdot x_n^{\beta_n}$ there is a polynomial map $\rho_\beta \colon (\mathbb{C}^n)^t \to \Lambda^-$ such that*

(1) For any point of $(\mathbb{C}^n)^t$ (i.e. a family of t points $Z_1, \ldots, Z_t \in \mathbb{C}^n$) the polynomial $\rho_\beta(Z_1, \ldots, Z_t) \in \Lambda^-$ has the same q-jets at the points $Z_1, \ldots, Z_t$ as the function x^β;

(2) The map ρ_β is invariant under permutations of these t points, i.e. it descends to the map $\mathrm{Sym}^t \mathbb{C}^n \equiv (\mathbb{C}^n)^t / S(t) \to \Lambda^-$;

(3) For $|\beta| \geq (q+1)t$ the polynomials defining the map ρ_β do not have constant terms.

PROOF OF THE LEMMA. Set $s = t(q+1)$. For $|\beta| < s$ the statement of the lemma is obvious (it suffices to set $\rho_\beta((\mathbb{C}^n)^t) = x^\beta$); we prove it for $|\beta| \geq s$. It suffices to do this for every factor $x_i^{\beta_i}$ of the monomial x^β, and hence, we can restrict ourselves to the case $n = 1$. In this case β is a scalar, and Λ^- is the base of the following deformation of the monomial x^β:

$$\varphi_\lambda = x^\beta - \lambda_0 - \lambda_1 x - \cdots - \lambda_{s-1} x^{s-1}. \tag{19}$$

Let $Z = (Z_1, \ldots, Z_t)$ be a family of points in $\mathbb{C}^1$. By definition, if in (19) λ_i are the coefficients of the desired polynomial $\rho_\beta(Z) \in \Lambda^-$, that is, $\varphi_\lambda \equiv x^\beta - \rho_\beta(Z)$, then φ_λ must have zeros of multiplicity $q+1$ at the points $Z_1, \ldots, Z_t$. We construct a map ρ_β for which this is always true. Let

$$x^\beta - \rho_\beta(Z) = \tilde{\rho}(Z) \cdot \overline{\rho}_\beta(Z),$$

where $\tilde{\rho}(Z) = (x - Z_1)^{q+1} \cdot \dots \cdot (x - Z_t)^{q+1}$, and let $\overline{\rho}_\beta(Z)$ be a polynomial of degree $\beta - s$. Obviously, the coefficients of $\tilde{\rho}$ depend polynomially on

Z_i, and the coefficients of $\bar{p}_\beta$ can be expressed successively (with decreasing degree) in terms of the higher coefficients and coefficients of $\tilde{p}$, starting from the condition that the coefficients of degree $\beta, \beta - 1, \ldots, s$ in the product $\tilde{p} \cdot \bar{p}_\beta$ must be equal to $1, 0, \ldots, 0$, respectively. For example, the leading coefficient in $\bar{p}_\beta$ is equal to 1, the next (if $\beta > s$) is equal to $(q + 1)(Z_1 + \cdots + Z_t)$, etc. Finally, the coefficients of the product of two polynomials are polynomials in their coefficients, and the first statement of the lemma is proved. The second and third statements follow from the construction.

5.6.2. The map ρ_β defined in the lemma for an arbitrary monomial x^β can be linearly extended to a map $P \colon \Lambda^+ \times (\mathbb{C}^n)^t \to \Lambda^-$ linear in $\lambda \in \Lambda^+$ and invariant under the action of the group $S(t)$ on the space $(\mathbb{C}^n)^t$. By part (3) of the lemma, there is a $\nu > 0$ such that, for any point $Z \in (\mathbb{C}^n)^t$ for which the corresponding t points belong to the polydisk $\{x|\max|x_i| \leq \nu\}$, the map $P(\cdot, Z)$ has the norm at most 1. By selecting $\delta \leq \nu^{N-1}$ we guarantee that all critical points of the function $F(\cdot, \lambda)$, $\lambda \in D_\delta$, in the polydisk $\{x|\max|x_i| \leq 1\}$ lie in fact in the polydisk $\{x|\max|x_i| \leq \nu\}$. Fix such δ and construct a retraction promised in Theorem 5.5.1. Consider an arbitrary point of the set $L(F)$, that is, a pair $(\lambda \in \{\Sigma\}(F); t$ points $Z_1, \ldots, Z_t \in \mathbb{C}^n)$. Decompose λ into the sum of vectors $\lambda^- \in \Lambda^-$ and $\lambda^+ \in \Lambda^+$. Let $\lambda_1^- = P(\lambda^+, \text{points } Z_1, \ldots, Z_t) \in \Lambda^-$. Then $\|\lambda_1^-\| \leq \|\lambda^+\|$. The desired retraction of the point λ is now given by the family $\lambda + (1 - \tau)(\lambda_1^- - \lambda^+)$, $\tau \in [0, 1]$. This family depends continuously on the starting point λ of the space $L(F)$, since the point $\lambda_1^- \in \Lambda^-$ is obtained from it by a sequence of continuous maps. Theorem 5.5.1 is proved.

5.7. Proof of Theorem 5.5.2. Define the spaces $\{L\}_\infty^-(F)$ and $L_\infty^-(F)$ similarly to the spaces $\{L\}^-(F)$ and $L^-(F)$ but without restrictions on $\|\lambda\|$ and without the condition that the critical points are close to 0. For example, if $L = (\mathfrak{M}^2, \ldots, \mathfrak{M}^2)$, then $\{L\}_\infty^-(F)$ is the set of all functions f_λ of the form (4), with $\lambda_\alpha = 0$ for $|\alpha| \geq (q+1)t$, that have t critical points with critical value 0 anywhere in $\mathbb{C}^n$. It is easy to see that $\{L\}^-(F) = \{L\}_\infty^-(F) \cap D_\delta$: indeed, any function f_λ of the form (4) with $\lambda \in \Lambda^-$ and $\|\lambda\| < \delta$ can have critical points only near 0, since for $|\alpha| < (q+1)t$, $|\operatorname{grad} x^\alpha| = o(|\operatorname{grad} f|)$ as $|x| \to \infty$. It suffices to prove that the imbedding $L^-(F) \to L_\infty^-(F)$ and the projection $L_\infty^-(F) \to [L]$ are homotopy equivalences. The latter assertion follows from any interpolation theorem. To prove the former assertion we consider the following action of the group of positive numbers $\mathbb{R}_+$ on the set $\{L\}_\infty^-(F)$: the number τ maps a polynomial $f_\lambda(\cdot)$ to $\tau^{-N} f_\lambda(\tau \cdot)$. This action lifts to a smooth action of $\mathbb{R}_+$ on $L_\infty^-(F)$, and it remains to observe that the intersection of every orbit of this action with $L^-(F)$ is an initial segment of the orbit.

5.8. Proof of the theorem from 3.4.6. Let us assume first that the versal deformation F has the special form (4) with indices α described in Theorem 5.4. Then the assertion of the theorem about homotopy equivalence is a special case of Theorem 5.4, and the smoothness of the subset $\tilde{\Sigma}^t \subset D_\delta \times \mathbb{C}^n(t)$, $t \leq c$, follows from the interpolation theorem: for any collection of t points in $\mathbb{C}^n$ the condition that the function $F(\cdot, \lambda)$ has singularities with critical value 0 at these points provides $t(n+1)$ linear restrictions on Λ, and all these restrictions are independent and depend smoothly on the collection of points. For any deformation of the form (4′) smoothness follows from the fact that the discriminants of all miniversal deformations of the singularity are diffeomorphic to each other, and the homotopy equivalence follows from the fact that the set of deformations of the form (4′) is path connected. □

§6. Stable cohomology of complements of caustics, and other generalizations

6.1. The proof of Theorem 3 from the introduction to this chapter is completely analogous to the proof of Theorem 1 above; the appearance of the space $\Sigma^{2n}\Lambda(n)$ in this theorem can be explained as follows.

LEMMA. *For any point $a \in \mathbb{C}^n$ the space of all possible 2-jets of functions $(\mathbb{C}^n, a) \to \mathbb{C}$ without degenerate (non-Morse) singularity at the point a is homotopy equivalent to the space $\Sigma^{2n}\Lambda(n)$.*

Proof is elementary ($\Lambda(n)$ is the homotopy type of the space of all non-degenerate quadratic forms in $\mathbb{C}^n$). □

6.2. The Smale-Hirsch principle for stable complements of strata of singularities. Theorems 1 and 3 from the introduction to this chapter are special cases of the following general fact.

Let σ be an arbitrary complex algebraic subset in the space $J_0^q(\mathbb{C}^n, \mathbb{C})$ of q-jets of holomorphic functions $(\mathbb{C}^n, 0) \to \mathbb{C}$ invariant under the action of the group of local diffeomorphisms $(\mathbb{C}^n, 0) \to (\mathbb{C}^n, 0)$ and under the multiplication of functions by nonzero constants. For any holomorphic deformation $F: (\mathbb{C}^n \times \mathbb{C}^l, 0) \to (\mathbb{C}, 0)$ denote by $\{\sigma\}$ the set of values of the parameter $\lambda \in \mathbb{C}^l$ such that the function $F(\cdot, \lambda)$ has a critical point of type σ near 0.

THEOREM. *The stable cohomology ring of complements of the sets $\{\sigma\}$ in the bases of deformations of isolated singularities of functions $(\mathbb{C}^n, 0) \to (\mathbb{C}, 0)$ is isomorphic to the ring $H^*(\Omega^{2n}(J_0^q(\mathbb{C}^n, \mathbb{C})\backslash\sigma))$. If the complex codimension of the set σ is greater than 1, then all stable homotopy groups π_i of complements of the sets $\{\sigma\}$ are isomorphic to the corresponding groups $\pi_{i+2n}(J_0^q(\mathbb{C}^n, \mathbb{C})\backslash\sigma)$.*

Proof of the first statement is the same as before, and the second statement follows from the first, the Whitehead theorem, and the equality (3) from Appendix 3. □

§7. Complements of resultants of polynomial systems in $\mathbb{C}^1$

In this section we prove Theorem 4 from the introduction to this chapter.

7.1. Resolutions of resultants. For any natural $d_1, \ldots, d_k$ we construct a resolution $\sigma_{d_1, \ldots, d_k}$ of the resultant $\sum_{d_1, \ldots, d_k}$ that is a simpler space with the same closed homology. This construction repeats the construction of the resolution σ of the discriminant of a singularity from §3.3 of this chapter.

Namely, we:

(1) for each natural t, consider a set $\widetilde{\Sigma}^t_{d_1, \ldots, d_k}$ consisting of pairs of the form (t distinct points in $\mathbb{C}^1$, polynomial system of the form (1) having the common roots at all these t points);

(2) fix a map $i: \mathbb{C}^1 \to \mathbb{R}^N$, $N \gg \min(d_i)$, in general position;

(3) define subsets $\sigma^t \subset \mathbb{R}^N \times \mathbb{C}^{d_1 + \cdots + d_k}$ consisting of pairs of the form (point of the simplex in $\mathbb{R}^N$ spanned by the points $i(x_1), \ldots, i(x_t)$, polynomial system in $\mathbb{C}^{d_1 + \cdots + d_k}$ with roots at the points $x_1, \ldots, x_t$);

(4) define the set $\sigma_{d_1, \ldots, d_k} \subset \mathbb{R}^N \times \mathbb{C}^{d_1 + \cdots + d_k}$ as the union of all sets σ^t.

Similarly, to investigate the set $\sum(m, k)$ of polynomials (2) with a root of multiplicity $\geq k$ we define a set $\sigma(m, k) \subset \mathbb{R}^N \times \mathbb{C}^m$ in the same way as $\sigma_{d_1, \ldots, d_k}$ with the only difference being that we take polynomials of the form (2) with roots of multiplicity k at the points $x_1, \ldots, x_t$ instead of polynomial systems (1) with common roots at these points; this is a special case of the construction from §3.3 of this chapter.

LEMMA. $\overline{H}_*(\sigma_{d_1, \ldots, d_k}) \simeq \overline{H}_*(\sum_{d_1, \ldots, d_k})$, $\overline{H}_*(\sigma(m, k)) \simeq \overline{H}_*(\sum(m, k))$. □

On the other hand, the one-point compactifications of these four spaces are simply-connected, and hence, the homology equivalences from the lemma extend to homotopy equivalences of the compactifications.

Both sets $\sigma_{d_1, \ldots, d_k}$ and $\sigma(m, k)$ carry standard filtrations: the ith term of the filtration consists of points of all sets σ^t with $t \leq i$.

NOTATION. If k symbols d_i are equal, $d_1 = \cdots = d_k = d$, then we use the notation $k|d$ for the multiindex $(d_1, \ldots, d_k)$.

7.2. THEOREM. *The homology spectral sequences $E^r_{p,q}(k|d)$ and $E^r_{p,q}(k \cdot d, k)$ converging to the closed homology of the spaces $\sigma_{k|d}$ and $\sigma(k \cdot d, k)$ and generated by the standard filtrations of these spaces are isomorphic starting with the E^1 term.*

The proof is given in 7.3 and 7.4.

7.3. For any p, q the groups $E^1_{p,q}(k|d)$ and $E^1_{p,q}(k \cdot d, k)$ are isomorphic; indeed, by construction they coincide with the $(p+q)$-dimensional closed homology groups of the spaces $\sigma_{k|d}^p - \sigma_{k|d}^{p-1}$ and $\sigma^p(k \cdot d, k) - \sigma^{p-1}(k \cdot d, k)$. Each of these spaces is the space of a fiber bundle which is a fibered product of two fiber bundles with common base (the configuration space $\mathbb{C}^1(p)$), common first factor (the bundle with the $(p-1)$-dimensional open simplex as a fiber) and second factors that are affine complex $k(d-p)$-dimensional bundles (whose fiber in the first case consists of polynomial systems (1) having roots at the given p points, and in the second case consists of polynomials (2) with $m = k \cdot d$ having roots of multiplicity k at the same points). Hence, by the Thom and Poincaré isomorphisms, both groups are isomorphic to $H^{2k(d-p)+2p-q-1}(\mathbb{C}^1(p), \pm\mathbb{Z})$ for $p \leq d$ and are trivial for $p > d$.

It remains to prove that this isomorphism is compatible with the other differentials of the spectral sequences. Consider the standard resolution $\sigma_{k \cdot d, k \cdot d-1, \ldots, k \cdot d-k+1}$ of the resultant $\Sigma_{k \cdot d, k \cdot d-1, \ldots, k \cdot d-k+1}$ constructed in 7.1, and in particular, the dth term $F_d \sigma_{k \cdot d, k \cdot d-1, \ldots, k \cdot d-k+1}$ of its standard filtration. Let $\mathscr{E}_{p,q}^r[d]$ be the spectral sequence converging to the closed homology of $F_d \sigma_{k \cdot d, k \cdot d-1, \ldots, k \cdot d-k+1}$ and generated by the filtration of this space by the sets $F_i \sigma_{k \cdot d, k \cdot d-1, \ldots, k \cdot d-k+1}$, $i = 1, 2, \ldots, d$.

7.4. PROPOSITION. *There exist bidegree* $(0; k(k-1)(2d-1))$ *isomorphisms of the spectral sequences* $E_{p,q}^r(k|d)$ *and* $E_{p,q}^r(k \cdot d, k)$ *onto the spectral sequence* $\mathscr{E}_{p,q}^r[d]$.

Let us construct these isomorphisms. Define a map $j: \mathbb{C}^{k \cdot d} \to \mathbb{C}^{k \cdot d + k \cdot d - 1 + \cdots + k \cdot d - k + 1}$ by sending a polynomial F of the form (2) with $m = k \cdot d$ to the system of k polynomials $F, F'/m, \ldots, F^{(k-1)}/(m \cdot (m-1) \cdots (m-k))$. On the space $\mathbb{C}^{d + \cdots + d}$ of systems (1) with $d_1 = \cdots = d_k$ define an arbitrary smooth function $M: \mathbb{C}^{d + \cdots + d} \to \mathbb{R}_+$ whose value on the system $\Phi = (f_1, \ldots, f_k)$ exceeds by at least 1 the maximum of the absolute values of all the roots of either of the f_i. Let us define an imbedding $l: \mathbb{C}^{d + \cdots + d} \to \mathbb{C}^{k \cdot d + k \cdot d - 1 + \cdots + k \cdot d - k + 1}$ by sending the system $\Phi = (f_1, \ldots, f_k)$ to the system $(f_1 \cdot (x - M(\Phi))^{d(k-1)}, f_2 \cdot (x - 2M(\Phi))^{d(k-1)-1}, \ldots, f_k \cdot (x - k \cdot M(\Phi))^{(d-1)(k-1)})$.

The imbeddings j and l map the sets $\Sigma(k \cdot d, k)$ and $\Sigma_{k|d}$ to $\Sigma_{k \cdot d, k \cdot d - 1, \ldots, k \cdot d - k + 1}$, and can be lifted in the obvious way to imbeddings of the resolutions

$$\tilde{j}: \sigma(k \cdot d, k) \to F_d \sigma_{k \cdot d, k \cdot d - 1, \ldots, k \cdot d - k + 1}, \tilde{l}: \sigma_{k|d} \to F_d \sigma_{k \cdot d, k \cdot d - 1, \ldots, k \cdot d - k + 1}.$$

LEMMA. *The images of the imbeddings $\tilde{j}$ and $\tilde{l}$ have neighborhoods in $F_d \sigma_{k \cdot d, k \cdot d - 1, \ldots, k \cdot d - k + 1}$ whose one-point compactifications are homotopy equivalent to the $(k(k-1)(2d-1))$-fold suspensions of the one-point compactifications of the spaces $\sigma(k \cdot d, k)$ and $\sigma_{k|d}$, as well as to the one-point compactification of the whole space $F_d \sigma_{k \cdot d, k \cdot d - 1, \ldots, k \cdot d - k + 1}$. These homotopy equiv-*

alences are realized by filtration-preserving maps, and hence, define isomorphisms $\mathscr{E}^1_{p,q}[d] \to E^r_{p,q-k(k-1)(2d-1)}(k \cdot d, k)$, $\mathscr{E}^1_{p,q}[d] \to E^r_{p,q-k(k-1)(2d-1)}(k|d)$. (In particular, the $(k(k-1)(2d-1))$-fold suspensions of the spaces $\sigma(k \cdot d, k)$ and $\sigma_{k|d}$ are homotopy equivalent.)

The construction of this homotopy equivalence for the space $\tilde{j}(\sigma(k \cdot d, k))$ repeats (with certain simplifications) the proof of Proposition 1 in 4.5 of Chapter III and, for the space $\tilde{l}(\sigma_{k|d})$, follows easily from the construction of spaces $\sigma_{k|d}$, $\sigma_{k \cdot d, k \cdot d-1, \ldots, k \cdot d-k+1}$ (moreover, in the last case the desired neighborhood of the space $\tilde{l}(\sigma_{k|d})$ in $F_d \sigma_{k \cdot d, k \cdot d-1, \ldots, k \cdot d-k+1}$ is homeomorphic to the direct product $\sigma_{k|d} \times \mathbb{R}^{k(k-1)(2d-1)}$). □

This lemma terminates the proof of Theorem 7.2.

Now, Theorem 4 from the introduction to this chapter follows from the fact that the Spanier-Whitehead duals of stable homotopy equivalent spaces are stable homotopy equivalent, see [W].

CHAPTER V

Cohomology of the Space of Knots

In this chapter we construct a series of numerical invariants of the regular isotopy type of oriented knots. These invariants are defined as linking indices in the space of smooth maps $S^1 \to S^3$ with appropriate closed hypersurfaces belonging to the discriminant of this space, i.e., to the set of maps which have self-intersections or singularities. The simplest features of this construction are outlined in the introduction to the book; the reader is advised to look through that part of the introduction before passing to the more concrete description below.

0.1. Noncompact knots. For convenience in computations we consider not ordinary knots but so-called noncompact knots, i.e., nonsingular imbeddings $\mathbb{R}^1 \to \mathbb{R}^3$ that asymptotically approach a fixed linearly imbedded line in $\mathbb{R}^3$ at infinity; see §1 below. The space of all (including singular) smooth maps $\mathbb{R}^1 \to \mathbb{R}^3$ with such an asymptotic behavior will be denoted by K; this space is homotopically trivial. We denote by Σ the discriminant of this space, that is, the set of maps with singularities or self-intersections.

It is easy to see that the components of the space of noncompact knots $K - \Sigma$ are in one-to-one correspondence with the isotopy classes of ordinary knots $S^1 \to S^3$. In particular, the numerical invariants of knots can be considered as elements of the group $H^0(K - \Sigma)$. We study this group simultaneously with the other groups $H^i(K - \Sigma)$, $i \geq 0$.

0.2. Geometric realization of the spectral sequence. Our spectral sequence generalizes the one from §5 of Chapter III, which converges to the cohomology of iterated loop spaces.

Invariants obtained using our spectral sequence form a group with a natural increasing filtration: sums of invariants coming from the cells $E_\infty^{-i,i}$ with $i \leq k$ have degree $\leq k$. We can visualize this filtration as follows.

Any nonsingular part of the discriminant Σ consists of maps with precisely one point of transversal intersection. Such a part has an invariantly defined transversal orientation, that is, a way to name one of the adjacent components of the space of knots positive and the other negative. Any numerical invariant of knots α associates to such a part its *index* defined as the jump of the value of α when passing through this part in the positive direction.

FIGURE 24

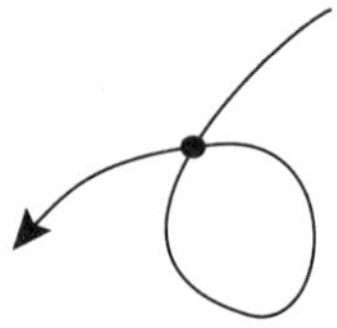

FIGURE 25

Consider a nonsingular self-intersection point of Σ (i.e., an immersion $\mathbb{R}^1 \to \mathbb{R}^3$ with two distinct points of transversal self-intersection). Near such a point the discriminant is represented by two irreducible components, each containing two nonsingular parts; see Figure 24. To such a point (and the part of self-intersection of the discriminant containing it) we assign a second order index defined as the difference of the indices of two smooth parts of any irreducible components of Σ near this point. (This number does not depend on the choice of the component since the sum of jumps of the invariant α along any circle is equal to 0.) Similar higher order indices correspond to nonsingular parts of arbitrary multiple self-intersections of the discriminant.

The invariant α has filtration k if and only if all corresponding indices of order $> k$ vanish.

EXAMPLE. There is no nontrivial invariant of order 1. Indeed, any invariant of the first order defines the same indices for all nonsingular parts of the discriminant. But, for the part containing the map in Figure 25, this index is always 0: two close knots separated by this part are equivalent.

0.3. Coding and computation of invariants. Two nonsingular points of a k-fold self-intersection of the discriminant are called *related* if it is possible to come from the first point to the second, always staying in the set of the nonsingular k-fold self-intersection of the discriminant and only several times passing transversely through the set of the nonsingular $(k+1)$-fold self-intersection. It is easy to see that for any k the number of classes of related points (the number of families) is equal to the number of unordered decompositions of the set of $2k$ elements into pairs: these classes are uniquely determined by the order of pairs of points on the line glued together by the corresponding maps $\mathbb{R}^1 \to \mathbb{R}^3$. For example, for $k = 2$ such decompositionsare shown in Figure 26, and the corresponding maps of the line are shown in Figure 27.

INTRODUCTION

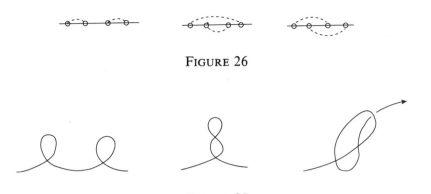

FIGURE 26

FIGURE 27

Any invariant of order k (= filtration k) is given by the following data. For any $i = 1, \ldots, k$ and for any class of related i-fold self-intersections of the discriminant we fix one map $\mathbb{R}^1 \to \mathbb{R}^3$ in this class and define its order i index.

The value of this invariant at any knot is computed using the following algorithm. We connect our knot and the trivial knot by a homotopy in general position in K. During this homotopy we intersect the discriminant in nonsingular points several times. The value of the invariant at the initial knot is equal to the sum of the indices of these points taken with sign + or − depending on whether we intersect Σ in the positive or negative direction. These indices are in turn computed as follows. Each of these points can be connected by a path in general position in Σ with the fixed map in the same (unique) class of related points. The desired index is equal to the sum of the index of this fixed map (which is known: if this map is the one in Figure 25, then the index is always 0) and the indices (of second order) of all parts of the set of self-intersections of the discriminant met along this path, taken with weight 1 or −1 depending on the direction in which we intersect them. The computation of these indices can be similarly reduced to the indices of order 3, etc. The entire procedure has k branching steps, where k is the filtration of the invariant. To compute the indices of points of k-fold self-intersection of Σ we do not need to draw paths: the index of order k is defined by the class of related points.

EXAMPLE. The simplest invariant and its value on the trefoil knot. This invariant has filtration 2. The corresponding index for the map in Figure 25 is equal to 0, and the indices of order 2 of the three maps in Figure 27 are 0, 0, and 1, respectively. Let us compute the value of this invariant on the trefoil knot (Figure 28, a, next page). This knot can be connected to the trivial knot by a path that intersects the discriminant only once at a point shown in Figure 28, b, and it remains to compute the index of this point. In turn, it can be connected with the simplest self-intersecting curve in Figure 28, e, by

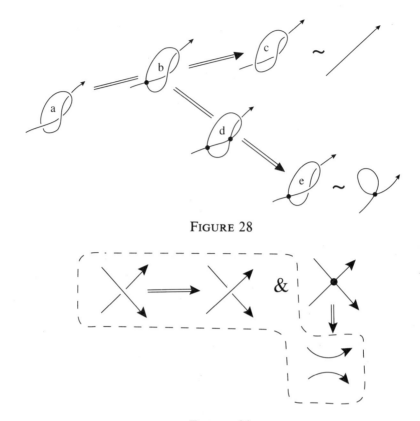

FIGURE 28

FIGURE 29

a path that intersects the set of double self-intersections of the discriminant (Figure 28, d) only once at a point of index 1. Finally, the value of our invariant equals $0 + (-1)[0 + (-1) \cdot 1] = 1$; here the first zero is the value of the invariant on the trivial knot, the first coefficient -1 corresponds to the direction of the intersection with Σ at the point in Figure 28, b, the second zero is the index of the map of Figure 28, e, the second coefficient -1 corresponds to the arrow between Figures 28, b and e, and, finally, 1 is the index of order 2 of the curve d.

REMARK. The sequence of local modifications in our algorithm becomes the sequence from the algorithm for computing Conway-Jones polynomials if after each modification we also "split" the obtained self-intersecting curve: the local reduction from the Conway algorithm is part of Figure 29 enclosed by the dotted line. Of course, during such splitting we lose some information. But in computing the simplest invariant (of filtration 2) this information is inessential, since the corresponding index of the unsplit self-intersecting curve can be recovered from its splitting.

0.4. The groups $E_1^{-i,i}$. For any i and any coefficient domain G the group $E_1^{-i,i}$ of our spectral sequence is naturally isomorphic to the kernel of the homomorphism $h_i \colon X_i \to Y_i$ defined below.

DEFINITION. An *[i]-configuration* is a collection of $2i$ distinct points on the oriented line $\mathbb{R}^1$ divided into i pairs in an arbitrary way. An *⟨i⟩-configuration* is a collection of $2i-1$ points divided into $i-2$ pairs and one triple. Two $[i]$- or $\langle i \rangle$-configurations are *equivalent* if one can be mapped onto another by an orientation preserving diffeomorphism of $\mathbb{R}^1$.

The group X_i is the direct sum of the groups X_i^0 and X_i^1 that are defined as free G-modules whose generators are in one-to-one correspondence with various $[i]$-configurations ($\langle i \rangle$-configurations, respectively) considered up to equivalence.

EXAMPLES. 1) $X_1^1 = 0$. The group X_1^0 is isomorphic to G and is generated by the configuration of two points in $\mathbb{R}^1$.

2) $X_2 \equiv X_2^0 \oplus X_2^1 \cong G^3 \oplus G$: the group X_2^1 is generated by any triple of points in $\mathbb{R}^1$, the generators of the group X_2^0 are shown in Figure 26.

To each $[i]$-configuration or $\langle i \rangle$-configuration corresponds an affine subspace of codimension $3i$ in K: this subspace consists of maps $\mathbb{R}^1 \to \mathbb{R}^3$ gluing together all points in each pair or triple.

Now we describe the group Y_i. It is defined as a direct sum $Y_i^1 \oplus Y_i^*$. The group Y_i^1 is isomorphic to the direct sum of three copies of the group X_i^1; a generator of Y_i^1 is an $\langle i \rangle$-configuration in which one of the points in the triple is marked. Such an $\langle i \rangle$-configuration with a marked point will be called an $\langle \tilde{i} \rangle$-configuration. The group Y_i^* for $i \geq 2$ is isomorphic to the direct sum of $2i-1$ copies of the group X_{i-1}^0: the standard generator of the group Y_i^* is (up to equivalence) an $[i-1]$-configuration with one added "singular" point $*$ that does not coincide with any of the $2i-2$ points of this $[i-1]$-configuration. This augmented $[i-1]$-configuration will be called an i^*-configuration. The group Y_1^* is defined as a one-dimensional G-module with the generator corresponding to the unique point $* \in \mathbb{R}^1$.

EXAMPLES. 1) $Y_2^* = G^3$; see Figure 30. 2) $Y_2^1 = G^3$; all generators of this group are represented in the upper row in Figure 31 (next page).

To each i^*-configuration corresponds a subspace of codimension $3i$ in K consisting of maps $\mathbb{R}^1 \to \mathbb{R}^3$ that glue together all points in each pair in the definition of configuration and have a singularity (zero of the derivative) at the point $*$.

The homomorphism $h_i \colon X_i \to Y_i$ is illustrated by Figure 31. Here and below the triples of points (in the definition of $\langle i \rangle$-configuration) are distinguished by triples of dotted curves connecting all points of these triples. If one of the points in a triple is marked (so that the corresponding $\langle i \rangle$-configuration

FIGURE 30

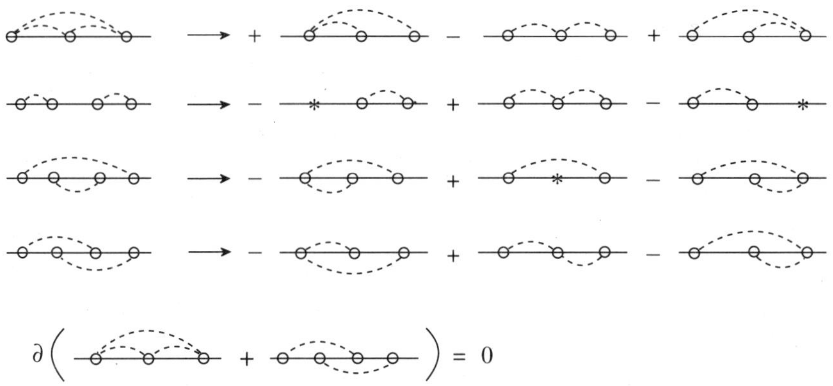

FIGURE 31

becomes an $\langle \tilde{i} \rangle$-configuration), then in the picture of this triple we omit the curve connecting two unmarked points.

Now we describe the homomorphism h_i. Let α be a generator in X_i^1, that is, an $\langle i \rangle$-configuration. Its image $h_i(\alpha)$ is a linear combination of three generators in Y_i^1 corresponding to the three $\langle \tilde{i} \rangle$-configurations geometrically coinciding with α. The $\langle \tilde{i} \rangle$-configuration in which the middle point of the triple is marked has coefficient -1 in this combination, and the two $\langle \tilde{i} \rangle$-configurations where the marked points are at the ends have the coefficient 1.

Now let α be a generator in X_i^0, i.e. an $[i]$-configuration. Then $h_i(\alpha)$ is a linear combination of $2i-1$ generators of the group Y_i that correspond to all finite segments of the line $\mathbb{R}^1$ bounded by points of the configuration α. Each such generator is represented by an i^*- or $\langle \tilde{i} \rangle$-configuration consisting of $2i-1$ points and coinciding with the initial configuration α everywhere outside the corresponding segment: the ends of this segment are replaced by the midpoint of the segment. If these endpoints belong to one pair of the configuration α, then the new point is a point of type $*$, and we get an i^*-configuration, i.e. a generator of the module Y_i^*. If the endpoints belong to different pairs, then these two pairs are replaced by a triple in which the marked point is the new one; this defines an $\langle \tilde{i} \rangle$-configuration, i.e. a generator of Y_i^1. Each of these $2i-1$ generators is present in the linear combination $h_i(\alpha)$ with coefficient $(-1)^{i+m-1}$, where m is the number (in the increasing order in $\mathbb{R}^1$) of the larger of the two points being glued.

We have completely defined the homomorphisms $h_i : X_i \to Y_i$. For any $i \geq 1$ and any coefficient group the group $E_1^{-i,i}$ of the basic spectral sequence is naturally isomorphic to the kernel of the corresponding homomorphism h_i.

EXAMPLES. Suppose $i = 1$. Then $X_1 \cong Y_1 \cong G$, and the homomorphism h_1 is an isomorphism; cf. the example at the end of 0.2.

Let $i = 2$. Then $X_2 = X_2^0 + X_2^1 \cong G^3 \oplus G$; $Y_2 = Y_2^1 \oplus Y_2^* \cong G^3 \oplus G^3$. The action of the differential h_2 is shown in Figure 31. In particular, it turns out that the sum of the first and the fourth generators of the group X_2 in this figure belongs to the kernel of h_2 (and so there exists a knot invariant corresponding to this sum). This is precisely the second-order invariant considered above, which distinguishes the trefoil knot from the line.

0.5. Agreement of basis invariants for symmetric knots. The system of invariants corresponding to basis elements of the graded group $\bigoplus_i E_\infty^{-i,i}$ admits a standard transformation group: for example, to each invariant of order i we can add any invariants of lower orders. Using these transformations we can insure the following self-symmetry property of invariants.

THEOREM. *Suppose the coefficient group of the basic spectral sequence possesses division into* 2. *Then all stable invariants corresponding to the terms* $E_\infty^{-i,i}$ *of this sequence can be chosen so that they take equal values on any two symmetric knots for even i and opposite values for odd i.*

0.6. Outline of the chapter. In §2 we construct the basic spectral sequence computing the cohomology of complements of discriminants in the space $\Gamma^d \subset K$ of a finite-dimensional approximation to the knot space. This spectral sequence stabilizes as $d \to \infty$. In §3 we prove that its terms $E_1^{p,q}$ (even nonstable ones) vanish for $p + q < 0$, and compute the term E_1 of the stable spectral sequence. In §4 we describe an algorithm to compute the terms $E_r^{-i,i}$, $r = 2, 3, \ldots, \infty$, of the stable sequence and the corresponding knot invariants, and also an algorithm to compute the value of such an invariant (given by the corresponding output file of the first algorithm) on any tame knot. In §5 we present the first results of these computations: the five simplest invariants (of orders 2, 3, and 4) and their values on tabular knots realizable by diagrams with ≤ 7 overlaps.

§1. Definitions and notation

1.1. Noncompact knots. A *knot* is an oriented C^1-submanifold in S^3 diffeomorphic to the circle. Two knots are said to be equivalent if they can be mapped onto each other (preserving orientation) by some C^1-isotopy of S^3. Similarly we define the equivalence of knots in $\mathbb{R}^3$.

1.1.1. PROPOSITION. *The obvious imbedding $\mathbb{R}^3 \to S^3$ defines a one-to-one correspondence between the classes of equivalent knots in $\mathbb{R}^3$ and in S^3. Two knots in S^3 are equivalent if and only if the corresponding imbeddings $S^1 \to S^3$ are homotopic in the class of imbeddings (that is, smooth maps that define diffeomorphisms of S^1 with their image).* □

Fix some tangent direction to S^3 at the base point $* \in S^3$.

DEFINITION. A smooth map $S^1 \to S^3$ is called *normal* if it sends the base point $* \in S^1$ (and only this point) to the base point $* \in S^3$, and the derivative of the map at the base point has the fixed direction (in particular, does not vanish). A knot is called *normal* if it is the image of a normal map.

Classification of normal knots is equivalent to classification of all knots. This is a consequence of the following elementary fact.

1.1.2. LEMMA. *Any class of equivalent knots in S^3 contains a normal knot. Any two normal knots are equivalent if and only if the corresponding normal imbeddings $S^1 \to S^3$ are homotopic in the class of normal imbeddings.* □

1.1.3. LEMMA. *The space of all normal maps $S^1 \to S^3$ is homotopy trivial (in particular, it contracts onto the inclusion of the great circle corresponding to the fixed direction at the base point $* \in S^3$).*

Indeed, the maps $\mathbb{R}^1 \to \mathbb{R}^3$ defined as restrictions of normal maps to $\mathbb{R}^1 = S^1 - *$ form a convex subset in the affine space of all smooth maps $\mathbb{R}^1 \to \mathbb{R}^3$. □

The space of normal maps $S^1 \to S^3$ is denoted by K. The set of maps of class K that are not imbeddings is called the *discriminant* and denoted by Σ.

It is convenient to represent normal maps as maps $\mathbb{R}^1 \to \mathbb{R}^3$ that go off to infinity in the fixed asymptotic direction. (Without loss of generality we will assume that this direction is given by the vector $(1, 1, 1)$ in the standard linear coordinates x, y, z in $\mathbb{R}^3$.) This representation explains the next definition.

DEFINITION. A *noncompact knot* is a regular C^1-submanifold in $\mathbb{R}^3$ diffeomorphic to $\mathbb{R}^1$ whose closure in the one-point compactification S^3 of $\mathbb{R}^3$ is a normal knot. Two noncompact knots are *equivalent* if the corresponding normal knots in S^3 are given by imbeddings that are homotopic in the space of (normal) imbeddings.

1.2. Approximation of knots. In this section we define a system of finite-dimensional subspaces of the space K of normal maps such that any (noncompact) knot is equivalent to some knot realizable by a point of some of these subspaces. Namely, consider the space of various polynomials of the form

$$t^{d+1} + a_1 t^d + \cdots + a_d t. \tag{1}$$

Let $\tilde{\Gamma}^d$ be the space of maps $\mathbb{R}^1 \to \mathbb{R}^3$ given by a triple of polynomials:

$$(x, y, z)(t) = (P_1(t), P_2(t), P_3(t)), \tag{2}$$

where all P_i have the form (1). Suppose d is even, then ([7]) $\tilde{\Gamma}^d$ is a subspace of the space K of normal maps; it has a natural affine space structure. This subspace is not in general position with respect to the discriminant. For example, it contains maps with a continuum of self-intersections: such are all

---
([7]) For a suitable choice of the identification $\mathbb{R}^1 \to S^1 - \{*\}$.

the maps of the form (2) in which $P_1 = P_2 = P_3$ is an arbitrary nonmonotonic polynomial of the form (1).

Move the space $\widetilde{\Gamma}^d$ so that it achieves general position with respect to the discriminant. To do this we note that for any odd $w > 1$ the space $\widetilde{\Gamma}^d$ can be considered as a subspace of $\widetilde{\Gamma}^{w(d+1)-1}$: the standard imbedding $I_w : \widetilde{\Gamma}^d \to \widetilde{\Gamma}^{w(d+1)-1}$ is given by a parameter change $t = s^w + s$.

For example, set $w = 3$. An arbitrarily small move of the affine subspace $I_3(\widetilde{\Gamma}^d) \subset \widetilde{\Gamma}^{3d+2}$ can put it in general position with respect to the discriminant $\Sigma \cap \widetilde{\Gamma}^{3d+2}$. The resulting perturbed space will be denoted by Γ^d.

By the Weierstrass approximation theorem any noncompact knot in $\mathbb{R}^3$ is equivalent to a knot given by a map of class Γ^d with sufficiently large d; moreover, any cycle in $K - \Sigma$ is homologous to cycles lying in the subspaces $\Gamma^d - \Sigma$ with sufficiently large d. Below we consider the cohomology of $\Gamma^d - \Sigma$ and its behavior as $d \to \infty$.

Here is the first reduction: by the Alexander duality theorem

$$\widetilde{H}^i(\Gamma^d \setminus \Sigma) \cong \overline{H}_{3d-i-1}(\Sigma \cap \Gamma^d), \tag{3}$$

where $\widetilde{H}^*$ is the cohomology reduced modulo the fundamental cocycle, and $\overline{H}_*$ is the closed homology, i.e. homology of one-point compactification modulo the compactifying point.

Thus it remains to study the closed homology of the discriminant $\Sigma \cap \Gamma^d$.

§2. Basic spectral sequence

2.1. Configurations of singularities and self-intersections. For any even d the discriminant $\Sigma \cap \Gamma^d$ is a stratified semialgebraic variety consisting of various affine linear manifolds that we now describe.

2.1.1. DEFINITION. Let A be an arbitrary finite sequence of integers of the form $\{a_1 \geq a_2 \geq \cdots \geq a_g\}$, $a_g \geq 2$. Denote by $|A|$ the number $\sum a_i$, and by $\#A$ the cardinality of this sequence (denoted by g in the line above). An A-configuration is any set of $|A|$ distinct points on the line $\mathbb{R}^1$ divided into $\#A$ groups of cardinality $a_1, \ldots, a_{\#A}$, respectively.

Also let b be a nonnegative integer. An (A, b)-configuration is a pair consisting of an A-configuration and an additional set of b distinct points in $\mathbb{R}^1$; here we do not exclude the possibility that some of these b points may coincide with points of the A-configuration.

EXAMPLE. The $[i]$-, $\langle i \rangle$-, and i^*-configurations defined in 0.4 are (in our present notation) (A, b)-configurations, where, respectively, $A = (\underbrace{2, \ldots, 2}_{i})$, $b = 0$; $A = (3, \underbrace{2, \ldots, 2}_{i-2})$, $b = 0$; $A = (\underbrace{2, \ldots, 2}_{i-1})$, $b = 1$, and all $2i - 1$ points in the configuration are distinct.

Two (A, b)-configurations are *equivalent* (or belong to the same class) if they can be mapped onto each other by an orientation preserving diffeomorphism of the line $\mathbb{R}^1$.

The dimension of the space of configurations equivalent to the given one equals the number of geometrically distinct points in the configuration; in particular, it does not exceed $|A| + b$.

A smooth map $\theta: \mathbb{R}^1 \to \mathbb{R}^3$ *respects* an (A, b)-configuration if it has singularities at all b marked points of this configuration and sends all points from each of the groups of cardinalities $a_1, \ldots, a_{\#A}$ to one point in $\mathbb{R}^3$ (in general these $\#A$ points are distinct).

The set of maps $\mathbb{R}^1 \to \mathbb{R}^3$ respecting a given configuration is an affine subspace in the space K (or in the space of all smooth maps); its codimension is $3(|A| - \#A + b)$. The intersection of this space with the space Γ^d defined in §1 has, in the case of generic Γ^d, the same codimension in Γ^d.

For any (A, b)-configuration J denote by $\chi(\Gamma^d, J)$ the set of maps of class Γ^d respecting J.

2.1.2. LEMMA. *For almost any choice of the space Γ^d and any (A, b)-configuration J the following statements are valid.*

(A) *For almost any (A, b)-configuration J' equivalent to J, $\chi(\Gamma^d, J')$ is an affine subspace of codimension $3(|A| - \#A + b)$ in Γ^d (in particular, it is empty if $|A| - \#A + b > d$).*

(B) *Let $|A| - \#A + b = k \leq d$. Then in the set of all configurations equivalent to J the subset of configurations J' such that $\chi(\Gamma^d, J') = \varnothing$ has codimension $\geq 3d - 3k + 1$, and the set of configurations J' such that the codimension of $\chi(\Gamma^d, J')$ in Γ^d is $3k - i$, $i \geq 1$, is a subset of codimension $\geq i(3d - 3k + i + 1)$. In particular, for $k \leq (3d + 1)/5$ the codimension of any set $\chi(\Gamma^d, J)$ defined by any (A, b)-configuration J with $|A| - \#A + b = k$ is precisely $3k$.*

(C) *Let $|A| - \#A + b = k > d$. Then in the set of configurations equivalent to J the set of configurations J' such that $\dim \chi(\Gamma^d, J') = l \geq 0$ has codimension at least $(l + 1)(3k - 3d + l)$. In particular, the set of J' such that $\chi(\Gamma^d, J') \neq \varnothing$, has codimension $\geq 3(k - d)$ and is empty for $k > 3d$.*

This is a standard statement about properties of "general position": it is easy to see that planes Γ^d that do not satisfy these conditions form a semialgebraic subset of positive codimension in the space of all $3d$-dimensional planes in $\tilde{\Gamma}^{3d+2}$. □

Below we assume that our space Γ^d satisfies all conditions of this lemma.

2.2. Generating collections.

2.2.1. Let $L \subset \mathbb{R}^1$ be an arbitrary finite set. A *generating collection* of the set L is an unordered collection of distinct pairs of points $(t_1 \neq t'_1), \ldots, (t_l \neq t'_l)$, $t_j, t'_j \in \mathbb{R}^1$, such that for all maps $\theta: \mathbb{R}^1 \to \mathbb{R}^3$ the following statements are equivalent:

(a) the map θ glues together all points of L,

(b) for any $j = 1, \ldots, l$, the map θ glues together the points t_j and t'_j.

Obviously, a collection of pairs $T = \{(t_1, t'_1), \ldots, (t_l, t'_l)\}$ is a generating collection for L if and only if all t_j, t'_j belong to L and for any two points of L there is a sequence of pairs in T connecting these points.

2.2.2. DEFINITION. The set of pairs $T = \{(t_1, t'_1), \ldots, (t_l, t'_l)\}$ is a *generating collection* for an A-configuration J if it is a union of generating collections for all $\#A$ sets forming the configuration A.

Obviously, the number of pairs in generating collections for a given A-configuration can vary from $\sum(a_i - 1) \equiv |A| - \#A$ to $\sum a_i(a_i - 1)/2$; these two bounds coincide if and only if all a_i are equal to 2.

2.2.3. A collection (T, V), where $T = \{(t_1, t'_1), \ldots, (t_l, t'_l)\}$ and V is exactly the set of points $v_1, \ldots, v_b \in \mathbb{R}^1$ is a *generating collection* for the (A, b)-configuration J if T is a generating collection for the corresponding A-configuration, and V is exactly the set of b points making the A-configuration an (A, b)-configuration. Two collections (T, V), (T_1, V_1) are *equivalent* if they are mapped onto each other by some orientation preserving diffeomorphism $\mathbb{R}^1 \to \mathbb{R}^1$. It is clear that equivalent collections generate equivalent configurations; the converse is not true in general.

2.3. Resolution of a discriminant. Here we construct an auxiliary space σ whose closed homology coincides with that of $\Gamma^d \cap \Sigma$. This construction generalizes the construction of similar sets in Chapters III and IV, for example, of the set $\{\Sigma_k\}$ from §III.3.3.

2.3.1. Denote by Ψ the set of unordered pairs of points in $\mathbb{R}^1$ (possibly coinciding): $\Psi = \mathbb{R}^2/\delta$, where δ is the involution $(t, t') \to (t', t)$. If we discard the diagonal $\{t = t'\}$ from Ψ, then we obtain the configuration space $\mathbb{R}^1(2)$; we will denote it by $\overset{\circ}{\Psi}$. By definition, the map $\Psi \to \mathbb{R}^N$ is the map $\mathbb{R}^2 \to \mathbb{R}^N$ invariant under reflection with respect to the diagonal $\{t = t'\}$.

2.3.2. LEMMA. *If N is sufficiently large then there exists a polynomial imbedding $\lambda \colon \Psi \to \mathbb{R}^N$ such that the images of any $2\binom{3d}{2} \equiv 3d(3d-1)$ points of the space Ψ are affinely independent in $\mathbb{R}^N$ (i.e. do not lie in the same $2\binom{3d}{2} - 2$-dimensional affine subspace).* □

2.3.3. Fix such an imbedding λ. For any (A, b)-configuration J consider the set of pairs of the form $(\theta, (T, V))$, where θ is any map $\mathbb{R}^1 \to \mathbb{R}^3$ of class Γ^d that respects the configuration J, and

$$(T, V) = \{(t_1, t'_1), \ldots, (t_l, t'_l), v_1, \ldots, v_b\}$$

is any generating collection of this configuration. To any such pair we associate a convex hull of $l + b$ points in the space $\Gamma^d \times \mathbb{R}^N$ which project to Γ^d

into one point θ, and project to $\mathbb{R}^N$ into the points $\lambda(t_1, t_1'), \ldots, \lambda(t_l, t_l')$, $\lambda(v_1, v_1), \ldots, \lambda(v_b, v_b)$. The last statement of Lemma 2.1.2 implies that the number $l+b$ of these points does not exceed $\binom{3d}{2}$. Hence, by the choice of the imbedding λ, this convex hull is a $(l+b-1)$-dimensional simplex; moreover, for any two such simplices in $\Gamma^d \times \mathbb{R}^N$ corresponding to distinct pairs $(\theta, (T, V))$ their projections to $\mathbb{R}^N$ cannot intersect in their interior points.

Define the set $\sigma \subset \Gamma^d \times \mathbb{R}^N$ as the union of such simplices over all pairs $(\theta, (T, V))$, all (A, b)-configurations J, and all collections (A, b).

These simplices will be called *standard simplices* of the space σ, and their projections into $\mathbb{R}^N$ are called standard simplices in $\mathbb{R}^N$.

REMARK. Of course, we get an equivalent definition of the set σ if take the union not of all standard simplices but just maximal $((\sum \binom{a_i}{2}) + b - 1)$-dimensional) simplices for all (A, b)-configurations J and all maps θ that respect them.

2.3.4. LEMMA. (A) *The set σ is semialgebraic.*

(B) *The obvious map $\sigma \to \Sigma \cap \Gamma^d$ given by the projection $\Gamma^d \times \mathbb{R}^N \to \Gamma^d$ defines an isomorphism*

$$\overline{H}_*(\sigma) \to \overline{H}_*(\Sigma \cap \Gamma^d). \qquad (4)$$

Part (A) follows directly from the construction and the Tarski-Seidenberg theorem; part (B) is proved similarly to Lemma 1 in §III.3.3. □

2.4. Filtration in the resolution of a discriminant. The set σ splits naturally into affine planes. Indeed, restrict the obvious projection $\pi: \Gamma^d \times \mathbb{R}^N \to \mathbb{R}^N$ to σ. The preimage of any point of the space $\mathbb{R}^N$ under this map is an affine space (possibly empty) whose projection to Γ^d is the set of maps $\theta: \mathbb{R}^1 \to \mathbb{R}^3$ satisfying several conditions of the form "the map θ has a singularity at the point v" or "θ glues the points t and t' together." Any set of conditions of this form defines a subspace $\chi(\Gamma^d, J)$ in Γ^d corresponding to an (A, b)-configuration J; see 2.1.

Define an increasing filtration $\sigma_1 \subset \sigma_2 \subset \cdots \subset \sigma_{3d} = \sigma$ on the set σ by assigning to σ_i the union of all subspaces in σ whose projections to Γ are subspaces $\chi(\Gamma^d, J)$ for some (A, b)-configurations J with $|A| - \#A + b \leq i$.

Consider the homology spectral sequence $\{E_{p,q}^r(d)\}$ converging to the group $\overline{H}_*(\sigma)$ and generated by this filtration. Make it a cohomological spectral sequence by renaming the term $E_{p,q}^r(d)$ by $E_r^{-p, 3d-1-q}(d)$. This is the spectral sequence announced in the introduction; to distinguish it from the sequences that appear below we will call it the *basic spectral sequence*. By Alexander duality in Γ^d it converges to the group $H^*(\Gamma^d \setminus \Sigma)$.

By construction, this spectral sequence lies in the region $\{p, q | p < 0\}$, and its term $E_\infty^{p,q}$ can be nonzero only if $p + q \geq 0$.

§2. BASIC SPECTRAL SEQUENCE

2.5. Fundamental properties of the basic spectral sequence.

THEOREM. *For any choice of the space Γ^d satisfying the assertions of Lemma* 2.1.2:

(A) *If the term $E_1^{p,q}(d)$ of the basic spectral sequence computing the cohomology of $\Gamma^d \setminus \Sigma$ with any coefficients does not vanish, then p, q satisfy the following conditions*:

(i) $p + q \geq 0$,
(ii) $p \geq -3d$,
(iii) $p \leq -2$.

(B) *For $i \leq (3d+1)/5$ and any coefficient domain the group $E_1^{-i,i}(d)$ is isomorphic to the group $\operatorname{Ker} h_i$ described in* 0.4; *in particular, it does not depend on d.*

(C) *For any even number $d' > d$ and any space $\Gamma^{d'}$ satisfying the assertions of Lemma* 2.1.2 *the corresponding basic spectral sequence $\{E_r^{p,q}(d')\}$ coincides with the sequence $\{E_r^{p,q}(d)\}$ for the following p, q, and r*:

(i) *for $r = 1$ and p, q in the "d-stable" region $\{p, q | p + q \geq 0, -p \leq (3d+1)/5\}$ (see Figure* 4);

(ii) *for any $r > 1$ and p, q such that for any $r' < r$ there is no differential acting into the cell $E_{r'}^{p,q}$ from the unstable region.*

COROLLARY 1. *For Γ^d, $\Gamma^{d'}$ in general position, $E_r^{-i,i}(d) = E_r^{-i,i}(d')$ for any $r = 1, 2, \ldots, \infty$ and even $d' > d \geq (5i-1)/3$.*

COROLLARY 2. *For any i the inclusion homomorphism $\overline{H}_{3d-1}(\sigma_i) \to \overline{H}_{3d-1}(\sigma)$ is a monomorphism.*

The restriction (ii) of part (A) of the theorem follows from Lemma 2.1.2, and the restriction (iii) follows from the explicit form of the column $\{p = -1\}$ of the term $E_0^{p,q}$. The restriction (i) and part (B) of the theorem will be proved in §3. Part (B) for $r = 1$ follows from the cellular structure of the spaces $\sigma_i - \sigma_{i-1}$, which we shall exhibit in 3.2, and for other r is proved similarly to Theorem 4.2.4 of Chapter III. In the most important case $p + q = 0$ all terms $E_r^{-q,q}$ are computed using an explicit algorithm that we shall present in §4 (and that does not depend on d for sufficiently large d).

REMARK. In the statement of the theorem it is possible to weaken the restrictions on $i \equiv -p$ and q. Namely, part (B) of the theorem is true for any $i \leq d - 1$, and the d-stable region from part (C) can be defined as the intersection of $\{p + q \geq 0\}$ with the union of $\{p \geq -(3d+1)/5\}$ and $\{q < 3d + 4p\}$. Also Corollary 1 is true for all $i \leq d - 1$.

2.6. The stable spectral sequence and stable invariants.
Part (C) of Theorem 2.5 allows us to define the *stable spectral sequence* $\{\underleftarrow{E}_r^{p,q}\}$: its term $\underleftarrow{E}_r^{p,q}$ is equal to the common term $E_r^{p,q}(d)$ of all spectral sequences corresponding to sufficiently large d.

CONJECTURE. For any p, q there is d_0 such that for all (even) $d > d_0$ and all r, $E_r^{p,q}(d) = E_r^{p,q}{\underset{\leftarrow}{}}$.

For example, Corollary 1 of the previous theorem says that this conjecture is true for $p + q = 0$.

The isomorphism $E_\infty^{-i,i}(d) \cong E_\infty^{-i,i}(d')$ from this corollary agrees with the map in homology: for $d' > d \geq (3i+1)/5$ there is a canonical isomorphism

$$\overline{H}_{3d-1}(\sigma_i(d)) \cong \overline{H}_{3d'-1}(\sigma_i(d')). \qquad (5)$$

This isomorphism agrees with the Alexander duality isomorphism: let θ and θ' be equivalent noncompact knots in K, and $\theta \in \Gamma^d \backslash \Sigma$, $\theta' \in \Gamma'^d \backslash \Sigma$, then the elements of groups (5) corresponding to each other by isomorphism (5) have the same values on these knots.

All these results (including the structure of the isomorphism (5)) follow directly from the algorithm for computing terms $E_r^{-i,i}$ of our spectral sequence; see §§3 and 4.

DEFINITION. A *stable invariant* of order i is an element of the group $\overline{H}_{3d-1}(\sigma_i(d))$, $d \geq (5i-1)/3$, whose image under the obvious map of this group into $\overline{H}_{3d-1}(\sigma_i(d) - \sigma_{i-1}(d))$ is nontrivial. Two elements of such groups corresponding to different d define the same invariant of order i if they correspond to each other via isomorphism (5).

REMARK. M. Kontsevich proved that the stable spectral sequence degenerates at the E_1 term (at least in the case of rational coefficients): $E_\infty^{p,q} = E_1^{p,q}$. See [Kontsevich$_1$], [BN$_2$].

§3. The term E_1 of the basic spectral sequence

By definition, the group $E_1^{-i,q}$ of the basic spectral sequence is isomorphic to the $(3d - 1 + i - q)$-dimensional closed homology group of the space $\sigma_i - \sigma_{i-1}$. In this section we start the computation of these groups. In particular, we prove that $E_1^{-i,q} = 0$ for $q - i < 0$; for stable values of i (i.e. for $i \leq (3d+1)/5$) we present a CW-complex structure on the one-point compactification of the space $\sigma_i - \sigma_{i-1}$ and explicitly describe all differentials of the corresponding chain complex. The cells of this complex correspond to various equivalence classes of generating collections in $\mathbb{R}^1$ such that the corresponding (A, b)-configurations satisfy $|A| - \#A + b = i$.

3.1. Decomposition of the space $\sigma_i - \sigma_{i-1}$ by the types of configurations. Let J be an arbitrary (A, b)-configuration. A *J-block* in σ is the union of the interior points of all standard simplices in $\Gamma^d \times \mathbb{R}^N$ such that the points $(t_1, t'_1), \ldots, (t_l, t'_l), v_1, \ldots, v_b$ form a generating collection of an (A, b)-configuration equivalent to J. Obviously, the space $\sigma_i - \sigma_{i-1}$ is the union of J-blocks corresponding to various (A, b)-configurations J such that $|A| - \#A + b = i$ (and the blocks corresponding to nonequivalent configurations do not intersect).

§3. THE TERM E_1 OF THE BASIC SPECTRAL SEQUENCE

3.1.1. DEFINITION. An (A, b)-configuration is *noncomplicated* if either $A = \{2, \ldots, 2\}$, $b = 0$, or $A = \{3, 2, \ldots, 2\}$, $b = 0$, or $A = \{2, \ldots, 2\}$, $b = 1$, and the last point v_1 of this configuration does not coincide with any of the $2\#A$ points forming the corresponding A-configuration.

The union of J-blocks over various complicated (A, b)-configurations with $|A| - \#A + b = i$ is a closed subset in $\sigma_i - \sigma_{i-1}$. Denote this union by S_i.

3.1.2. THEOREM. *For any* i, $\overline{H}_k(S_i) = 0$ *if* $k \geq 3d - 2$. *In particular, the factorization map*

$$\overline{H}_k(\sigma_i - \sigma_{i-1}) \to \overline{H}_k((\sigma_i - \sigma_{i-1}) - S_i) \qquad (6)$$

is an epimorphism for $k \geq 3d - 2$ *and isomorphism for* $k \geq 3d - 1$.

Proof is given in 3.4–3.7 below.

It follows from this theorem that in order to compute the 0-dimensional cohomology of the space $\Gamma^d \backslash \Sigma$ (that is, the $(3d - 1)$-dimensional closed homology of σ) it suffices to consider only J-blocks with noncomplicated J. By Lemma 2.1.2 the dimension of such blocks is at most $3d - 1$. This immediately implies restriction (i) of part (A) of Theorem 2.5. Part (B) of the theorem follows directly from the form of these blocks; see 3.2.4 and 3.3 for the argument.

3.2. Cellular decomposition of spaces $\sigma_i - \sigma_{i-1}$ for stable values of i.

3.2.1. Let $i \leq (3d - 1)/5$. By Lemma 2.1.2, in this case any J-block in $\sigma_i - \sigma_{i-1}$ is the total space of a locally trivial bundle whose base is the set of all (A, b)-configurations equivalent to J, and the fiber is the product of a $3(d - i)$-dimensional affine space and the union of interior points of various standard simplices in $\mathbb{R}^N$ defined by all generating collections of the configuration J. The base of this bundle is obviously homeomorphic to an open cell, and hence, its total space is homeomorphic to the product of the above three spaces. Decompose the last factor into open cells corresponding to various generating collections for J and consisting of the interior points of the corresponding standard simplices in $\mathbb{R}^N$. Such a decomposition of the factor induces a decomposition of the whole product, that is, our J-block. This decomposition is called the *canonical decomposition of the J-block*.

The *canonical decomposition of the space* $\sigma_i - \sigma_{i-1}$ is the decomposition into various cells of canonical decompositions of all J-blocks in $\sigma_i - \sigma_{i-1}$.

3.2.2. PROPOSITION. *For* $i \leq (3d + 1)/5$ *the decomposition of the one-point compactification of the space* $\sigma_i - \sigma_{i-1}$ *into cells of the canonical decomposition and the added point defines a structure of a finite CW-complex on this compactification.* □

3.2.3. We describe the corresponding chain complex with coefficients in $\mathbb{Z}_2$. Its generator, i.e. an arbitrary pair of the form $((A, b)$-configuration J

with $|A| - \#A + b = i$ (considered up to equivalence), generating collection of this configuration), will be shown as a set of $|A| + b$ points on the line (see Figure 32; cf. Figures 26, 30, and 31). Points in the A-configuration are shown as small circles, and b singular points as stars. Circles forming a pair from the generating collection are connected by a dotted arc; some stars can coincide geometrically with circles. The dimension of such a generator is equal to the sum of four numbers: $3(d-i)-1$, b, the number of dotted arcs, and the number $\rho(J)$ of geometrically distinct points in the configuration (it equals the dimension of the space of configurations equivalent to J). This dimension can be greater than $3d - 1$: for example, this occurs in the case of the (A, b)-configuration with $A = \{4\}$, $b = 0$ when the generating collection consists of all six possible pairs of four points forming this configuration.

The boundary operator maps this generator into the sum of similar generators that can be obtained from the initial one by the following two methods.

(1) Generators in the same J-block. Remove any dotted arc from the picture of our generator. If its end points remain to be connected by a sequence of dotted arcs, then the generator depicted by the picture thus obtained appears in the boundary of the initial one with coefficient 1 (mod 2). For example, in Figure 32 we can remove any of the three arcs connecting points 1, 3, and 4 (counting from the left).

(2) Generators obtained when the initial configuration degenerates. Consider any of $\rho(J) - 1$ segments of the line between adjacent points of the configuration J. To each such segment (except for the segments of the prohibited type, see below) we associate a generator whose picture coincides with the initial one outside this segment, and this segment contracts to a point which inherits all the arcs connecting the ends of the segment to other points of the configuration. The type of the obtained point (and the possibility of contraction) depends on the types of the endpoints of the segment. There are six possible types; see Figure 33.

FIGURE 32

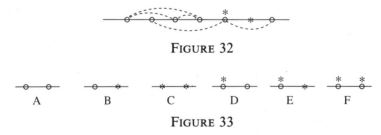

FIGURE 33

A. Type A splits into five subtypes. A1: if both points are connected by arcs to a third point, then the contraction is prohibited. A2: if the points are connected to each other and none of them is connected to any third point, then we get a point of type $*$. A3: if the points are connected to each other and at least one of them is connected to other points (but we do not have the subtype A1), then we get the point $\overset{*}{\circ}$. A4: if these two points are connected by sequences of ≥ 3 arcs only, then we get a point of type $\circ$. A5: if the points are not connected, then we also get $\circ$.

B. The obtained point has type $\overset{*}{\circ}$.

C, E, and F: the contraction is prohibited.

D. If these two points are connected directly or by a sequence of two arcs, then the contraction is prohibited; if they are connected by a sequence of ≥ 3 arcs or not connected at all, then the obtained point has type $\overset{*}{\circ}$.

All generators of our complex obtained by method (1) or (2) occur in the boundary of the initial cell with coefficient 1 (mod 2); all the other generators occur with coefficient 0. This gives a complete description of the boundary operator of the $\mathbb{Z}_2$-chain complex corresponding to the canonical CW-complex structure on the one-point compactification of the space $\sigma_i - \sigma_{i-1}$. A similar integral complex will be described in 3.3.

3.2.4. EXAMPLE. Let us compute the group $\overline{H}_k(\sigma_i - (\sigma_{i-1} \cup S_i))$ (see formula (5)) with coefficients in $\mathbb{Z}_2$. The definition of S_i shows that this group coincides with the k-dimensional homology group of the (quotient) chain complex whose generators are the cells of the canonical decomposition of $\sigma_i - \sigma_{i-1}$ corresponding to J-blocks with noncomplicated J, and the differentials are defined by the incidence coefficients of these cells. These cells belong to one of the following types.

1. If J is an (A, b)-configuration with $A = \{2, \ldots, 2\}$, $b = 0$, then the J-block consists of just one cell of dimension $3d - 1$.

2. If $A = \{2, \ldots, 2\}$, $b = 1$, then the J-block consists of just one cell of dimension $3d - 2$.

3. If $A = \{3, 2, \ldots, 2\}$, $b = 0$, then the J-block consists of one cell of dimension $3d - 1$ and three cells of dimension $3d - 2$; these four cells correspond to four possible generating collections of the three-element set.

In defining the homomorphism $h_i: X_i \to Y_i$ in 0.4 we have introduced two families of generators of the group X_i: $[i]$- and $\langle i \rangle$-configurations, and two families of generators of the group Y_i: i^*- and $\langle \tilde{i} \rangle$-configurations. In our present terminology, these are respectively the cells of type 1, $(3d - 1)$-dimensional cells of type 3, cells of type 2, and $(3d - 2)$-dimensional cells of type 3. The homomorphism h_i is precisely the homomorphism given by the incidence coefficients of these cells. Finally, we get the following

PROPOSITION. *For $i \leq (3d + 1)/5$, the group $\overline{H}_k(\sigma_i - (\sigma_{i-1} \cup S_i); \mathbb{Z}_2)$ is trivial if $k > 3d - 1$, and is isomorphic to the kernel of the operator h_i (mod 2) if $k = 3d - 1$.*

Part (B) of Theorem 2.5 in the $\mathbb{Z}_2$-case immediately follows from this proposition and Theorem 3.1.2 (which we prove in 3.4–3.7 below).

3.3. The integral spectral sequence. The chain complex described in 3.2 computes the column $\underleftarrow{E}^{-i, *}$ of the stable spectral sequence with coefficients in $\mathbb{Z}_2$. Now we describe a similar complex for a similar integral spectral sequence. This complex has the same system of generators (corresponding to the cells of the canonical cellular decomposition of $\sigma_i - \sigma_{i-1}$), but now these generators are given orientations, and their incidence coefficients equal ± 1, depending on the choice of these orientations. Fix such a system of orientations.

3.3.1. Consider an arbitrary cell of the canonical decomposition of $\sigma_i - \sigma_{i-1}$; it is defined by (a) some (A, b)-configuration

$$J = \{M_1, \ldots, M_{\#A}, v_1, \ldots, v_b\},$$

where each M_j is a subset of cardinality a_j in

$$\mathbb{R}^1 \colon M_j = \{t_{j,1} < \cdots < t_{j,a_j}\},$$

and (b) some generating collection $(T, V) = \{(t_1, t'_1), \ldots, (t_l, t'_l), v_1, \ldots, v_b\}$ of the configuration J.

The orientation of the cell consists of

(a) an orientation of the simplex in $\mathbb{R}^N$ spanned by the points

$$\lambda(t_1, t'_1), \ldots, \lambda(t_l, t'_l), \lambda(v_1, v_1), \ldots, \lambda(v_b, v_b); \qquad (7)$$

(b) an orientation of the set of configurations equivalent to J;

(c) an orientation of the subspace $\chi(\Gamma^d, J) \subset \Gamma^d$ consisting of maps that respect J.

To define these orientations, we order the sets M_j according to decrease of the corresponding numbers a_j, and for equal a_j, according to increase of their smallest elements $t_{j,1}$.

3.3.2. An orientation of a simplex is the order of its vertices, that is, the points (7). Order them as follows: first put the points $\lambda(t_{1,\alpha}, t_{1,\beta})$ according to lexicographic order of the pairs $(t_{1,\alpha}, t_{1,\beta})$ of the collection T generating the set M_1: the pair $(t_{1,\alpha}, t_{1,\beta})$ appears before $(t_{1,\gamma}, t_{1,\delta})$ if either $\min(t_{1,\alpha}, t_{1,\beta}) < \min(t_{1,\gamma}, t_{1,\delta})$ or these two numbers are equal and $t_{1,\alpha} + t_{1,\beta} < t_{1,\gamma} + t_{1,\delta}$. Then put the similarly ordered pairs from the generating collections for $M_2, \ldots, M_{\#A}$ and, finally, the points $\lambda(v_j, v_j)$ in the order of increasing v_j.

3.3.3. To orient the set of configurations equivalent to J list all $\rho(J)$ points defining this configuration in increasing order. Then define a positive tangent frame at the point J of this space as follows: its first vector is tangent to the line consisting of configurations with all $\rho(J)$ points except the first the same as in J, and the first one increases; the second vector is defined by increasing the second point, etc.

3.3.4. Let x, y, z be the linear coordinates in $\mathbb{R}^3$. The configuration J defines $3(|A| - \#A + b)$ linear functions on the space Γ^d: for any map

§3. THE TERM E_1 OF THE BASIC SPECTRAL SEQUENCE 171

$\theta \colon \mathbb{R}^1 \to \mathbb{R}^3$, $\theta(t) = (x(t), y(t), z(t))$, these functions take the values

$$x(t_{1,2}) - x(t_{1,1}), \quad y(t_{1,2}) - y(t_{1,1}), \quad z(t_{1,2}) - z(t_{1,1}),$$
$$x(t_{1,3}) - x(t_{1,2}), \quad y(t_{1,3}) - y(t_{1,2}), \quad z(t_{1,3}) - z(t_{1,2}),$$
$$\cdots\cdots\cdots\cdots\cdots\cdots\cdots\cdots\cdots\cdots\cdots\cdots\cdots\cdots\cdots\cdots$$
$$x(t_{1,a_1}) - x(t_{1,a_1-1}), \quad y(t_{1,a_1}) - y(t_{1,a_1-1}), \quad z(t_{1,a_1}) - z(t_{1,a_1-1}),$$
$$x(t_{2,2}) - x(t_{2,1}), \ldots, \tag{8}$$
$$\cdots\cdots\cdots\cdots\cdots\cdots\cdots\cdots\cdots$$
$$x(t_{\#A, a_{\#A}}) - x(t_{\#A, a_{\#A}-1}), \ldots,$$
$$\partial x/\partial t|_{v_1}, \quad \partial y/\partial t|_{v_1}, \quad \partial z/\partial t|_{v_1},$$
$$\cdots\cdots\cdots\cdots\cdots\cdots\cdots\cdots\cdots$$
$$\partial x/\partial t|_{v_b}, \quad \partial y/\partial t|_{v_b}, \quad \partial z/\partial t|_{v_b}.$$

If the number $i \equiv |A| - \#A + b$ is in the stable region $i \leq (3d+1)/5$, then these functions are linearly independent. The space $\chi(\Gamma^d, J)$ consisting of the maps of class Γ^d that respect the configuration J is precisely the zero set of all these functions. Define a coorientation of $\chi(\Gamma^d, J)$ in Γ^d (that is, an orientation of its orthogonal complement) using the exterior $3i$-form equal to the exterior product of the linear forms (8) taken in the order indicated in (8):

$$d(x(t_{1,2}) - x(t_{1,1})) \wedge \cdots \wedge d(\partial z/\partial t|_{v_b}).$$

Suppose that we have fixed an orientation in Γ^d. Then there is a uniquely defined orientation on $\chi(\Gamma^d, J)$ such that a frame in Γ^d is positively oriented if its first $3i$ vectors are normal to $\chi(\Gamma^d, J)$ and determine the above coorientation, and the last $3(d-i)$ vectors are tangent to $\chi(\Gamma^d, J)$ and are positive with respect to the orientation of $\chi(\Gamma^d, J)$ we have defined. The exact form of the orientation of Γ^d is not important: by changing it we simultaneously change the orientations of all cells, so that the incidence coefficients do not change. Let us describe these incidence coefficients.

3.3.5. Suppose two cells of the canonical decomposition of $\sigma_i - \sigma_{i-1}$ belong to the same J-block, and the generating collection used in the definition of one of them is obtained from the generating collection of the other by re-

a   b

FIGURE 34

moving one pair. Then the incidence coefficient of these cells equals $(-1)^{j-1}$, where j is the number of the removed pair in the enumeration from 3.3.2.

3.3.6. If the smaller of these incident cells is obtained by a degeneration of the configuration defining the larger one (see 3.2.3), then the incidence coefficient depends on the type of this degeneration. In cases A2, A3 from 3.2.3, this coefficient equals $(-1)^{\mu-\nu+\Delta+m}\zeta$, where μ is the number of the point $\lambda(t_j, t'_j)$ in the list (7) of vertices of the simplex for the initial cell corresponding to the contracted pair (t_j, t'_j), ν is the number of the new point $\lambda(v_k, v_k)$ in the similar list for the obtained configuration, Δ is the dimension of these simplices for both cells (that is, the number $l + b - 1$ from (7)), m is the number of the larger of two points being glued among the points in the initial configuration (see 3.3.3), and ζ is the coefficient responsible for agreement of orientations of the spaces $\chi(\Gamma^d, \cdot)$ for the initial and the obtained configurations. (In turn, ζ is computed as follows. Suppose the points $t_j < t'_j$ of the contracted pair are in the list (8) under the names $t_{s,\alpha}$, $t_{s,\beta}$, $\alpha < \beta$. Then $\zeta = (-1)^{\varkappa - \pi}$, where $\varkappa$ and π are respectively the number of the row $\{x(t_{s,\beta}) - x(t_{s,\beta-1}), \ldots\}$ in the list (8) for the initial configuration and the number of the row $\{\partial x/\partial t|_{v_k}, \ldots\}$ in the list (8) for the obtained configuration.)

In cases A4, A5 the incidence coefficient equals $(-1)^{\Delta + m} \cdot \varepsilon \cdot \zeta$, where Δ, m, and ζ are the same as before, and ε is the parity of the renumeration of the corresponding dotted arcs before and after gluing. For example, the gluing shown in Figure 34a gives $\varepsilon = \zeta = -1$, and the gluing in Figure 34b gives $\varepsilon = \zeta = 1$. In general, if the initial configuration has the type $(A, b) = ((2, \ldots, 2), 0)$, then always $\varepsilon = \zeta$.

In case B the incidence coefficient is $(-1)^{\Delta + m}$.

In the nonprohibited subcases of case D the rule is the same as in cases A4, A5.

EXAMPLE. For $i = 2$ the set $\sigma_i - \sigma_{i-1}$ has four, six, and three cells of dimensions $3d - 1$, $3d - 2$, and $3d - 3$, respectively. The incidence coefficients of these cells are computed by the rules 3.3.5 and 3.3.6; these computations give the following result.

PROPOSITION. *The complex* $(E_0^{-2,*}, d_0)$ *of the stable spectral sequence is acyclic in all dimensions except* $q = 2$, *and* $E_1^{-2,2} = G$. □

The remaining part of §3 is devoted to the proof of Theorem 3.1.2.

Again let i be an arbitrary positive integer.

3.4. An auxiliary filtration in the set $\sigma_i - \sigma_{i-1}$. Recall the notation $\rho(J)$ for the dimension of the space of configurations equivalent to J.

Define the *auxiliary filtration* $\varnothing = F_{[i/2]} \subset F_{[i/2]+1} \subset \cdots \subset F_{2i} \equiv \sigma_i - \sigma_{i-1}$ on the set $\sigma_i - \sigma_{i-1}$ by setting F_α to be the union of all J-blocks with $\rho(J) \leq \alpha$.

3.4.1. LEMMA. *The subset* $S_i \subset \sigma_i - \sigma_{i-1}$ *defined in* 3.1 *is precisely the term* F_{2i-2} *of the auxiliary filtration.* □

Consider the *auxiliary spectral sequence* $\mathscr{E}^\gamma_{\alpha,\beta}$ converging to the group $\overline{H}_*(\sigma_i - \sigma_{i-1})$ and generated by our auxiliary filtration. By definition, its term $\mathscr{E}^1_{\alpha,\beta}$ is isomorphic to the group $\overline{H}_{\alpha+\beta}(F_\alpha - F_{\alpha-1})$.

3.4.2. LEMMA. *Any connected component of the set $F_\alpha - F_{\alpha-1}$ belongs to some J-block for an (A, b)-configuration J such that $|A| - \#A + b = i$, $\rho(J) = \alpha$. In particular, the group $\mathscr{E}^1_{\alpha,\beta}$ splits into the direct sum of the groups $\overline{H}_{\alpha+\beta}(J\text{-block})$ over all such (A, b)-configurations J (one from each equivalence class).* □

To prove Theorem 3.1.2 it remains to prove that if the configuration J is complicated, then $\overline{H}_k(J\text{-block}) = 0$ for $k \geq 3d - 2$.

3.5. The bundle structure in J-blocks. By construction, any J-block is the space of a locally trivial bundle whose base is the set of all pairs of the form ((A, b)-configuration equivalent to J, map $\theta \in \Gamma^d$ respecting this configuration), and the typical fiber is the union of the interior points of various standard simplices in $\mathbb{R}^N$ defined by all generating collections of the configuration J. Lemma 2.1.2 immediately implies that the dimension of the base of this bundle is at most $3d - 3i + \rho(J)$. If the configuration J is complicated, then the last number does not exceed $3d - i - 2$, and it remains to prove that the one-point compactification of the fiber of the bundle is acyclic in dimensions exceeding $i - 1$. We study this fiber in detail.

Obviously, the dimension of this fiber is $\sum \binom{a_l}{2} + b - 1$; a dense set in this fiber is the open simplex spanned by all the points $\lambda(t, \tau)$, $\lambda(v, v)$ such that the points t, τ are in one of the $\#A$ groups of points defining the A-part of the configuration J, and v is one of the points defining its b-part. In addition to this open simplex, the fiber contains some of its (nonclosed) faces. Enumerating these faces leads to the following formalism.

3.6. The generating complex of (A, b)-configurations.

3.6.1. To each set M of cardinality $a < \infty$ corresponds its *generating complex* $K(M)$, defined as follows.

Consider a simplex of dimension $\binom{a}{2} - 1$ whose vertices correspond to the two-element subsets of our set M. A face of this simplex is said to be *generating* if any two elements of M can be connected by a chain consisting of pairs corresponding to the vertices of this face. Obviously, the dimension of the generating face can lie between $a - 2$ and $\binom{a}{2} - 1$.

The obvious triangulation of our simplex defines an (acyclic) chain complex of dimension $\binom{a}{2} - 1$; the generating complex $K(M)$ is defined as the quotient complex by the subcomplex spanned by all nongenerating faces.

EXAMPLE. Let M consist of three elements $\varkappa, \lambda$, and μ. Then the complex $K(M)$ consists of one nonclosed triangle (whose vertices are the pairs $(\varkappa, \lambda)$, (λ, μ), and $(\mu, \varkappa)$) and three open segments (its sides).

Now let J be an (A, b)-configuration. Its generating complex is the join([8]) of $\#A$ generating complexes corresponding to all sets that form the

---

([8]) The definition of the join of two chain complexes is an obvious formalization of the definition of the join of CW-complexes (see Appendix 1): the join of k chain complexes is

A-configuration, and b acyclic (in positive dimensions) complexes corresponding to b singular points. Of course, the generating complexes for all (A, b)-configurations with the same A, b are isomorphic to each other.

3.6.2. TAUTOLOGICAL LEMMA. *For any (A, b)-configuration J the closed homology group of the fiber of the bundle from 3.5 is isomorphic to the homology group of the generating complex of (A, b)-configurations.* □

It remains to prove that the generating complex of any complicated (A, b)-configuration is acyclic in dimensions greater than $i-1$. But $i = |A|-\#A+b$, so (by the Künneth formula) it suffices to prove the following statement.

3.6.3. THEOREM. *For any a-element set M, $H_l(K(M)) = 0$ for $l \neq a-2$, and $H_{a-2}(K(M)) = G^{(a-1)!}$.*

3.7. Proof of Theorem 3.6.3. Let $\Sigma(M)$ be a subcomplex of the $(\binom{a}{2} - 1)$-dimensional simplex consisting of all nongenerating faces. Since the simplex is acyclic, $H_l(K(M)) \cong \tilde{H}_{l-1}(\Sigma(M))$, and it remains to prove that the complex $\Sigma(M)$ is acyclic in dimensions other than 0 and $a - 3$, and $\tilde{H}_{a-3}(\Sigma(M)) \cong G^{(a-1)!}$.

3.7.1. *Auxiliary construction: the complex of incomplete decompositions.* Consider the simplex $\nabla(M)$ with $2^{a-1} - 1$ vertices corresponding to the various decompositions of the set M into two nonempty subsets. Call a face of this simplex *incomplete* if there is a pair of points in M which is not separated by any of the decompositions corresponding to the vertices of this face. Denote by $\Omega(M)$ the union of all incomplete faces of the simplex $\nabla(M)$.

3.7.2. THEOREM. *The complexes $\Sigma(M)$ and $\Omega(M)$ are homotopy equivalent.*

PROOF. We construct a cellular subcomplex $\sigma(M) \subset \Sigma(M) \times \Omega(M)$ that is homotopy equivalent to both $\Sigma(M)$ and $\Omega(M)$. Namely, consider an arbitrary decomposition of the set M into k nonempty subsets, $2 \leq k \leq a - 1$. To this decomposition correspond a simplex in $\Sigma(M)$ whose vertices are various pairs of points not separated by this decomposition, and a simplex in $\Omega(M)$ whose vertices are decompositions of M into pairs of subsets, each being the union of several parts of our decomposition. Consider the product of these two simplices in $\Sigma(M) \times \Omega(M)$. Denote by $\sigma(M)$ the union of all such products over all nontrivial decompositions of M.

PROPOSITION. *The obvious projections of $\sigma(M)$ to $\Sigma(M)$ and to $\Omega(M)$ are homotopy equivalences.*

---

their tensor product increased in dimension by $k - 1$. In particular, the generating complex of the (A, b)-configuration defined here coincides with the tensor product of $\#A$ generating complexes for sets of cardinality $a_1, \ldots, a_{\#A}$ shifted in dimension by $\#A + b - 1$.

Indeed, over any open simplex in $\Sigma(M)$ or $\Omega(M)$ such a projection is the trivial bundle with simplex as fiber; repeating the proof of Lemma 1 in §III.3.3 we see that these projections are homology equivalences. It can easily be verified that for $a > 4$ the three complexes are simply connected, and it remains to use the Whitehead theorem; the case $a \leq 4$ is trivial. □

3.7.3. It remains to prove the following four statements.

LEMMA A. *The complex $\Sigma(M)$ is acyclic in dimensions lower than $a - 3$.*

LEMMA B. *The complex $\Omega(M)$ is acyclic in dimensions greater than $a - 3$.*

LEMMA C. *The G-module $H_{a-3}(\Omega(M), G)$ is free.*

LEMMA D. *The Euler characteristic of the complex $\Omega(M)$ is equal to $1 - (-1)^a (a-1)!$.*

Lemma A follows from the fact that for $i < a - 2$ all i-dimensional faces of the $(\binom{a}{2} - 1)$-dimensional simplex belong to the subcomplex $\Sigma(M)$.

Define an increasing filtration $F_2 \Omega(M) \subset \cdots \subset F_{a-1} \Omega(M) = \Omega(M)$ in $\Omega(M)$ by associating to $F_p \Omega(M)$ the union of all simplices corresponding to decompositions of the set M into k subsets, $k \leq p$, in the proof of Theorem 3.7.2. Let $\mathscr{E}^r_{p,q} \to H_{p+q}(\Omega(M))$ be the spectral sequence generated by this filtration; by definition, $\mathscr{E}^1_{p,q} \cong H_{p+q}(F_p \Omega(M), F_{p-1} \Omega(M))$.

PROPOSITION. *For any $p = 2, \ldots, a-1$ the group $\mathscr{E}^1_{p,q}$ splits into the direct sum of subgroups corresponding to various decompositions of M into p nonempty sets, and each of these subgroups is isomorphic to $\widetilde{H}_{p+q-1}(\Omega(\{p\}))$, where $\{p\}$ is a set of p elements.*

Indeed, consider any such decomposition and the corresponding simplex in $\Omega(M)$. This simplex has $2^{p-1} - 1$ vertices corresponding to decompositions of M into nonempty pairs of subsets that are unions of several parts of the initial decompositions; it is naturally identified with the simplex $\nabla(\{p\})$. A face of our simplex belongs to $F_{p-1}(\Omega(M))$ if and only if the corresponding face in $\nabla(\{p\})$ is incomplete; a face that does not belong to $F_{p-1}(\Omega(M))$ cannot be a face of a simplex in $\Omega(M)$ corresponding to another decomposition of M into p parts. This implies the proposition.

Let us now assume that Lemma B is proved for all complexes $\Omega(\{k\})$, $k < a$. Then the proposition implies that the term $\mathscr{E}^1_{p,q}$ of our spectral sequence vanishes for $q > -2$; moreover $\mathscr{E}^1_{p,q} = 0$ for $p \geq a$ since $F_{a-1} \Omega(a) = \Omega(a)$. This also proves Lemma B for an a-element set M.

Lemma C follows from Lemma B and the Universal Coefficient Theorem.

Lemma D will be proved by induction on a. The induction hypothesis and our proof of Lemma B imply that the required Euler characteristic equals $\sum_{k=2}^{a-1} (-1)^k (k-1)! \langle a?k \rangle$, where $\langle a?k \rangle$ is the number of unordered decompositions of the a-element set into k nonempty subsets. Since

$\langle a?1\rangle = \langle a?a\rangle = 1$, this implies that Lemma D is equivalent to the equality $\sum_{k=1}^{a}(-1)^{k}(k-1)!\langle a?k\rangle = 0$. This equality follows from the obvious combinatorial identity $\langle a?k\rangle = k\langle a-1?k\rangle + \langle a-1?k-1\rangle$, and Theorem 3.6.3 is proved completely. This also completes the proof of Theorem 3.1.2 and part (A) of Theorem 2.

REMARK. Another proof of Lemma B follows from the Goresky-MacPherson formula (or, more exactly, from its Alexander dual formula (37) in Section III.6). Indeed, consider the space $\mathbb{R}^a$ with coordinates $x_1, \ldots, x_a$ and a plane arrangement in it, formed by all hyperplanes with equations $x_i = x_j$, $i \neq j$. Then our complex $\Sigma(M)$ coincides with the complex $K(\Lambda)$ which corresponds in the Goresky-MacPherson formula to the "deepest" stratum Λ of this arrangement; that is, to the line $x_1 = \cdots = x_a$. According to this formula, the i-dimensional homology of the complex $\Sigma(M)$ enters as a direct summand in the $i+1$-dimensional homology of the one-point compactification of that arrangement. Since this compactification is an $(a-1)$-dimensional complex, our lemma follows.

§4. Algorithms for computing the invariants and their values

In this section we describe two algorithms ready to be programmed.

The first algorithm computes stable invariants of the ith order for any $i > 0$ (in particular, the terms $\underleftarrow{E}_r^{-i,i}$, $r = 2, 3, \ldots, \infty$, of the basic spectral sequence converging to the cohomology of the space $\Gamma^d \setminus \Sigma$, $d \geq (5i-1)/3$).

The second algorithm for each such invariant (given by the corresponding output data of the first algorithm) computes its value on an arbitrary noncompact knot $\theta \in K$ encoded by a finite diagram (see [CF], [Lick], [FCh]).

REMARK (added in second edition). Recently, Ted Stanford of Columbia University realized both these algorithms and using them calculated all invariants of order ≤ 7 as well as their values on all table knots with ≤ 10 overlaps.

4.1. Truncated spectral sequence.

For any $i > 0$ and any coefficient group G define the groups $X_i = X_i(G)$ and $Y_i = Y_i(G)$ as free G-modules whose generators are $[i]$- and $\langle i \rangle$-configurations ($\langle \tilde{i} \rangle$- and i^*-configurations, respectively) described in 0.4. To all these configurations correspond cells of the canonical decomposition of $\sigma_i - \sigma_{i-1}$; the incidence coefficients of these cells described in 3.3 define a homomorphism $h_i = h_i(G) : X_i \to Y_i$.

By Theorem 3.1.2, the terms $\underleftarrow{E}_1^{-i,i}$ and $\underleftarrow{E}_1^{-i,i+1}$ of the basic stable spectral sequence are isomorphic to the kernel of the homomorphism h_i and to a subgroup of the group $Y_i/h_i(X_i)$, respectively.

Replace by zeros all terms $\underleftarrow{E}_r^{p,q}$ of the stable spectral sequence with $p + q \geq 2$. The *truncated spectral sequence* $\widetilde{E}_r^{p,q}$ thus obtained has the following properties:

(i) $\widetilde{E}_r^{p,q} = 0$ for $p \geq 0$ or $p+q \notin \{0, 1\}$; $\widetilde{E}_0^{-i,i} \equiv X_i$, $\widetilde{E}_0^{-i,i+1} \equiv Y_i$;

§4. ALGORITHMS FOR COMPUTING THE INVARIANTS AND THEIR VALUES   177

(ii) for any $i, r \geq 1$ the term $\underleftarrow{E}_r^{-i,i}$ of the stable spectral sequence is canonically isomorphic to the corresponding term of the truncated sequence, and the term $\underleftarrow{E}_r^{-i,i+1}$ is canonically isomorphic to a subgroup of the group $\widetilde{E}_r^{-i,i+1}$.

Below we describe an algorithm for computing this truncated spectral sequence $\{\widetilde{E}_r^{p,q}\}$.

4.2. Actuality table. Any invariant of ith order is defined by an *actuality table*: this table is filled in for each (generating) element of the group $\widetilde{E}_1^{-i,i}$ during the computation of the spectral sequence (and the process may terminate if the corresponding differential of this element happens to be nontrivial (^{9})). Here is the description of this table.

The empty actuality tables are the same for all invariants of ith order. Each such table consists of i levels. The lth level consists of $2l!/2^l l!$ cells in bijective correspondence with various $[l]$-configurations considered up to equivalence. In each such cell there is drawn (or encoded) an immersed connected curve in $\mathbb{R}^3$ that coincides with a standard imbedded line outside some compact set, respects the corresponding $[l]$-configuration, and has no "unnecessary" self-intersections except the l intersections determined by this configuration; another condition is that at each of the l self-intersection points the tangents to its branches are noncollinear. For example, the first level consists of one cell: it can contain the curve of Figure 25. The second level contains three cells; they can contain curves of Figure 27. In the general case, I leave the choice of such curves to the programmer. (^{10})

The table must be filled in from the top starting with the ith level. In each cell we place a number (or, in general, an element of the coefficient group of the cohomology under consideration). This number is called the *actuality index* of the corresponding self-intersecting curve.

For the cells of the upper level these indices are defined as follows. The initial element of the group $E_1^{-i,i}$ for which we are trying to construct the actuality table is a formal linear combination of $[i]$-configurations and $\langle i \rangle$-configurations; see 0.4. The coefficient of an $[i]$-configuration is the actuality index placed in the corresponding cell of the ith level.

The cells of the $(i-1)$th level are filled in during the computation of the differential $d_1 : \widetilde{E}_1 \to \widetilde{E}_1$ of the truncated spectral sequence; the cells of the

---

(^{9}) It follows from a recent theorem of M. Kontsevich that this cannot happen: all elements of $\widetilde{E}_1^{-i,i}$ can be extended to elements of $\widetilde{E}_\infty^{-i,i}$.

(^{10}) With one exception though. Suppose that a pair of points defining our configuration is not separated by points of other pairs. Then the loop in $\mathbb{R}^3$ that is the image of the segment between the points of this pair must be unlinked with the other parts of the curve (i.e. our curve can be continuously deformed in such a way that all the intermediate curves respect some $[l]$-configurations equivalent to the initial one and have no extra self-intersections, and at the end of this deformation our self-intersection degenerates into a singular point; see Figure 35). All pictures in Figures 27 and 25 satisfy this condition.

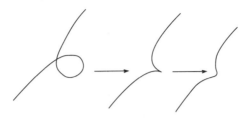

FIGURE 35

$(i-2)$th level are filled in during the calculation of d_2, etc. We describe this process in 4.5, but here we show how to use the table.

4.3. Algorithm for computing the values of invariants.

4.3.1. *Coorientation of the discriminant.* The hypersurface $\Sigma \subset \Gamma^d$ has the following standard transversal orientation at nonsingular points. A nonsingular point $\theta \in \Sigma$ is a map $\mathbb{R}^1 \to \mathbb{R}^3$ that glues together some points $t < t' \in \mathbb{R}^1$ so that the derivatives of θ at these points are not collinear; see Figure 25. For any point $\theta_1 \in \Gamma^d$ near θ define a number $w(\theta_1)$ as the determinant of the triple of vectors $\partial\theta_1/\partial\tau|_t$, $\partial\theta_1/\partial\tau|_{t'}$, $(\theta_1(t') - \theta_1(t))$ with respect to the standard coordinates x, y, z. Obviously, $w(\theta) = 0$, and the hypersurfaces Σ and $\{\theta_1 | w(\theta_1) = 0\}$ have the same tangent planes at the point θ. The vector attached at the point θ and transversal to the set Σ is called positive if the derivative of the function $w(\cdot)$ in the direction of this vector is positive, and negative if the derivative is negative.

This rule defines a coorientation of the hypersurface Σ at all its nonsingular points and, moreover, a coorientation of any nonsingular local branch of Σ at points of self-intersection of Σ.

This coorientation is related to the orientation of the manifold σ_1 defined in 3.3 as follows. Suppose there is a fixed orientation of the space Γ^d; then it is possible to orient the surface Σ near any of its nonsingular points as the boundary of the positive (with respect to our coorientation) part of its complement. But near nonsingular points the projection $\sigma_1 \to \Sigma$ is a local diffeomorphism, and hence, we get an orientation of σ_1. This orientation always coincides with the orientation defined in 3.3.3–3.3.4 (which, by the way, also depends on the choice of orientation for Γ^d).

4.3.2. Given a noncompact knot $\theta \colon \mathbb{R}^1 \to \mathbb{R}^3$, $\theta \in \Gamma^d$, where d is sufficiently large, and an invariant γ of order i defined by its actuality table. The computation of the value of the invariant γ on the knot θ consists of i steps.

The first step. Connect an unknotted imbedding $\mathbb{R}^1 \to \mathbb{R}^3$ with the knot θ by a path in general position in the space of normal immersions $\mathbb{R}^1 \to \mathbb{R}^3$ (i.e. immersions with a fixed asymptotics at infinity, see §1). This path

§4. ALGORITHMS FOR COMPUTING THE INVARIANTS AND THEIR VALUES

has several transversal intersections with the discriminant Σ at nonsingular points of the latter. These intersection points $\theta_1, \ldots, \theta_s$ are immersions $\mathbb{R}^1 \to \mathbb{R}^3$ with one self-intersection. The value of the invariant γ on the knot θ is equal to the sum of the actuality indices of these points θ_j (all remaining steps of the algorithm are devoted to computing these indices), taken with coefficients equal to either $+1$ or -1 depending on whether we intersect Σ in the positive or negative direction.

The second step. We start by computing the actuality indices for each of the points θ_j. By a reparametrization of $\mathbb{R}^1$ we make the map θ_j glue together the same pair of points t_1, t_1' as the model map in the first level of the actuality table. For this model map the actuality index is already known (and written in the table); if this model map satisfies the requirements from footnote (10) in 4.2 (for example, if this is a curve in Figure 25), then this index is 0. Let $K(t_1, t_1')$ be the space of all normal maps $\mathbb{R}^1 \to \mathbb{R}^3$ gluing points t_1, t_1' together. Connect the model map with the map θ_j by an arbitrary path in general position in the space $K(t_1, t_1')$. This path contains several maps $\theta_{j,1}, \ldots, \theta_{j,s_j}$, each of which glues together another pair of points t_2, t_2' different from t_1, t_1'. The required actuality index θ_j is equal to the index of the model map plus the sum of indices of the maps $\theta_{j,1}, \ldots, \theta_{j,s_j}$ (determined in the future steps of the algorithm) taken with coefficients $\varepsilon \cdot (-1)^{n(t_2)+n(t_2')}$, where ε is equal to $+1$ or -1 depending on whether we intersect the discriminant in the positive or negative direction, and $n(t_2)$ and $n(t_2')$ are the numbers of the points t_2, t_2' in the collection $\{t_1, t_1', t_2, t_2'\}$ arranged in increasing order.

At the lth step of the algorithm ($l < i$) we have to determine the actuality indices of some set of maps $\theta_{j_1, \ldots, j_{l-1}}$, each of which glues together $l-1$ nonintersecting pairs of points in $\mathbb{R}^1$. We do this for each of these maps separately. Find on the $(l-1)$th level of the actuality table a cell corresponding to an $[l-1]$-configuration respected by the map $\theta_{j_1, \ldots, j_{l-1}}$. Reparametrize the line $\mathbb{R}^1$ so that this map and the map shown in this cell respect the same configuration. The subsequent construction is the same as before: we consider a generic homotopy of the model immersed curve onto the curve $\theta_{j_1, \ldots, j_{l-1}}$ in the space $K(t_1, t_1') \cap \cdots \cap K(t_{l-1}, t_{l-1}')$; the actuality indices of the points $\theta_{j_1, \ldots, j_l}$ are taken with coefficients $\varepsilon \cdot (-1)^{n(t_l)+n(t_l')+l}$, where ε is defined as before, and $n(t_l)$, $n(t_l')$ are the indices of points t_l, t_l' among the points $\{t_1, t_1', \ldots, t_{l-1}', t_l, t_l'\}$ listed in increasing order.

Finally, the actuality index of any map $\theta_{j_1, \ldots, j_i}$ computed at the last ith step is equal to the index written in the table in the ith level cell that corresponds to the $[i]$-configuration respected by this map. (This is compatible with the scheme described above if we add to the table the $(i+1)$th level where only the zero actuality indices are written.)

To complete the actuality table we need the following object.

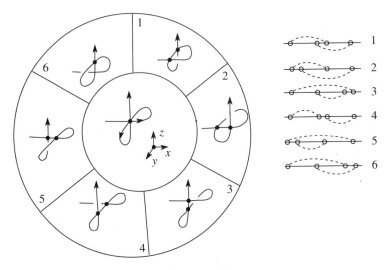

FIGURE 36

4.4. The *extended actuality table* looks the same as the basic one defined in 4.2 but contains some additional cells corresponding to various $\langle l \rangle$-configurations, $l \leq i$; before filling in the tables we fix for each such cell an immersed curve in $\mathbb{R}^3$ that respects the corresponding $\langle l \rangle$-configuration and has no unnecessary self-intersections. Here at each point of self-intersection (at triple as well as at double ones) the tangent directions to its branches must be linearly independent in $\mathbb{R}^3$ (see Figure 36); moreover, at each triple point, where the corresponding map θ glues together three points $t < t' < t''$, the frame $(\partial\theta/\partial\tau|_t, \partial\theta/\partial\tau|_{t'}, \partial\theta/\partial\tau|_{t''})$ must be positively oriented.

EXAMPLE. The first level does not contain new cells, and the second contains just one new cell, in which we put an immersed curve with a triple point as in the center of Figure 36 and without any other self-intersections.

We complete the extended actuality table for each (generating) element γ_0 of the group $\widetilde{E}_1^{-i,i}$: the data that are not contained in the basic table are not necessary to compute the values of invariants, but in completing the lth level of the basic table it is necessary, in general, to know the $(l+1)$th level not only of the basic table but also of the extended one.

The upper (ith) level of the extended table is completed as before: in the cell corresponding to some $\langle i \rangle$-configuration we write the coefficient of this configuration in the element $\gamma_0 \in \widetilde{E}_1^{-i,i} \subset X_i$ we are considering.

4.5. The differential d_1 of the truncated spectral sequence and completion of the $(i-1)$th level of the actuality table.

NOTATION. We denote by Ξ_l the set $(\sigma_l - \sigma_{l-1}) - S_l$ (see (3.1)), i.e. the union of all J-blocks of the space $\sigma_l - \sigma_{l-1}$ over all noncomplicated configurations J.

By construction, $\overline{H}_k(\Xi_i) = \widetilde{E}_1^{-i,3d-1-k+i}$; in particular, this group is trivial for k not equal to $3d-1$ and $3d-2$.

Let γ_0 be an arbitrary nonzero element of the group $\widetilde{E}_1^{-i,i} \equiv \overline{H}_{3d-1}(\Xi_i)$. Let us compute the corresponding element $d_1 \gamma_0 \in \widetilde{E}_1^{-i+1,i}$.

4.5.1. *The geometric boundary of* γ_0. By construction, γ_0 is a linear combination of oriented cells of the canonical decomposition of Ξ_i corresponding to $[i]$- and $\langle i \rangle$-configurations, and the boundary of this linear combination in Ξ_i is equal to 0. We describe its boundary in Ξ_{i-1}.

The boundary in Ξ_{i-1} of any of its cells is defined as follows. If this cell C corresponds to an $[i]$-configuration J, $C = C(J)$, then its boundary in Ξ_{i-1} contains i components corresponding to all possible pairs of points defining the configuration J. Consider any such pair; let it be the mth pair with respect to the natural order on pairs (see 3.3.2). Remove this pair from the configuration J, denote the remaining $[i-1]$-configuration by J_m, and consider the cell of the canonical decomposition of the space Ξ_{i-1} corresponding to it. By construction this cell is a subset in $\Gamma^d \times \mathbb{R}^N$ consisting of the pairs of the form (θ, u), where u is a point of a nonclosed $(i-2)$-dimensional simplex in $\mathbb{R}^N$ corresponding to a certain $[i-1]$-configuration $\widetilde{J}_m$ equivalent to J_m, and θ is a map of class Γ^d that respects this configuration. Consider a subset of this cell consisting of (θ, u) such that θ also glues together such a pair of points not in $\widetilde{J}_m$ that, adding this pair of points to $\widetilde{J}_m$, we get an $[i]$-configuration equivalent to J. This subset is a hypersurface in our cell; in general, it has singularities and self-intersections. Such a hypersurface is called a singular hypersurface of the *first type*.

The boundary of the cell $C(J)$ can also contain other surfaces lying in the cells corresponding to $\langle i-1 \rangle$-configurations. Namely, suppose that among the i characteristic pairs of the initial $[i]$-configuration J there is a triple of pairs (t_1, t_1'), (t_2, t_2'), (t_3, t_3') such that the points t_1 and t_2', t_2 and t_3', t_3 and t_1' are next to each other in $\mathbb{R}^1$. Consider the $\langle i-1 \rangle$-configuration $J!$ obtained from J by replacing these three pairs by one triple consisting of the centers of the corresponding intervals. In the cell $C(J!)$ consider the hypersurface consisting of pairs (θ, u) such that the derivatives of the map θ at the points of the characteristic triple are coplanar. Such hypersurfaces will be called singular surfaces of the *2nd type*. The boundary of the initial cell $C = C(J)$ in Ξ_{i-1} is the union of surfaces of the 1st and 2nd types, taken with suitable orientations.

If the cell C corresponds to an $\langle i \rangle$-configuration then its boundary in Ξ_{i-1} consists of $i-2$ hypersurfaces of the *3rd type* corresponding to all possible pairs of points used in the definition of this configuration (not in its triple) and lying in the cells corresponding to $\langle i-1 \rangle$-configurations obtained from the initial one by forgetting the pair.

Finally, the geometric boundary $\partial_1 \gamma_0$ of our cycle γ_0 in Ξ_{i-1} is defined by linearity; since γ_0 is a cycle in Ξ_i, $\partial_1 \gamma_0$ is a cycle in Ξ_{i-1}.[11] In order to find the element $d_1 \gamma_0$ we have to construct a cycle in Ξ_{i-1} homologous to $\partial_1 \gamma_0$ and realized by a linear combination of $(3d-2)$-dimensional cells of the canonical decomposition of Ξ_{i-1}. This is done in 4.5.2 and 4.5.3 below.

4.5.2. *The homological boundary of γ_0.* For any $[l]$- (or $\langle l \rangle$-)configuration J denote by $\Sigma(J)$ the union of all singular hypersurfaces of the 1st (respectively, 2nd and 3rd) order in the corresponding cell $C(J)$ of the space Ξ_l. Denote by $\Sigma(l)$ the union of sets $\Sigma(J)$ over all $(3d-1)$-dimensional cells in Ξ_l. By construction, the support of the chain $\partial_1 \gamma_0$ belongs to $\Sigma(i-1)$.

DEFINITION. An *elementary component* in Ξ_l is any connected component of the complement of $\Sigma(l)$ in any $(3d-1)$-dimensional cell of the canonical decomposition of Ξ_l. The standard orientation of such a component is given by the orientation of the corresponding cell; see 3.3.

Let us try to span the cycle $\partial_1 \gamma_0 \subset \Xi_{i-1}$ by a chain in Ξ_{i-1} consisting of elementary components (taken with appropriate multiplicities).

DEFINITION. A chain in Ξ_{i-1} given by a linear combination of elementary components is called *compatible with the cycle $\partial_1 \gamma_0$* if in the intersection with any open $(3d-1)$-dimensional cell C of the canonical decomposition of Ξ_{i-1} the boundary of this chain is equal to $\partial_1 \gamma_0$ (that is, under the restriction homomorphism $\overline{H}_*(\Xi_{i-1}) \to \overline{H}_*(C)$ the images of the element $\partial_1 \gamma_0$ and the boundary of this chain coincide).

DEFINITION. The *main component* of the complement of $\Sigma(l)$ in a $(3d-1)$-dimensional cell $C \subset \Xi_l$ is the component that contains the curve defined in the corresponding cell of the actuality table.

LEMMA. *The group of chains in Ξ_{i-1} compatible with $\partial_1 \gamma_0$ is isomorphic (noncanonically) to the group X_{i-1}.*

Indeed, the coefficient in this chain of the main component of the complement of $\Sigma(i-1)$ in the cell $C(J)$ uniquely determines the coefficients of all other components in this cell. Namely, in passing from one such component to another through a smooth region in $\Sigma(i-1)$ of the 1st or 3rd type in the positive direction (for the definition of which, see 4.3.1) this coefficient increases by the product of two numbers $I \cdot \chi$. Here I is the actuality index of any curve in this region either with i self-intersections or with $i-2$ self-intersections and one triple point (the value of this index can be extracted from the ith level of the actuality table), and χ is defined as follows. To each point of the cell C corresponds a map $\theta: \mathbb{R}^1 \to \mathbb{R}^3$ that respects a

---

[11] This conclusion requires a justification, since the union of the sets S_l is not closed in σ. But the intersection of the closure of the set S_i with Ξ_{i-1} has dimension ≥ 2 in Ξ_{i-1}, and our argument is correct.

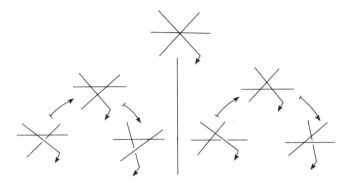

FIGURE 37

certain $[i-1]$- or $\langle i-1\rangle$-configuration depending on the type of this cell. At a nonsingular point of the surface $\Sigma(i-1)$ of the 1st or 3rd type this map glues together one more pair of points τ, τ', i.e. it respects a certain $[i]$-(respectively, $\langle i\rangle$-)configuration. Let $n(\tau)$ and $n(\tau')$ be the numbers of these points in the list of all $2i$ $(2i-1)$ points defining this configuration and taken in increasing order. The number χ equals $(-1)^{n(\tau)+n(\tau')+i}$.

The computation of the jump of the coefficient in passing through a surface of the 2nd type in an $\langle i-1\rangle$-cell reduces to passages through surfaces of the 1st type. Namely, let $C(J)$ be a cell in Ξ_{i-1} corresponding to the $\langle i-1\rangle$-configuration J, and let (θ, u) be a generic point of a surface of the 2nd type in $C(J)$. Then all three derivatives of θ at the points of the glued triple do not vanish and are not pairwise collinear. Such a map θ can be perturbed in two ways so that the resulting maps θ_1 and θ_2 preserve all double points remote from our triple point, and this triple point splits into three points of self-intersection; see Figure 37. The passage through the surface $\Sigma(i-1)$ at the point (θ, u) can be realized by a one-parameter family of maps θ^t, $t \in (-\varepsilon, \varepsilon)$, preserving two local branches of the curve $\theta(\mathbb{R}^1)$ and turning the third so that its tangent always lies in a fixed plane orthogonal to the plane spanned by the tangents to the first two branches. At the same time deform the perturbed map θ_1 (and θ_2) so that only the third local branch of the curve $\theta_1^t(\mathbb{R}^1)$ gets changed, namely, it must always intersect one of the first two branches at the same point, and its tangent at this point must repeat the movement of the tangent to the corresponding branch of θ^t; see Figure 37. Then all maps θ_1^t respect a certain $[i-1]$-configuration J_1, and this entire deformation realizes a passage in the corresponding cell $C(J_1) \subset \Xi_{i-1}$ through a singular surface of the 1st type. Let X_1 be the corresponding jump of the coefficient of the chain in the cell $C(J_1)$ compatible with $\partial_1 \gamma_0$; define X_2 similarly, using the map θ_2. The jump in the cell $C(J)$ in moving along the path $\{\theta^t, u\}$ is equal to the sum $X_1 + X_2$. □

Choose a chain in Ξ_{i-1}, compatible with $\partial_1 \gamma_0$, so that the main component of any $(3d-1)$-dimensional cell is included in this chain with coefficient 0. By construction, the boundary of this (and any other) compatible chain is the sum of the cycle $\partial_1 \gamma_0$ and a certain linear combination of $(3d-2)$-dimensional cells of the canonical decomposition of the space Ξ_{i-1}. Denote this linear combination by $\tilde{d}_1 \gamma_0$. This is an element of the group $Y_{i-1} \equiv \tilde{E}_0^{-i+1,i}$. It depends on the choice of the compatible chain (i.e. the choice of the curve in the actuality table), but the class of this element in the group $\tilde{E}_1^{-i+1,i} \equiv Y_{i-1}/h_{i-1}(X_{i-1})$ is defined invariantly. This is precisely $d_1 \gamma_0$.

4.5.3. *Algorithm for the computation of $\tilde{d}_1 \gamma_0$.* Take an arbitrary $(3d-2)$-dimensional cell $\tilde{C}$ of the canonical decomposition of Ξ_{i-1} corresponding to a $\langle \widetilde{i-1} \rangle$-configuration and compute the coefficient of this cell in $\tilde{d}_1 \gamma_0$. Consider the corresponding $\langle i-1 \rangle$-configuration and the cell corresponding to it in the $(i-1)$th level of the extended actuality table. Consider the immersed curve in $\mathbb{R}^3$ given in this cell. There are exactly six different ways to perturb this curve so that the resulting curve respects some $[i-1]$-configuration; then its triple point splits into two double points; see Figure 36. Choose two ways from these six such that the dotted arcs of the obtained $[i-1]$-configurations corresponding to these two pairs arise from the dotted arcs in the picture of the initial $\langle \widetilde{i-1} \rangle$-configuration. For example, if the cell $\tilde{C}$ is given by the $\langle \widetilde{i-1} \rangle$-configuration shown as the first summand of the right side of the upper row in Figure 31 (the second and third summands, respectively), then the ways 2 and 5 (respectively, 1 and 4, 3 and 6) correspond to it in Figure 36. These two ways define two immersed curves in $\mathbb{R}^3$ that respect two distinct $[i-1]$-configurations and have no extra self-intersections (hence belong to certain components of the complement of $\Sigma(i-1)$ in the corresponding cells of the canonical decomposition of Ξ_{i-1}). Let us compute the coefficients of these components in our compatible chain: to do this, connect the two obtained curves by paths in general position in the corresponding cells with the model ones and sum the numbers indicated in the proof of the lemma from 4.5.2 over all intersection points of these paths with $\Sigma(i-1)$. Finally, the desired coefficient that defines the multiplicity of the cell $\tilde{C}$ in $\tilde{d}_1 \gamma_0$ is the sum of the two coefficients above multiplied by the incidence coefficients of the cell $\tilde{C}$ and the corresponding two $[i-1]$-cells.

Now let $\tilde{C}$ be a $(3d-2)$-dimensional cell in Ξ_{i-1} corresponding to a certain i^*-configuration. If the programmer did not neglect footnote 10 (see 4.2), then this cell occurs in $\tilde{d}_1 \gamma_0$ with coefficient zero.

4.5.4. *Filling the $(i-1)$th level of the actuality table.* Suppose we have already computed the elements $\tilde{d}_1(\cdot) \in Y_{i-1}$ for all generators of the group $\tilde{E}_1^{-i,i}$, and hence, the homomorphism $\tilde{d}_1 : \tilde{E}_1^{-i,i} \to Y_{i-1}$. Define the group $\tilde{E}_2^{-i,i}$ as the subgroup of $\tilde{E}_1^{-i,i}$ consisting of elements sent by this homomorphism to elements of the subgroup $h_{i-1}(X_{i-1}) \subset Y_{i-1}$. Choose a new system of generators in $\widetilde{E}_1^{-i,i}$ so that it contains a system of generators for $\widetilde{E}_2^{-i,i}$.

For any such generator $\gamma_0 \in \widetilde{E}_2^{-i,i}$ consider the corresponding element $\tilde{d}_1 \gamma_0$ and an arbitrary element $\alpha \in X_{i-1}$ such that $h_{i-1}(\alpha) + \tilde{d}_1 \gamma_0 = 0$; here α is a linear combination of cells corresponding to $[i-1]$-configurations and $\langle i-1 \rangle$-configurations. Then the geometric boundary $\partial_1 \gamma_0$ of the element γ_0 is spanned in Ξ_{i-1} by a compatible chain in which each main component of the complement of $\Sigma(i-1)$ in each cell is present with the same coefficient this cell has in the linear combination α. This is the coefficient that must be put into the place in the $(i-1)$th level of the extended actuality table corresponding to this cell.

The generators of the group $\widetilde{E}_1^{-i,i}$ that are not in $\widetilde{E}_2^{-i,i}$ will not be used in further computations of the invariants of the ith order; nevertheless we should not forget them. For all such generators we keep the corresponding values $\tilde{d}_1(\cdot)$; they will be used in the computation of the homomorphisms $d_2(\widetilde{E}_2^{-i+1,i+1})$, $d_3(\widetilde{E}_3^{-i-2,i+2})$, etc.

4.6. The differential d_2 and subsequent differentials are computed using the same scheme as for $\tilde{d}_1$. Point out the major steps of this scheme.

Suppose we have already computed the subgroups $\widetilde{E}_r^{-j,j} \subset \widetilde{E}_{r-1}^{-j,j} \subset \cdots \subset \widetilde{E}_0^{-j,j} \equiv X_j$ for all $j \leq i$; in particular, $\widetilde{E}_r^{-i,i}$ is the kernel of the homomorphism

$$d_{r-1}: \widetilde{E}_{r-1}^{-i,i} \to \widetilde{E}_{r-1}^{-i+r-1,i-r+2}.$$

For the given generator γ_{r-1} of this group the ith, $(i-1)$th, ..., and $(i-r+1)$th levels in the extended actuality table are already filled out. Let us compute $d_r(\gamma_{r-1})$.

Define a trial compatible chain in Ξ_{i-r}: this is a linear combination of elementary components of the complement of $\Sigma(i-r)$ in the open $(3d-1)$-dimensional cells; the main component occurs in this combination with coefficient 0, whereas the coefficients of any pair of adjacent components differ by an actuality index (taken with proper sign) of the component in Ξ_{i-r+1} corresponding to any part of their common boundary; see the proof of the lemma from 4.5.2 (at the previous stage we learned how to compute such indices).

The boundary of this compatible chain is the sum of the geometric boundary $\partial_r(\gamma_{r-1})$ and a certain cycle consisting of $(3d-2)$-dimensional canonical cells. This cycle is denoted by $\tilde{d}_r(\gamma_{r-1})$; we denote by $d_r(\gamma_{r-1})$ the class of this cycle in the quotient group

$$Y_{i-r}/\{h_{i-r}(X_{i-r}), \tilde{d}_1(\widetilde{E}_1^{-i+r-1,i-r+1}), \ldots, \tilde{d}_{r-1}(\widetilde{E}_{r-1}^{-i+1,i-1})\}. \quad (9)$$

The kernel of the obtained map d_r of the group $\widetilde{E}_r^{-i,i}$ in (9) is denoted by $\widetilde{E}_{r+1}^{-i,i}$.

Choose a new system of generators in the group X_i in such a way that the generators outside $\widetilde{E}_r^{-i,i}$ do not change, whereas those of $\widetilde{E}_r^{-i,i}$ are changed to satisfy the requirement that a subset of them forms at the same time a system of generators for $\widetilde{E}_{r+1}^{-i,i}$. This change defines an obvious transformation of the filled out levels of the actuality tables of these generators: for example, the data in such a table for the new generator $\gamma + \gamma'$ is obtained by componentwise summation of the corresponding data in the tables for γ and γ'.

For each generator γ_r of the group $\widetilde{E}_{r+1}^{-i,i}$ we choose an arbitrary element $\alpha = (\alpha_{i-r}, \alpha_{i-r+1}, \ldots, \alpha_{i-1}) \in X_{i-r} \oplus \widetilde{E}_1^{-i+r-1, i-r+1} \oplus \cdots \oplus \widetilde{E}_{r-1}^{-i+1, i-1}$ such that

$$-\tilde{d}_r(\gamma_r) = h_{i-r}(\alpha_{i-r}) + \tilde{d}_1(\alpha_{i-r+1}) + \cdots + \tilde{d}_{r-1}(\alpha_{i-1});$$

the existence of such an element follows from the definition of $\widetilde{E}_{r+1}^{-i,i}$; see (9). Add componentwise the actuality indices written in the corresponding cells of the ith, $(i-1)$th, $\ldots$, $(i-r)$th levels of the actuality tables of the elements $\alpha_{i-r}, \ldots, \alpha_{i-1}, \gamma_r$ (here we assume that at the $(i-r)$th level of γ_r there are only zeros, and for any element $\alpha_j \in X_j$ the actuality table has all the levels greater than j and all the cells of these levels also contain only zeros). The table thus obtained is declared to be the actuality table of the element γ_r; its $(i-r)$th level is now also complete.

4.7. Proof of the theorem from 0.5. Choose a plane in $\mathbb{R}^3$ containing the given direction $\partial/\partial x + \partial/\partial y + \partial/\partial z$. Reflection with respect to this plane acts on the space of noncompact knots K and preserves the discriminant Σ. We may choose the space Γ^d so that it is invariant with respect to this action (although to be able to do it we may need to move the space $I_5(\widetilde{\Gamma}^d) \subset \widetilde{\Gamma}^{5d+4}$ instead of the space $I_3(\widetilde{\Gamma}^d) \subset \widetilde{\Gamma}^{3d+2}$; see 1.2). Then this reflection also acts on the set $\sigma = \sigma(d)$ and respects its filtration by the sets σ_i. The induced action of this map on the group $\overline{H}_*(\sigma_i - \sigma_{i-1}) \sim \widetilde{E}_1^{-i,*}$ is the identity if i is even, and coincides with multiplication by -1 if i is odd: this follows easily from our definition of orientation of canonical cells in $\sigma_i - \sigma_{i-1}$; see §3.5.

Thus, by replacing each invariant of the ith order by its half-sum (for i even) or half-difference (for i odd) with its mirror image under this map we obtain an invariant of the same order defining the same element in the group $E_1^{-i,i}$.

This is the desired invariant. □

§5. The simplest invariants and their values for tabular knots

5.1. Theorem. *For any coefficient domain G, in our stable spectral sequence, $E_\infty^{-1,1} = 0$, $E_\infty^{-2,2} = G$, $E_\infty^{-3,3} = G$, $E_\infty^{-4,4} = G^3$.*

Proof consists in straightforward computations. □

§5. THE SIMPLEST INVARIANTS AND THEIR VALUES FOR TABULAR KNOTS

NOTE ADDED IN SECOND EDITION. D. Bar-Natan calculated all the groups $E_1^{-i,i}$, $i \leq 9$ (which coincide with the corresponding groups $E_\infty^{-i,i}$ by the above-mentioned theorem of Kontsevich). Namely, the group of all finite-order invariants has a natural structure of a free filtered algebra; the numbers of its multiplicative generators of dimensions 1, 2, 3, 4, 5, 6, 7, 8, 9 equal 0, 1, 1, 2, 3, 5, 8, 12, 18, respectively.

The corresponding basis invariants in the cases $G = \mathbb{Q}$ (or $\mathbb{Z}[1/2]$) are presented in 5.2; their values for tabular simple knots with ≤ 7 overlaps and for the two simplest nonsimple knots are given in Figure 5 in square brackets (first we give the value of the invariant of the second order, then of the third, and then the values of three invariants of the fourth order). In order to get a basis in the space of invariants of order ≤ 4 over $\mathbb{Z}$, it suffices to add to the second of the given invariants of the fourth order the basis invariants of the 2nd and 3rd order with any half-integer coefficients; however, the basis thus obtained will not satisfy the self-symmetry condition from 0.5.

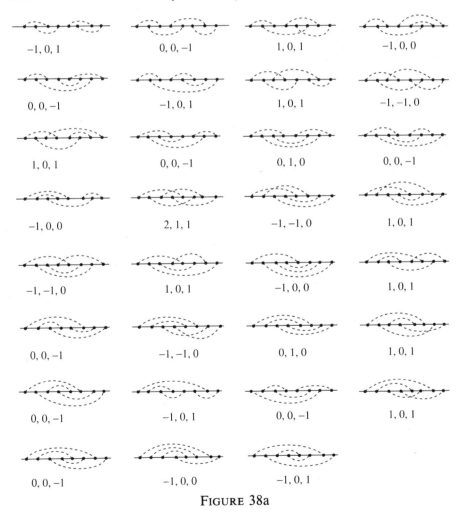

FIGURE 38a

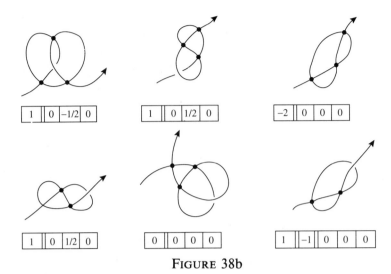

FIGURE 38b

5.2. The fourth level of the actuality tables for the three invariants of the fourth order is shown in Figure 38a; in Figure 38b (next page) we give the third level of the tables for invariants of the 3rd and 4th orders and the second level for all five invariants. Here we omit the cells corresponding to [*l*]-configurations, one of the characteristic pairs of which is not separated by points of other pairs (see footnote 10 in 4.2): in such cells we always write 0 independently of the encoded invariant.

For example, the notation $\boxed{1 \| 0 \| -1/2 \| 0}$ under the first picture of the upper level in Figure 38b means that the corresponding actuality index defined by the basis invariant of 3rd order is equal to 1, and the indices of the three basis invariants of 4th order are 0, $-1/2$, and 0, respectively. The indices of the invariants of orders 2, 3, and 4 for the only picture of the 2nd level are 1, -1, and $(0, 0, 0)$, respectively.

§6. Conjectures, problems, and additional remarks

6.1. Stabilization conjecture. The classes $\underleftarrow{E}_\infty^{-i,i}$, $i \geq 2$, of the stable spectral sequence give a complete system of invariants (that is, the zero-dimensional cohomology classes corresponding to them distinguish any two nonequivalent knots).

Obviously, any two nonequivalent knots differ by suitable elements encoded in the terms $E_\infty^{-i,i}(d)$, $i = i(d)$, of our spectral sequences converging to the cohomology of the spaces $\Gamma^d \backslash \Sigma$ for d such that both knots can be approximated by polynomial maps of degree d; the conjecture above says that for sufficiently large d such terms lie in the stable region $i(d) \leq (3d+1)/5$.

If the conjecture is true, then we can define the following integer-valued functions I and I': in order to distinguish between any pair of knots realized by diagrams with $\leq s$ overlaps (realized by polynomial maps of degree $\leq d$, respectively) it suffices to compute and compare their invariants of order $\leq I(s)$ (order $\leq I'(d)$, respectively). Problem: estimate the asymptotics of the functions I and I'.

6.2. Problem. Study the terms $E_r^{-i,i}(d)$ in the unstable region. This problem is closely related to the conjecture 6.1 but may have an independent significance.

6.3. Problem. Compute the stable spectral sequence as far as possible. The computation of the next several terms might give a hint about the general form of the term $\underleftarrow{E}_\infty^{-i,i}$.

Here are two trial problems:

Is it true that the stable $\mathbb{Z}_2$-spectral sequence $\{\underleftarrow{E}_r^{p,q}\langle\mathbb{Z}_2\rangle\}$ is equal to $\{\underleftarrow{E}_r^{p,q}\langle\mathbb{Z}\rangle\} \otimes \mathbb{Z}_2$?

Can the trefoil knot and its mirror image be distinguished by a stable $\mathbb{Z}_2$-invariant?

6.4. Problem. Compute higher degree terms (corresponding to $p+q > 0$) of the stable spectral sequence. What is the geometrical realization of the corresponding cohomology classes of the space of knots? Is the conjecture from 2.6 true?

6.5. Our spectral sequence can be trivially generalized to the case of links in $\mathbb{R}^3$ and imbeddings $S^1 \to \mathbb{R}^k$, $k \geq 4$. In the latter case there is no problem with verifying convergence: the corresponding stable spectral sequence converges to the cohomology of the space of nonsingular imbeddings (and its support has the structure shown in Figure 22 with $\tan\alpha > 1$). ([12])

We describe a more general case where our method applies.

Let M be a smooth manifold. Consider the space $J_2^l(M, \mathbb{R}^k)$ of various bijets of maps $M \to \mathbb{R}^k$ (i.e. unordered pairs of l-jets of maps $M \to \mathbb{R}^k$ applied at two distinct points of M). Let Ξ be a closed semialgebraic subset in $J_2^l(M, \mathbb{R}^k)$ invariant under the obvious action of the diffeomorphism group of M. The set Ξ is called *transitive* if for any smooth map $\varphi: M \to \mathbb{R}^k$ the fact that the bijets of φ at a pair of points $a, b \in M$ and a pair of points b, c belong to Ξ implies the fact that the bijet of φ at the pair a, c also belongs to Ξ. The most important examples are: the sets of bijets corresponding to self-intersections and points of self-tangency of $\varphi(M)$ are transitive, and the set corresponding to the points of nontransversality of two branches of a two-dimensional immersed surface in $\mathbb{R}^4$ is nontransitive.

([12]) Kontsevich proved, that in this case as well our spectral sequence (with rational coefficients) degenerates at the E_1 term. In principle, this gives a complete determination of the rational cohomology of the spaces of knots in $\mathbb{R}^k$, $k > 3$.

If the codimension of a transitive subset Ξ in $J_2^l(M, \mathbb{R}^k)$ is at least $2 \dim M + 1$, then it is possible to construct a spectral sequence similar to our sequence and providing invariants of maps $M \to \mathbb{R}^k$ with no multisingularities of type Ξ (or monosingularities obtained by their degeneration). If this codimension is at least $2 \dim M + 2$, then the set of such maps is connected and has no invariants, but our spectral sequence converges to its cohomology in all dimensions (and there is no problem with its convergence since the term $\underleftarrow{E}_1 = \{\underleftarrow{E}_1^{p,q}\}$ of this sequence has only a finite number of nonzero cells on any diagonal of the form $p + q = \mathrm{const}$).

6.6. Obviously, if the knot is symmetric, then all our invariants of odd order vanish on it. Is the converse true?

6.7. When we were describing the algorithms for computing our invariants and their values in §4 we did not formalize one step: two immersions $\mathbb{R}^1 \to \mathbb{R}^3$, each with l self-intersections with the same alternation of glued points (given, for example, by diagrams of their projections to a plane), are connected by a path in the space of immersions with l self-intersections, and we list all immersions with $l + 1$ self-intersections encountered along this path. The realization of an algorithm creating this list is a simple problem in combinatorics and programming, but how can one make this algorithm as efficient as possible?

§7. A guide to the recent results on the finite-order invariants

In the last two years, the theory of invariants described in this chapter has been intensively developed. For different aspects of it, see the expository articles [Ar_{19}], [Birman_2], [BN_2], [Vogel], [Sossinsky].

Dror Bar-Natan [BN_1] and, independently, Mikhail Gussarov, proved that all the coefficients in the Conway polynomial of links are finite-order invariants: more exactly, the coefficient a_i of the Conway polynomial

$$\nabla_K(t) \equiv \sum_i a_i t^i$$

is an invariant of order $\leq i$.

Joan Birman and Xiao-Song Lin then proved that all coefficients in the Jones, HOMFLY, and Kauffman polynomials can be reduced to ours. Indeed, the two-variable HOMFLY polynomial (see [FHYLMO]) has an equivalent representation as an infinite series of polynomials $\mathscr{H}_{n,\,\mathrm{a\,knot}}(t)$ in one variable t, indexed by the integer numbers $n \neq -1$ and determined by the inductive formulae

$$\mathscr{H}_{n,\,\mathrm{the\,unknot}} \equiv 1,$$

$$t^{(n+1)/2}\mathscr{H}_{n,K_+}(t) - t^{-(n+1)/2}\mathscr{H}_{n,K_-}(t) = (t^{1/2} - t^{-1/2})\mathscr{H}_{n,K_0}(t)$$

for every three links K_+, K_-, K_0, which coincide outside a disc, and behave in this disc as three configurations shown on the left, middle, and right lower pictures of Figure 29, see [J_3]. For instance, the original Jones polynomial (see [J_1]) is just the polynomial $\mathscr{H}_{-3,\cdot\cdot}$

The substitution $t = e^x$ defines a function

$$W_{n,K}(x) \equiv \mathscr{H}_{n,K}(e^x).$$

This function is regular close to zero; consider its Taylor decomposition

$$W_{n,K}(x) = \sum_{i=0}^{\infty} w_{n,K,i} x^i. \tag{1}$$

THEOREM (see [BL]). *For any $n \neq -1$, the coefficient $w_{n,\cdot,i}$ in the power series* (1) *is an invariant of order $\leq i$.*

A similar representation is obtained in [BL] for the Kauffman polynomial [Ka], and in [Lin^1] for all other polynomial knot invariants arising from the quantum groups (see [Tur]); for a different proof of this result see [Birman_2].

Bar-Natan (see [BN_1, BN_2]) indicated how to construct the elements of the groups $E_1^{-i,i}$ of our stable spectral sequence starting from an arbitrary irreducible representation of a simple Lie algebra.

Maxim Kontsevich (see [Kontsevich_1], [BN_2]) then proved that the stable spectral sequence with complex coefficients degenerates at the term E_1, $E_1^{p,q} \otimes \mathbb{C} = E_\infty^{p,q} \otimes \mathbb{C}$, and hence the Bar-Natan's construction can be extended to a construction of knot invariants of order i.

Moreover, Kontsevich constructed an explicit integral representation of the finite-order invariants: he wrote a universal integral formula which for any element α of the group $E_1^{-i,i}$ determines an invariant of order i whose class modulo the group of invariants of order $i - 1$ coincides with α.

One more result of Kontsevich asserts that the variant of our spectral sequence, which calculates the (complex) cohomology of the space of knots in $\mathbb{R}^n$, $n \geq 4$, also degenerates in the term E_1. The chain complexes $(E_0^{-i,*}; d^0)$ calculating the terms E_1 of these spectral sequences look very similar to the ones from §3 (in the case of odd n, these complexes, for all i, are obviously isomorphic to these from §3 up to a shift of dimensions, and in the case of even n their cells are in an obvious one-to-one correspondence with these from §3, but the incidence coefficients are different). Hence, we get in principle (modulo the calculation of the homology of given finite complexes) a complete determination of the rational cohomology groups of the spaces of knots in multidimensional spaces.

Sergey Piunikhin ([Piunikhin_3]) and, independently, Pierre Cartier ([Cartier]) gave a combinatorial realization of our invariants which also starts from elements of the groups $E_1^{-i,i}$: their construction is much more convenient for precise calculations and is based essentially on the result of [Drinfeld_2].

Lin and Bar-Natan proved independently that all Milnor's higher linking numbers (=the Milnor μ-invariants) are of finite type, and that the finite-order invariants of the homotopy type of links are strong enough to separate all classes of homotopy equivalence.

Lin investigated the finite-type invariants of links in any two-connected three-dimensional manifold M: he proved by geometrical methods that any standard imbedding $\mathbb{R}^3 \to M$ induces an isomorphism between the groups of these invariants in M and in $\mathbb{R}^3$. (This follows also immediately from a comparison theorem of corresponding spectral sequences constructed as in §2.)

Bar-Natan calculated (using a computer) the numbers of knot invariants of order i for all $i \le 9$, these numbers can be described as follows. By a theorem of Kontsevich, the ring of finite-order invariants over $\mathbb{C}$ admits a natural structure of a graded (by the orders) Hopf algebra, in particular it is a free polynomial algebra (see [MM]). According to Bar-Natan's calculations, the numbers of multiplicative generators of orders 1, 2, 3, 4, 5, 6, 7, 8, 9 of this algebra are equal to 0, 1, 1, 2, 3, 5, 8, 12, 18, respectively. It follows from these calculations, that all the invariants of order ≤ 9 can be obtained from the quantum groups.

However, the general behavior of this sequence is unknown yet; for some upper bound of it see [ChD].

Ted Stanford realized both algorithms described in §4 and using them calculated all (integer) invariants of orders ≤ 7 and their values on all knots with ≤ 10 crossings.

Joan Birman and Ted Stanford (see [Stanford], [Birman$_2$]) also generalized the construction of finite-order invariants of knots to the invariants of spatial graphs.

N. Nekrasov [Nekrasov] applied our spectral sequence to a different problem of singularity theory, also concerning the discriminants of multisingularities: the investigation of the stable cohomology ring of the complement of function bifurcation varieties of isolated singularities.

§8. A Morse theory on the space of knots

In [O'Hara], a continuous functional on the space of knots in Euclidean three-space was defined, which is bounded from below and tends to infinity when the knots tend to the degeneration (here the knots are considered as smooth imbedded circles in $\mathbb{R}^3$). This functional was defined by an integral formula. There is another similar functional (see [Nabutovsky]), which is more geometrical (at least, the local constraints, which it stationary points must satisfy, seem to be more explicit).

Suppose that a knot K is imbedded in a homogeneous conducting media, and consider an instant perturbation at all the points of the knot. Then for small values t of time, the corresponding wave front is the boundary of the t-neighborhood of K and is smooth. Let $T(K)$ be the first instant at which the front fails to be a C^∞-manifold.

§8. A MORSE THEORY ON THE SPACE OF KNOTS

THEOREM ([Nabutovsky]). *The function $T(K)$ on the space of smooth knots of length 1 in $\mathbb{R}^3$ is continuous in the C^2 topology, bounded from above, and tends to 0 when the knot tends to a singular or self-intersecting curve in $\mathbb{R}^3$. For any $\varepsilon \geq 0$, there is only a finite number of ambient isotopy classes of knots with $T(\cdot) \geq \varepsilon$.*

The function $T(K)$ is not smooth: for instance, it has singularities of the $|x|$ type; but it is very probable that it is a topological Morse function in finite codimensions:

CONJECTURE. For any smooth generic finitedimensional submanifold in the space of knots, the restriction of the function $T(K)$ on this submanifold is topologically equivalent to a Morse function.

CHAPTER VI

Invariants of Ornaments

Denote by C_k the disjoint union of k circles.

DEFINITION. A *k-ornament* (or simply *ornament*) is a C^∞-smooth map $C_k \to \mathbb{R}^2$ such that the images of no three different circles intersect at the same point in $\mathbb{R}^2$. Two ornaments are *equivalent*, if the corresponding maps $C_k \to \mathbb{R}^2$ can be connected by a homotopy $C_k \times [0, 1] \to \mathbb{R}^2$ such that for any $t \in [0, 1]$ the corresponding map of $C_k \times t$ is an ornament. See Figures 39, 40.

(Similar objects were considered in [FT] under the name *doodles*: the only difference is that the doodles are collections of Jordan curves (without self-intersections). Of course, the invariants of ornaments are also invariants of doodles; conversely, the invariants introduced in [FT] can be easily generalized to invariants of ornaments, see §1.2).

In this chapter we construct a series of numerical invariants of equivalence classes of ornaments. Like the knot invariants from Chapter V, these invariants appear from the study of the *discriminant*, i.e., of the space of all maps $C_k \to \mathbb{R}^2$ which have forbidden triple intersections.

(We follow again the general strategy from [Ar$_4$]: replace the study of the soft, homogeneous space of nonsingular objects by the study of the complementary space of singular objects, which has usually a rich geometrical structure.)

Using the geometry of the discriminant, we construct a spectral sequence $E_r^{p,q}$ which calculates the cohomology groups of the space of ornaments; in particular, the groups $E_\infty^{-i,i}$ of this spectral sequence provide the invariants of ornaments.

Many of our invariants can be interpreted in absolutely classical terms, these invariants are described in §1.4.

As in Chapter V, the invariants coming from the cell $E^{-i,i}$ of this spectral sequence are called the invariants of order i, and all such invariants corresponding to different i are called the *finite-order invariants*.

This theory is a model version of a wide class of problems (stated in [FNRS] in a connection with the higher dimensional generalizations of the Chern-Simons theory) where a similar technique works: say, the next prob-

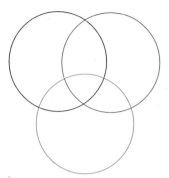

FIGURE 39

FIGURE 40

lem of this class is the classification of all maps of k two-spheres (or arbitrary fixed Riemannian surfaces) in $\mathbb{R}^3$ in such way that no four of them intersect at the same point. The construction of our invariants can be immediately generalized to these problems. However, in §§1.2, 1.3 we describe several other invariants of ornaments (due essentially to R. Fenn and P. Taylor) which seem to be specifically "one-dimensional". A refinement of these invariants, due to A. B. Merkov, is presented in the §9.

In §§2, 3 we describe the elementary characterization of finite-order invariants and show how to calculate the values which these invariants take on an ornament. In §§4–7 we construct and investigate the principal spectral sequence which provides such invariants. The first calculations are presented

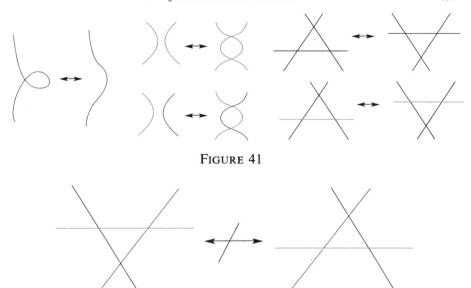

FIGURE 41

FIGURE 42

in §8. A large list of unsolved problems is given in §10.

About §9. All of the new results in §9 and a main part of the text are due to A. B. Merkov. I thank him very much for this work.

During my work on Chapter VI, I was partially supported by the American Mathematical Society fSU Aid Fund.

§1. Elementary theory

1.1. Reidemeister moves.

DEFINITION. An ornament $\varphi: C_k \to \mathbb{R}^2$ is *regular* if it is an immersion of C_k and all the multiple points of the image of C_k in $\mathbb{R}^2$ are double transversal intersection points only.

THEOREM 1. *Any ornament is equivalent to some regular ornament. Two regular ornaments are equivalent iff they can be transformed one into the other by a finite sequence of isotopies of* $\mathbb{R}^2$ *(which do not change the topological picture of the image of the ornament), and of local moves shown in Figure* 41 *(or obtained from these moves by recoloring the strands).*

(In other words, only the local move shown in Figure 42 is forbidden among the moves which can appear in a generic homotopy of a generic smooth map $C_k \to \mathbb{R}^2$.)

This theorem follows immediately from Thom's multijet transversality theorem, see [GG]. □

1.2. Fenn-Taylor invariants.

In [FT], Fenn and Taylor introduced an invariant of *doodles*, i.e., of collections of Jordan curves without triple intersections. This invariant can be easily generalized to an invariant of ornaments; this generalization is due to Merkov. This invariant is a collection of k elements of the free group with $k - 1$ generators considered up to cyclic permutations of symbols.

Indeed, let us go along one (say, the first) component of a regular ornament beginning with any point at which it does not intersect other components. Any intersection with some *other* (say, with the ith) component which we meet during this travel is encoded by the symbol a_{i-1} (resp., a_{i-1}^{-1}) if this new component intersects ours from right to left (resp., from left to right). The words obtained in such way are said to be equivalent if they can be reduced one to the other by some iteration of a cyclic permutation of letters and insertion (deletion) of pairs $a_i a_i^{-1}$ or $a_i^{-1} a_i$. Similar words can be defined by going along all other $k - 1$ components of the ornament.

THEOREM 2 (cf. [FT]). *Equivalent k-ornaments define equivalent collections of k words.*

PROOF. Obvious, by Theorem 1. □

The described invariants are essentially invariants of ornaments on a 2-sphere: they do not distinguish two ornaments in $\mathbb{R}^2$ which become equivalent ornaments in S^2 after obvious imbeddings $\mathbb{R}^2 \to S^2$.

For a refinement of these invariants see §9, which is independent on the intermediate §§2–8.

1.3. Reduction to the homotopy classification of links.

The classification of k-ornaments can in several ways be partially reduced to the homotopy classification of links. These reductions (which have been introduced also in [FT]) are numbered by the orientations of the complete graph with k vertices: given such an orientation, we assign a k-component link in $\mathbb{R}^3$ to any k-ornament in such a way that to the equivalent ornaments there correspond homotopy equivalent links.

Indeed, let us fix such an orientation. Make a link diagram from the (image of) our regular ornament in the following way: the ith string goes everywhere under the jth in the points of their crossings if the edge (ij) of the complete graph is oriented from the ith vertex to the jth. In the self-intersection points of the same component the over/undercrossings should be chosen in arbitrary way.

THEOREM 3. *If two ornaments are equivalent, then the links assigned to them by the above rule (based on arbitrary orientation of the complete graph) are homotopy equivalent.* □

Indeed, any Reidemeister move from Theorem 1 can be lifted to an admissible move of a link which preserves its homotopy class, and the resulting link diagram satisfies again the above rule on over/undercrossings.

In particular, the homotopy invariants of links provide invariants of ornaments. These invariants can be nontrivial: for instance, any cyclic orientation of the complete graph with 3 vertices makes the 3-ornament from Figure 39 the Borromean link.

1.4. Index-type invariants. Remember that any closed oriented immersed curve c in $\mathbb{R}^2$ defines an integer-valued function ind_c on its complement: for any point t of the complement, $\text{ind}_c(t)$ equals the rotation number of the vector (t, x) when x runs one time along c.

To any regular k-ornament there correspond $\binom{k}{2}$ functions $I_{i,j} = I_{i,j}(b_1, \ldots, b_k)$, $1 \le i < j \le k$, with integer values and arguments; these functions are invariant under the moves from Theorem 1, and hence, define invariants of ornaments.

Indeed, to any (transversal) intersection point x of the ith and jth curves we assign in the following way k integer numbers $b_1(x), \ldots, b_k(x)$ and a sign $\sigma(x)$.

If $l \ne i, j$, then $b_l(x)$ is just the number $\text{ind}_l(x)$, the index of x with respect to the lth curve. Now close to any regular point of the ith curve (in particular, to the intersection point x) the values of the corresponding function $\text{ind}_i(\cdot)$ take two neighboring integer values on different sides of the curve. Define the number $b_i(x)$ as the smallest of these values in the neighboring points to x. The number $b_j(x)$ is defined in the same way by means of ind_j. Finally, $\sigma(x)$ equals 1 if the tangent vectors of the ith and jth curves at the point x define a positive frame (with respect to a fixed orientation of $\mathbb{R}^2$) and equals -1 if this frame is negatively oriented.

Given a regular k-ornament and k integer numbers $b_1, \ldots, b_k$, define the number $I_{i,j}(b_1, \ldots, b_k)$ as the number of transversal intersection points x of the ith and jth curves of our ornament, such that $b_1(x) = b_1, \ldots, b_k(x) = b_k$ and $\sigma(x) = 1$, minus the number of similar points with $\sigma(x) = -1$.

THEOREM 4. *All the functions $I_{i,j}$, $1 \le i < j \le k$, are invariant under all the Reidemeister moves of Figure* 41.

PROOF. The proof is immediate. □

The functions $I_{i,j}$ are not independent. For instance, for $k = 3$ let us define the numbers $i_{1,2}, i_{2,3}, i_{3,1}$ as the sums

$$\sum_{b_1, b_2, b_3 = -\infty}^{\infty} b_3 I_{1,2}(b_1, b_2, b_3), \quad \sum_{b_1, b_2, b_3 = -\infty}^{\infty} b_1 I_{2,3}(b_1, b_2, b_3),$$

$$\sum_{b_1, b_2, b_3 = -\infty}^{\infty} -b_2 I_{1,3}(b_1, b_2, b_3),$$

respectively.

PROPOSITION 1. *The three numbers $i_{1,2}$, $i_{2,3}$, $i_{3,1}$ coincide.*

PROOF. Indeed, for the unlinked ornament all three numbers are equal to 0, and any forbidden move from Figure 42 simultaneously increases or decreases by 1 all three numbers. □

This number $i_{1,2}$ is called the *index* of the 3-ornament φ and is denoted by $i(\varphi)$.

More generally, for any $k \geq 3$ and any k-ornament φ define the *index* of φ, $i(\varphi)$, as the number

$$(1/3) \sum_{1 \leq i < j \leq k} \sum_{b_1, \ldots, b_k = -\infty}^{\infty} (b_1 + \cdots + b_{i-1} - b_{i+1} - \cdots - b_{j-1} + b_{j+1} + \cdots + b_k)$$
$$\times I_{i,j}(b_1, \ldots, b_k).$$

This number $i(\varphi)$ is always a natural number: again, any elementary surgery of Figure 42 decreases or increases the previous double sum by 3.

In a similar way, given a regular k-ornament, for any k integer nonnegative exponents $\beta_1, \ldots, \beta_k$ we can define the corresponding *momenta*

$$M_{i,j}(\beta_1, \ldots, \beta_k) = \sum_{b_1, \ldots, b_k = -\infty}^{\infty} b_1^{\beta_1} \cdots b_k^{\beta_k} \cdot I_{i,j}(b_1, \ldots, b_k).$$

It is natural to call the function $M_{i,j}$ the *Laplace transform* of $I_{i,j}$. Since all functions $I_{i,j}$ are finite, they can be reconstructed from their Laplace transforms. Here are some other relations on the indices $I_{i,j}$ and their momenta.

PROPOSITION 2. *For any $1 \leq i < j \leq k$ and any two values b_i and b_j, the sum*

$$\sum_{b_1, \ldots, b_{i-1}, b_{i+1}, \ldots, b_{j-1}, b_{j+1}, \ldots, b_k = -\infty}^{\infty} I_{i,j}(b_1, \ldots, b_k)$$

equals 0.

PROOF. The curves other than the ith and jth one actually do not participate in these sums; after they are removed, the statement becomes trivial. □

Here is an equivalent reformulation of this proposition in the terms of momenta.

PROPOSITION 2'. *If $\beta_l = 0$ for all l other than i or j, then $M_{i,j}(\beta_1, \ldots, \beta_k) = 0$.*

REMARK. The construction of invariants $I_{i,j}$ and $M_{i,j}$ can be immediately extended to the invariants which distinguish maps of collections of $(n-1)$-dimensional manifolds in $\mathbb{R}^n$, no $n+1$ of which intersect at the same point; the corresponding functions I and M in this case have n lower indices.

REMARK. I expect that there are many other elementary invariants of orna-

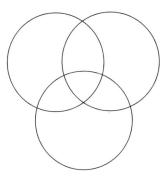

FIGURE 43

ments, and the spectral sequence of §4 can be considered as a regular method of guessing them: for instance, I guessed the invariants $I_{i,j}$ and $M_{i,j}$ after calculating the terms $E^{-i,i}$ of the sequence with $i = 2, 3, 4$.

1.5. Examples.

(A) The simplest picture of nontrivial ornament (see Figure 43) has 16 nonequivalent realizations, depending on the orientation and ordering (coloring) of circles. All of them can be distinguished by the functions $I_{i,j}$. The Fenn-Taylor invariants split these 16 ornaments into two groups, with 8 ornaments in each (and are constant on any of these two groups): indeed, all ornaments in any of these groups are equivalent as ornaments on a sphere.

(B) For the ornament from Figure 44, all invariants $I_{i,j}$ vanish. However, this ornament is nontrivial because its Fenn-Taylor invariant is nontrivial.

(C) For the ornament from Figure 45, both the Fenn-Taylor invariant and the invariants $I_{i,j}$ vanish. (This example is due to A. B. Merkov, who also proved the nontriviality of this ornament by the use of a refinement of the Fenn-Taylor invariant, see §9.) On the other hand, the link obtained from this ornament by the construction of §1.3 is homotopy nontrivial (this calculation

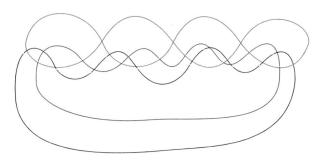

FIGURE 44

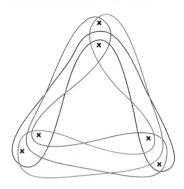

FIGURE 45

is due to X.-S. Lin), and hence, also proves the nontriviality of this ornament.

§2. Elementary definition of finite-order invariants

DEFINITION AND NOTATION. A *quasiornament* is any C^∞-smooth map $C_k \to \mathbb{R}^2$. The space of all k-quasiornaments is denoted by κ_k. The *discriminant* $\Sigma \subset \kappa_k$ is the space of all quasiornaments which are not ornaments, i.e., have forbidden triple points.

The discriminant is a singular subvariety in κ_k. Its regular points are the quasiornaments having only one forbidden triple point and such that three local branches of them at the triple point are smooth and pairwise nontangent; the singular points of Σ correspond to the quasiornaments with several or more complicated singularities. A natural stratification of the discriminant is provided by the classification of these singularities.

Any numerical invariant of ornaments can be expressed in the terms of the discriminant. Indeed, to each nonsingular piece of the discriminant (i.e., a connected component of the set of its nonsingular points) we can assign its *index*, i.e., the difference of values of this invariant on two neighboring topologically different ornaments, taken in an appropriate order.

The value of the invariant on any ornament can be reconstructed from this system of indices (under the assumption that any invariant takes zero value on the trivial ornament): to do this we connect our invariant with the trivial one by a generic homotopy in the space κ_k and count all the indices of all quasiornaments at which this homotopy intersects the discriminant.

Conversely, suppose that to each nonsingular component of the discriminant we have assigned a numerical index. In order for this collection of indices to define an invariant of ornaments, it should satisfy a homological condition: the sum of these components, taken with the corresponding coefficients (i.e., their indices) must have no boundary in the space of all quasiornaments. Enumerating such admissible collections is a problem in homology theory and can be solved by standard methods of this theory. A

§2. ELEMENTARY DEFINITION OF FINITE-ORDER INVARIANTS

partial solution to this problem is described in §§4–8; in the present section we give an elementary characterization of the invariants thus obtained.

DEFINITION. A *degree j standard singularity* of ornaments is a pair of the form (a quasiornament $\varphi \colon C_k \to \mathbb{R}^2$; a point $x \in \mathbb{R}^2$) such that $\varphi^{-1}(x)$ consists of exactly $j+1$ points $z_1, \ldots, z_{j+1}$, at least three of which belong to different components of C_k, the map φ close to all these points is an immersion, and the corresponding $j+1$ local branches of $\varphi(C_k)$ are pairwise nontangent at x. A quasiornament is called a *regular quasiornament of complexity i*, if all its forbidden points (i.e., the points, at which at least three different components meet) are standard singular points, and the sum of the degrees of these singularities equals i.

Given an invariant of ornaments, to any regular quasiornament of finite complexity there corresponds a collection of *characteristic numbers* which we define below; the invariant is *of order i* iff all such numbers corresponding to all quasiornaments of complexity $> i$ are equal to zero.

Let us define the characteristic numbers. Let φ be a regular quasiornament with m singular points $x_1, \ldots, x_m \in \mathbb{R}^2$, let the inverse images in C_k of these points be

$$z_{1,1}, \ldots, z_{1,j_1} \in \varphi^{-1}(x_1), \ldots, z_{m,1}, \ldots, z_{m,j_m} \in \varphi^{-1}(x_m).$$

A *degeneration mode* corresponding to the regular quasiornament φ is some arbitrary order of marking all these points $z_{1,1}, \ldots, z_{m,j_m}$, satisfying the following conditions: at any step we mark either some three points of this group $\varphi^{-1}(x_l)$, belonging to some three different components of C_k (if no point of the same group is already marked) or one point (if some three or more other points of the same group are already marked).

EXAMPLE. Suppose that our 3-quasiornament has one singular point at which 4 points of C_3 meet: two from the first component of C_3, one from the second and one from the third. (See Figure 46.) Then there are two different degeneration modes, see Figure 47. If our quasiornament has exactly one additional singular point, at which 3 points meet, then there are 6 different degeneration modes: for any case from Figure 47, we can mark the whole second group before, after, or in between the steps of marking the points from the first.

FIGURE 46

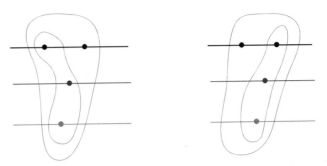

FIGURE 47

To any degeneration mode there corresponds a characteristic number of our invariant (and of the quasiornament φ). This number will be defined by the induction over the process of marking.

Base of induction. For a nonsingular ornament φ (and the empty degeneration mode) the characteristic number equals the value which our invariant takes on φ.

Suppose that the last step of degeneration consist in marking some three points z, z', z'' (and hence, the corresponding group $\varphi^{-1}(\cdot)$ consists of these three points only). Let $j < l < n$ be the numbers of components of C_k containing these points. Then we can move our map φ slightly in two ways such that all the critical points other than $\varphi(z)$ stay unmoved, and the triple point $\varphi(z)$ resolves in two ways shown on Figure 48. These two resolutions are not equal: we can always call one of them *positive*, and the other will be *negative*. Indeed, to any of these two pictures there correspond three integer numbers: the first of them is the index of the intersection point

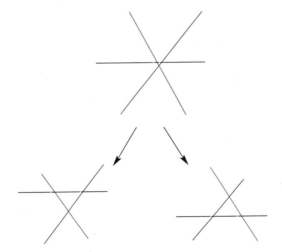

FIGURE 48

$*_{j,l}$ of the jth and lth curve with respect to the nth curve, multiplied by the sign $\sigma(*_{j,l})$ (see subsection 1.4); the second number is defined in a similar way by the point $*_{l,n}$ and sign $\sigma(*_{l,n})$, and the third by $*_{j,n}$ and the sign $-\sigma(*_{j,n})$. It is easy to prove that for some of our two resolutions all three of these numbers are one more than the corresponding numbers for the other; this resolution is called positive. Then the characteristic number which our invariant assigns to the quasinornament φ and the degeneration mode equals the characteristic number defined by the same invariant for the positive resolution and for the same degeneration mode without the last step (this value is already known by the conjecture of induction) minus the similar characteristic number of the negative resolution.

Now let the last step of degeneration consists in adding only one point $z_{l,j} \in \varphi^{-1}(x_l)$. In this case, we can partially resolve the quasiornament φ in two topologically different ways φ', φ'' such that the resting singularity glues together the same points $z_{l,j'}$ as φ, except only for this one. Indeed, we preserve our map φ everywhere outside a small neighborhood of $z_{l,j}$, and in this neighborhood change it in such a way that the corresponding local branch of the (image of) ornament translates parallel to itself to one side or to the other, see Figure 49. Again, one of these resolutions can be invariantly called positive and the other negative. To do this, take an arbitrary point in $\mathbb{R}^2$ not on the ornament and very close to $\varphi(z_{l,j})$ (much closer than any of the points $\varphi'(z_{l,j})$, $\varphi''(z_{l,j})$: for instance, the point $*$ in Figure 49). Then the positive resolution of φ is those of two maps φ', φ'', for which the index of the point $*$ with respect to the containing $z_{l,j}$ component of C_k is greater. Again, the characteristic number corresponding to φ (and our degeneration mode) equals the characteristic number of those of the resolutions φ', φ'' (and the previous degeneration mode) which is positive, minus the characteristic number of the negative resolution.

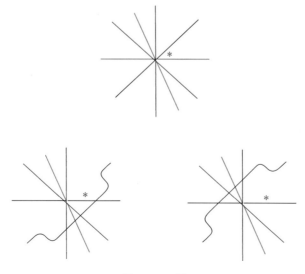

FIGURE 49

DEFINITION. An invariant of ornaments is *of order* i if all corresponding characteristic numbers for any regular quasiornament of complexity $> i$ vanish.

An equivalent definition will be given in §4.

REMARK. Of course, the characteristic numbers corresponding to different degeneration modes of the same regular quasiornament satisfy some natural relations; the study of these relations is closely related to the theory of arrangements, order complexes, etc., see §§5–7.

For instance, if two degeneration modes differ only by a permutation of markings, which for any $l = 1, \ldots, m$ preserves the order of marking the points in the group $\varphi^{-1}(x_l)$, then the corresponding characteristic numbers coincide.

THEOREM 5. *Any invariant* $M_{i,j}(\beta_1, \ldots, \beta_k)$ *from subsection* 1.4 *is an invariant of order* $\beta_1 + \cdots + \beta_k + 1$.

PROOF. The proof is immediate by induction over the process of degeneration. □

Let A be a finite series of integer numbers $A = (a_1 \geq a_2 \geq \cdots \geq a_m \geq 3)$. Denote by $|A|$ the number $a_1 + \cdots + a_m$ and by $\#A$ the number of elements a_l of the series (denoted in the previous line by m).

DEFINITION. An *A-configuration* is a collection of $|A|$ pairwise different points in C_k divided into groups of cardinalities $a_1, \ldots, a_{\#A}$ such that any group contains the points of at least three different components of C_k. Two A-configurations are *equivalent* if they can be transformed one into the other by a diffeomorphism $C_k \to C_k$ which preserves ordering and orientations of all components of C_k. A quasiornament $\varphi: C_k \to \mathbb{R}^2$ *respects* an A-configuration if it sends any of corresponding $\#A$ groups of points into one point in $\mathbb{R}^2$. φ *strictly respects* the A-configuration if, moreover, all these $\#A$ points in $\mathbb{R}^2$ are distinct, have no extra preimages than these $|A|$ points, and φ has no extra points in $\mathbb{R}^2$ at which there intersect images of three of more different components of C_k. A degeneration mode of an A-configuration is a degeneration mode of arbitrary quasiornament strictly respecting it.

Obviously, the space of all quasiornaments which respect a given A-configuration J is a linear subspace of codimension $2(|A| - \#A)$ in the space of all quasiornaments. We shall denote this subspace by $\chi(J)$. The set of all quasiornaments which strictly respect this configuration is an open dense subset in this subspace.

Let M be an invariant of ornaments, and let J be an A-configuration.

THEOREM 6. *If* M *is an invariant of order* i *and* $|A| - \#A = i$, *then for any regular quasiornament* φ *which strictly respects the A-configuration* J, *all characteristic numbers defined by* M *and* φ *depend on the configuration* J *only.*

§2. ELEMENTARY DEFINITION OF FINITE-ORDER INVARIANTS

PROOF. Any two regular quasiornaments φ, φ' which strictly respect the same A-configuration J, can be transformed one into the other by some homotopy $\varphi \colon C_k \times [0, 1] \to \mathbb{R}^2$, $\varphi \equiv \varphi(\cdot, 0)$, $\varphi' \equiv \varphi(\cdot, 1)$, such that

(1) any quasiornament $\varphi_t \equiv \varphi(\cdot, t)$, $t \in [0, 1]$, respects the configuration J, and

(2) for almost all $t \in [0, 1]$ it *strictly* respects this configuration, and only at a finite number of instants $t_l \in (0, 1)$ the topological picture of the set $\varphi(C_k)$ undergoes one of the following local surgeries:

   (a) one of the permitted surgeries from Figure 40 aside from the "bad" points of $\varphi(t)$,
   (b) the local move connecting two lower pictures in Figure 48,
   (c) the local move connecting two lower pictures in Figure 49,
   (d) the local move preserving all local branches of $\varphi(C_k)$ at all its singular points but one and changing the last branch as shown in Figure 50.

The invariance of the characteristic numbers under one of the surgeries of type (a) is obvious, and that under the surgeries (b) and (c) follows from the definition of invariants of order i (indeed, the difference of the characteristic numbers of the quasiornaments on the left and right sides of Figure 48 or 49 is just a characteristic number of the upper quasiornament on the same picture, and it is zero because the complexity of this quasiornament is greater than i). Finally, for surgery (d) the invariance follows easily from the definitions, and theorem is proved. □

Any equivalence of A-configurations establishes a one-to-one correspondence between their degeneration modes.

THEOREM 6'. *Any invariant of order i assigns equal characteristic numbers to any two equivalent A-configurations and their degeneration modes corresponding to each other via this equivalence.*

PROOF. The proof is obvious. □

COROLLARY. *For any natural k and i, the space of order i invariants of k-ornaments is finite-dimensional.*

Indeed, for any k there is only a finite number of equivalence classes of configurations of a given complexity, as well as of their degeneration modes. □

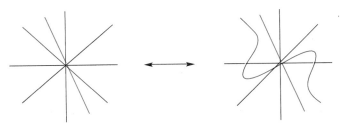

FIGURE 50

In §§4–8 we show how to calculate all such invariants.

§3. Coding the finite-order invariants and calculating their values on ornaments

Any invariant of order i can be encoded by its *actuality table* which we now describe.

This table has $i+1$ levels numbered by $0, 1, \ldots, i$. The lth level consists of several cells, which are in one-to-one correspondence with all possible pairs consisting of

(a) an equivalence class of A-configurations of complexity l,

(b) a degeneration mode of this A-configuration.

In each cell we put

(a) a picture (or a code) representing a "model" regular quasiornament, which respects some A-configuration from the corresponding equivalence class (this picture is the same for all invariants), and

(b) a number (called the *actuality index* corresponding to this picture) which is nothing but the characteristic number which our invariant and the degeneration mode corresponding to the cell assign to this quasiornament.

By the Theorem $6'$, we may not draw the pictures in the cells of the highest (ith) level of the table: indeed, the corresponding characteristic numbers depend only on the data indexing the cell.

For instance, the 0th level consists of the trivial (disjoint) ornament, and the corresponding characteristic number equals 0 (recall that we assume that all invariants take value zero on the trivial ornament). The 1st level is empty, because there are no configurations of complexity 1.

In order to calculate the value of our invariant on some ornament, we joint this ornament with the trivial one by a generic path in the space κ_k. This path has only a finite number of transversal intersections with the discriminant in its nonsingular points, i.e., in some regular quasiornaments with only one simplest triple point. The value of the invariant on our ornament equals the sum of characteristic numbers of these quasiornaments, taken with the signs depending on the direction in which we traverse the discriminant at the corresponding points (i.e., do we go from the negative side to the positive or in the opposite direction).

To calculate these characteristic numbers, we use an inductive process, the general (lth) step of which consists of the following.

Before this step, we have reduced our problem to the calculation of the characteristic numbers which our invariant assigns to several regular quasiornaments of complexity $\geq l$, taken together with certain degeneration modes of the configurations strictly respected by them.

To calculate such a number for some one quasiornament φ of this list, we choose in the actuality table the cell, corresponding to the (equivalence class of the) A-configuration, strictly respected by this quasiornament, and to the degeneration mode. (Without loss of generality, we shall assume that the quasiornament encoded in this cell respects exactly the same A-configuration as

φ.) Then we join these two quasiornaments by a generic path in the space of all quasiornaments which respect this A-configuration. For almost all points of the path, corresponding quasiornaments respect this configuration strictly, and at only a finite number of instants do they undergo one of the local surgeries shown in Figures 48, 49, 50. The surgery from Figure 50 should be left without attention. At the instant of any other surgery, we get a regular quasiornament which strictly respects some A'-configuration, whose complexity $|A'|-\#A'$ is strictly greater (by 2 in the case in Figure 48, and by 1 in the case in Figure 49) than that for the configuration A. Also a degeneration mode for this A'-configuration is well defined: it is obtained from the previous mode for an A-configuration by adding one more step: marking all additional points which belong to the A'-configuration but do not belong to the A-configuration. The characteristic number of the original quasiornament φ and the original degeneration mode equals the similar characteristic number for the table quasiornament respecting the same A-configuration and for the same degeneration mode, plus the sum of characteristic numbers for the regular quasiornaments (and the degeneration modes just defined) respecting more complicated configurations which we meet along the path: these characteristic numbers in the sum must be taken with the coefficient 1 (resp., -1) if the corresponding local surgery goes from the negative picture to the positive (see §2) (resp., from the positive picture to the negative).

Thus, we have reduced the calculation of characteristic numbers for some regular quasiornament respecting an A-configuration to that for several quasiornaments respecting certain A'-configurations, where the complexities of all A'-configurations are strictly greater than these for A. Since our invariant is of order i, this process stops when the complexities of all such configurations attain i.

§4. Discriminants and their resolutions

In this section, we begin the regular topological study of the discriminant $\Sigma \subset \kappa_k$; in particular, we present a method of calculation of all finite-order invariants (and show why this class of invariants is natural).

We construct a spectral sequence $E_r^{p,q}$ that calculates the cohomology of the space of ornaments $\kappa_k - \Sigma$. This construction is based on the natural stratification of the discriminant by the types of degeneration of quasiornaments. For $r \geq 1$ this spectral sequence lies in the domain $\{p, q | p < 0, p+q \geq 0\}$, see Figure 51. The invariants of ornaments correspond to the elements of the groups $E_\infty^{-i,i}$, $i \geq 1$. The invariants, which appear from our spectral sequence, are exactly the invariants of finite order described in §2: they are, in a sense, exactly the invariants which can be expressed in the terms of strata of finite codimension in the discriminant.

The description of invariants in terms of the discriminant, which was given

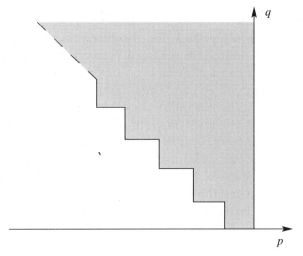

FIGURE 51

in §2, can be intuitively considered as a variant of the Alexander duality formula:

$$H^0(\kappa_k - \Sigma) \cong \overline{H}_{\infty-1}(\Sigma). \tag{1}$$

In this (strictly speaking, meaningless) formula and everywhere below $\overline{H}_*$ denotes the closed homology, i.e., the homology of the one-point compactification reduced modulo the compactifying point; ∞ is the notation for the dimension of κ_k, and the right part $\overline{H}_{\infty-1}(\Sigma)$ is just the group of all the above-described linear combinations of smooth pieces of Σ satisfying some homological concordance constraints, see the beginning of §2.

The spectral sequence, which we construct in this section, provides also the elements of the higher cohomology groups $H^i(\kappa_k - \Sigma)$, $i > 0$. Indeed, the formula (1) can be extended to the general Alexander duality formula

$$H^i(\kappa_k - \Sigma) \cong \overline{H}_{\infty-i-1}(\Sigma) \tag{2}$$

which can also be given an exact sense in the terms of the discriminant and its strata. To justify these formulas, we proceed as follows:

(a) choose an increasing sequence of affine subspaces Γ_k^d, $d \to \infty$, in κ_k such that any compact subset in κ_k can be approximated arbitrarily well by the points of appropriate subspace of this sequence;

(b) for any d, construct a spectral sequence $E_r^{p,q}(d) \to H^{p+q}(\Gamma_k^d \backslash \Sigma)$,

(c) prove the stabilization of these spectral sequences.

The groups $E_\infty^{p,q}(\infty)$ of the stabilized spectral sequence provide the elements of the group $H^{p+q}(\kappa_k - \Sigma)$ (as usual, these elements are well defined modulo the similar elements coming from the groups

$$E_\infty^{p+l,q-l}(\infty), \qquad l \geq 1).$$

§4. DISCRIMINANTS AND THEIR RESOLUTIONS

4.1. Finite-dimensional approximations of the space of ornaments. For any $d \geq 1$, define the subspace $\widetilde{\Gamma}_k^d \subset \kappa_k$ as the space of all maps $C_k \to \mathbb{R}^2$ given by $2k$ trigonometric polynomials of degree $\leq d$.

The space $\widetilde{\Gamma}_k^d$ is tautologically imbedded in all spaces $\widetilde{\Gamma}_k^{d'}$, $d' \geq d$.

The spaces $\widetilde{\Gamma}_k^d$ are in an intuitive sense "nongeneric" in κ_k: for instance, they contain "infinitely degenerated" constant maps of C_k: of course, this situation cannot arise in any generic finite-dimensional family of maps. For our approximation spaces Γ_k^d we use small perturbations of $\widetilde{\Gamma}_k^d$, which are generic in the following precise sense.

For any A-configuration $J \subset C_k$ and any affine subspace $\Gamma \subset \kappa_k$ denote by $\chi(\Gamma, J)$ the space $\Gamma \cap \chi(J)$, i.e., the space of all maps $\varphi \in \Gamma$ which respect J.

PROPOSITION 3. *In the space of all affine subspaces of fixed finite dimension in κ_k, a residual (in particular, dense) subset consists of such planes Γ that for any index $A = (a_1 \geq \cdots \geq a_{\#A} \geq 3)$ and any A-configuration $J \subset C_k$, the following assertions hold*:

(A) *for almost any A-configuration J' equivalent to J, the set $\chi(\Gamma, J')$ is an affine subspace of codimension $2(|A|-\#A)$ in Γ (in particular, it is empty if $2(|A|-\#A) > \dim \Gamma$).*

(B) *Denote the number $|A|-\#A$ by i and suppose that $2i \leq \dim \Gamma$. Then in the set of all configurations equivalent to J the subset of those configurations J', for which $\chi(\Gamma, J')$ is empty, is of codimension $\geq \dim \Gamma - 2i + 1$, while the set of J' such that the codimension of $\chi(\Gamma, J')$ in Γ equals $2i - l$, $l \geq 1$, is a subset of codimension $\geq l(\dim \Gamma - 2i + l + 1)$. In particular, when $i < 2(\dim \Gamma + 1)/7$, the codimension of any set $\chi(\Gamma, J)$ defined by any A-configuration J with $|A|-\#A = i$ is exactly equal to $2i$.*

(C) *Suppose that $|A|-\#A = i \geq \dim \Gamma / 2$. Then in the set of configurations equivalent to J the set of all configurations J' such that $\dim \chi(\Gamma, J') = l \geq 0$ is of codimension no less than $(l+1)(2i - \dim \Gamma + l)$ or is empty. In particular, the set of all J' such that $\chi(\Gamma, J')$ is nonempty, is of codimension $\geq 2i - \dim \Gamma$ and is empty when $\dim \Gamma < i$.*

This proposition follows from the transversality theorem, see [GG]. □

DEFINITION. The subspaces Γ in κ_k satisfying the conditions of Proposition 3 are called Σ-*nondegenerate* subspaces.

In what follows Γ_k^d denotes a Σ-nondegenerate affine subspace in κ_k, which is sufficiently close to the subspace $\widetilde{\Gamma}_k^d$ in the space of all subspaces of the same dimension.

By the Weierstrass approximation theorem, any quasiornament and any compact subset in the space of quasiornaments can be approximated arbitrarily close (in any prescribed C^m-topology) by the quasiornaments lying in appropriate space Γ_k^d.

4.2. Geometrical resolution of the discriminant.
Here we construct a resolution of the space $\Sigma \cap \Gamma_k^d$, i.e., a semialgebraic set σ together with proper projection $\sigma \to \Sigma \cap \Gamma_k^d$ such that the induced map $\overline{H}_*(\sigma) \to \overline{H}_*(\Sigma \cap \Gamma_k^d)$ is an isomorphism.

Denote by Ψ the disjoint union of all $\binom{k}{3}$ possible three-dimensional tori $T_{\alpha\beta\gamma}^3$, $1 \le \alpha < \beta < \gamma \le k$, which are the direct products of three different components of the manifold C_k.

Let N be a very large natural number, and let $\lambda\colon \Psi \to \mathbb{R}^N$ be a smooth embedding. Denote by σ_2 the subset in $\Gamma_k^d \times \mathbb{R}^N$ consisting of all possible pairs of the form

$$(\text{a map } \varphi \in \Gamma_k^d;\ \text{a point } \lambda(x, y, z) \in \mathbb{R}^N) \tag{3}$$

such that x, y, z are points of three different components of C_k and $\varphi(x) = \varphi(y) = \varphi(z)$.

PROPOSITION 4. *If N and d are sufficiently large, the map λ is generic and Γ_k^d satisfies conditions of Proposition 3, then σ_2 is a smooth manifold with a natural structure of the space of an orientable (and even stably trivial) $(\dim \Gamma_k^d - 4)$-dimensional affine bundle over Ψ, the projection in this bundle is defined by forgetting the first elements φ in the pairs (3).*

PROOF. This follows immediately from the construction. □

The obvious map $\sigma_2 \to \Sigma \cap \Gamma_k^d$ is a smooth normalization of $\Sigma \cap \Gamma_k^d$; its inverse image over a nonsingular point consists of only one point, while the inverse images of singular points can consist of several points.

For any point $\varphi \in \Sigma \cap \Gamma_k^d$, let us take all possible points $(x, y, z) \in \Psi$ such that $\varphi(x) = \varphi(y) = \varphi(z)$; let $\widetilde{\Delta}(\varphi)$ be the convex hull in $\mathbb{R}^N$ of all points $\lambda(x, y, z)$ where the point (x, y, z) satisfies this condition.

PROPOSITION 5. *Suppose that Γ_k^d satisfies the conditions of Proposition 3, N is sufficiently large, and the imbedding $\lambda\colon \Psi \to \mathbb{R}^N$ is generic. Then for any $\varphi \in \Sigma \cap \Gamma_k^d$, the polyhedron $\widetilde{\Delta}(\varphi)$ is a simplex whose vertices are all the points $\lambda(x, y, z)$ such that $\varphi(x) = \varphi(y) = \varphi(z)$.*

PROOF. This follows from Thom's multijet transversality theorem, see [GG]. □

In the sequel we shall assume that Γ_k^d, N, and λ satisfy the conditions of this proposition.

For any $\varphi \subset \Sigma \cap \Gamma_k^d$, denote by $\Delta(\varphi)$ the simplex $\varphi \times \widetilde{\Delta}(\varphi) \in \Gamma_k^d \times \mathbb{R}^N$ and by σ the union of all simplices $\Delta(\varphi)$ over all φ.

The topology of the space σ is induced from the ambient space $\Gamma_k^d \times \mathbb{R}^N$. The obvious projection $\Gamma_k^d \times \mathbb{R}^N \to \Gamma_k^d$ maps σ onto $\Sigma \cap \Gamma_k^d$.

PROPOSITION 6. *Suppose that the conditions of Propositions 5 are satisfied. Then the projection $\pi\colon \sigma \to \Sigma \cap \Gamma_k^d$ is proper and induces a homotopy equiv-*

alence of one-point compactifications of these two spaces, in particular, an isomorphism $\overline{H}_*(\sigma) \to \overline{H}_*(\Sigma \cap \Gamma_k^d)$.

Indeed, the properness follows from the construction; the induced map of compactifications is a piece-wise-algebraic map of semialgebraic compact sets with contractible fibers. This implies the assertion about the homotopy equivalence, see [D]. □

The space σ together with the projection π is called the *geometrical resolution* of $\Sigma \cap \Gamma_k^d$.

4.3. Filtration on the spaces of geometrical resolutions. Restrict on the space σ the obvious projection $\Gamma_k^d \times \mathbb{R}^N \to \mathbb{R}^N$. If N is sufficiently large and λ is generic, then the inverse image of any point θ in $\mathbb{R}^N$ under this projection is an affine subspace of the form $\chi(\Gamma_k^d, J) \times \theta$, where J is some A-configuration.

Define an increasing filtration $\sigma_1 \subset \sigma_2 \subset \cdots \subset \sigma_{\text{last}} = \sigma$ on the set σ by assigning to σ_i the union of all subspaces of the form $\chi(\Gamma_k^d, J) \times \theta$ where J is an A-configuration with $|A| - \#A \leq i$.

By Proposition 3, the number *last* of elements of the filtration is not greater than $\dim \Gamma_k^d$.

Consider the homology spectral sequence $E_{p,q}^r(d)$ converging to the group $\overline{H}_*(\sigma)$ and generated by this filtration.

By definition, the term $E_{p,q}^1$ of this spectral sequence equals $\overline{H}_{p+q}(\sigma_{-p} - \sigma_{-p-1})$.

PROPOSITION 7. *If i is in the stable range, $i < 2(\dim \Gamma_k^d + 1)/7$, then the term $\sigma_{-p} - \sigma_{-p-1}$ of our filtration is the space of an affine fibre bundle (whose fibers are the fibers of the projection $\Gamma_k^d \times \mathbb{R}^N \to \mathbb{R}^N$ restricted on σ); this bundle is orientable.*

PROOF. Indeed, the fact that this projection is an affine fibre bundle follows from Proposition 3. This bundle can be considered as a subbundle of the trivial bundle with the fiber Γ_k^d. The fibre of this subbundle at each point is distinguished in Γ_k^d by several conditions of type $\varphi(x) = \varphi(y)$, where the pairs of points x, y are defined by the point of the base. For stable i, all these conditions are linearly independent. Hence, the quotient bundle of our subbundle at each point splits into the direct sum of several examples of $\mathbb{R}^2$ with the canonical two-dimensional coordinate $\varphi(x) - \varphi(y)$ on each. This splitting is defined invariantly, up to permutations of summands. Such permutations do not change the orientation of the sum, and proposition is proved. □

Thus, by the Thom isomorphism, the closed homology group $\overline{H}_*$ of the space $\sigma_{-p} - \sigma_{-p-1}$ reduces to that of the base of this bundle, which can, in principle, be described in combinatorial terms. For the first calculations see §8 below.

Make our spectral sequence $E^r_{p,q}(d)$ a cohomological spectral sequence by renaming the term $E^r_{p,q}(d)$ by $E_r^{-p,\dim\Gamma_k^d-1-q}(d)$. This spectral sequence is called the *main spectral sequence*, and the previous one the *main homological spectral sequence*. By the Alexander duality theorem, the spectral sequence $E_r^{p,q}$ converges to $H^*(\Gamma_k^d\setminus\Sigma)$.

By the construction, this spectral sequence lies in the region $\{p, q | p < 0\}$, and its $E_\infty^{p,q}$ term can be nontrivial only if $p + q \geq 0$.

4.4. Main properties of the main spectral sequence.

THEOREM 7. *For any choice of the space Γ_k^d satisfying the conditions of Proposition 3;*

(A) *If the term $E_1^{p,q}(d)$ of our spectral sequence is nontrivial, then*
   (i) $p + q \geq 0$,
   (ii) $p \geq -\dim\Gamma_k^d$.

(B) *For any $d' \geq d$ and any space $\Gamma_k^{d'}$ satisfying the assertions of Proposition 3, the corresponding spectral sequence $E_r^{p,q}(d')$ coincides with $E_r^{p,q}(d)$ for following values of p, q and r:*
   (i) *for $r = 1$ and p, q in the "stable" region $\{p+q \geq 0, -p < 2(\dim\Gamma_k^d + 1)/7\}$,*
   (ii) *for any $r > 1$ and p, q such that the differentials d^t, $t < r$, do not act into the cell $E_t^{p,q}$ from the unstable region.*

COROLLARY. *For any i the inclusion homomorphism $\overline{H}_{\dim\Gamma_k^d-1}(\sigma_i) \to \overline{H}_{\dim\Gamma_k^d-1}(\sigma)$ is a monomorphism.* □

The restriction (ii) of part (A) of the theorem follows from Proposition 3. Part (i) will be proved in §6 (and, in a different way, in §7).

Let $s = 2(\dim\Gamma_k^d + 1)/7$, then the identity imbedding $\Gamma_k^d \to \Gamma_k^{d'}$ induces imbedding $\sigma_s(d) \to \sigma_s(d')$ of sth terms of the corresponding resolutions; this imbedding respects the above defined filtrations on both spaces.

LEMMA. *The last imbedding can be extended to a homotopy equivalence of the $4k(d' - d)$-fold suspension of the one-point compactification of $\sigma_s(d)$ onto the one-point compactification of $\sigma_s(d')$ such that for any $u < s$ the restriction of this homotopy equivalence on the $4k(d' - d)$ fold suspension of the compactification of $\sigma_u(d)$ is the homotopy equivalence of this space onto the compactification of $\sigma_u(d')$.*

The proof repeats that of Theorem 4.2.4 in Chapter III; this lemma implies part (B) of Theorem 7. □

4.5. The stable spectral sequence.
Part (B) of Theorem 7 allows us to define the *stable spectral sequence* $E_r^{p,q} \equiv E_r^{p,q}(\infty)$: its term $E_r^{p,q}$ is equal to the common term $E_r^{p,q}(d)$ of all spectral sequences corresponding to sufficiently large d.

PROPOSITION 8. *For any i, there exists $d = d(i)$ such that for all $d' \geq d$ and all r we have $E_r^{-i,i}(d') \cong E_r^{-i,i}(d)$, in particular, $E_r^{-i,i} \cong E_r^{-i,i}(d)$. Namely, it is sufficient to take $d > [(7i/2 - 1)2k - 1]/2$.*

The proposition follows immediately from the structure of the spectral sequence (see Theorem 7) and the stabilization properties of the strata of $\Sigma \cap \Gamma_k^d$ (see Proposition 3). □

The isomorphism $E_\infty^{-i,i}(d') \cong E_\infty^{-i,i}(d)$ from this proposition agrees with the map in homology: for $d' > d$, sufficiently large with respect to i, there is a canonical isomorphism

$$\overline{H}_{2k(2d+1)-1}(\sigma_i(d)) \cong \overline{H}_{2k(2d'+1)-1}(\sigma_i(d')). \tag{4}$$

This isomorphism is compatible with the Alexander duality isomorphism: let θ and θ' be equivalent ornaments in κ_k, and let $\theta \in \Gamma_k^d$, $\theta' \in \Gamma_k^{d'}$, then the elements of the groups in (4), corresponding to each other by the isomorphism in (4), have the same values on these ornaments.

DEFINITION. A *stable invariant of order i* is an element of the group $\overline{H}_{\dim \Gamma_k^d - 1}(\sigma_i(d))$, d sufficiently large. Two elements of such groups corresponding to different d define the same invariant of order i if they correspond to each other via isomorphism (4).

Any such stable invariant of finite order i can be regarded as a well-defined invariant of ornaments: this follows from the stability property and from the fact that any homotopy, which realizes the equivalence of two ornaments, can be approximated arbitrarily well by some homotopy, also avoiding the discriminant, in an appropriate space Γ_k^d.

DEFINITION. An invariant of ornaments is *of order i* if it is obtained by the above construction from some stable invariant of order i.

THEOREM 8. *This definition is equivalent to the definition from §2.*

A proof will be given in §7.

In §§5, 6 we begin the study of the term E_1 of the stable spectral sequence, and in §8 we present the results of the first calculations.

§5. Complexes of connected hypergraphs

Recall, that for i in stable range, $i < 2(\dim \Gamma_k^d + 1)/7$, the space $\sigma_i - \sigma_{i-1}$ is the space of an oriented affine bundle over a stratified variety; in particular, the closed homology group $\overline{H}_*(\sigma_i - \sigma_{i-1})$ reduces to a similar group of that base variety.

To study this last group (and to estimate the similar groups for i in the nonstable domain) we need some homological preliminaries.

Let θ be a finite set together with some subdivision of it into k subsets, $\theta = (\theta_1, \ldots, \theta_k)$. Denote by $\Delta[\theta]$ (or $\Delta[\theta_1, \ldots, \theta_k]$) the simplex, whose vertices correspond to all triplets of points in θ, belonging to different subsets $\theta_j, \theta_l, \theta_m, 1 \leq j < l < m \leq k$.

DEFINITION. A collection of vertices of the simplex $\Delta[\theta]$ is called *connecting* if any two points of θ can be joined by a chain of points, any two neighboring points in which belong to one triplet corresponding to some vertex of the collection. A face of $\Delta[\theta]$ is called connecting if the collection of its vertices is connecting.

The *connecting part* of the simplex $\Delta[\theta]$ is the union of interior points of all its connecting faces (including the simplex itself, if there are at least 3 nonempty sets θ_j).

Now, let $\Theta = (\theta^1, \ldots, \theta^{\#A})$ be a collection of $\#A$ sets θ^j, $j = 1, \ldots, \#A$, of cardinalities $a_1, \ldots, a_{\#A}$, any of which is divided into some subsets $\theta_1^j, \ldots, \theta_k^j$ (k is the same for all j).

Consider the join of the simplices $\Delta[\theta^j]$ over all $j = 1, \ldots, \#A$, i.e., the simplex whose vertices are all the vertices of these $\#A$ simplices. Denote this simplex by $\Delta[[\Theta]]$.

DEFINITION. A face of the simplex $\Delta[[\Theta]]$ is *essential* if for any $j = 1, \ldots, \#A$, the vertices of this face belonging to the simplex $\Delta[\theta^j]$ span a connecting face of this simplex. The *essential part* of the simplex $\Delta[[\Theta]]$ is the union of interior points of all its essential faces.

The union of all nonessential faces in the simplex $\Delta[[\Theta]]$ is obviously a subcomplex of its natural triangulation. Denote by $\Xi(\theta)$ the corresponding quotient complex. The (reduced modulo a point) homology group of this quotient complex is another realization of the closed homology group of the essential part of the simplex.

CONVENTION. From now on we consider homologies with coefficients in the field $\mathbb{R}$ only.

EXAMPLES. Let $\#A = 1$, so that $\Theta = (\theta^1)$, and let $k = 3$ so that θ^1 consists of 3 subsets of cardinalities $\alpha_1, \alpha_2, \alpha_3$.

(1) Let $\alpha_2 = \alpha_3 = 1$. Then the simplex $\Delta[[\Theta]]$ is $(\alpha_1 - 1)$-dimensional and has only one essential facet: the simplex itself. In particular, the group $\overline{H}_i(\Xi(\Theta))$ equals $\mathbb{R}$ if $i = \alpha_1 - 1$ and is trivial for all other i.

(2) Let $\alpha_1 = \alpha_2 = 2$, $\alpha_3 = 1$, θ_1 consists of points x_1 and x_2, θ_2 consists of points y_1 and y_2, and θ_3 has only one point z. Then the dimension of the simplex $\Delta[[\Theta]]$ equals 3, and its vertices are called

$$(x_1, y_1, z), (x_1, y_2, z), (x_2, y_1, z), (x_2, y_2, z),$$

see Figure 52. The essential faces of this simplex are the simplex itself, all its two-dimensional faces, and two edges $((x_1, y_1, z); (x_2, y_2, z))$ and $((x_1, y_2, z); (x_2, y_1, z))$. The group $\overline{H}_i(\Xi(\Theta))$ equals $\mathbb{R}$ for $i = 2$ and is trivial for other i. The group $\overline{H}_2(\Xi(\Theta))$ is generated by either of two chains $((x_2, y_1, z); (x_2, y_2, z); (x_1, y_1, z)) + ((x_1, y_1, z); (x_2, y_2, z); (x_1, y_2, z))$ or $((x_1, y_2, z); (x_2, y_1, z); (x_1, y_1, z)) + ((x_2, y_2, z); (x_2, y_1, z); (x_1, y_2, z))$; the sum of these chains is homologous to zero.

§5. COMPLEXES OF CONNECTED HYPERGRAPHS

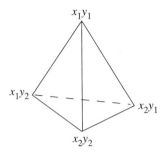

FIGURE 52

The case when Θ consists of more than one collection θ^j can be reduced to the case $\#A = 1$ by the following assertion.

PROPOSITION 9. *For any $\Theta = (\theta^1, \ldots, \theta^{\#A})$, there is natural isomorphism*

$$H_{*-\#A+1}(\Xi(\Theta)) = \bigotimes_{j=1}^{\#A} H_*(\theta^j).$$

PROOF. This follows from the formula for the homology of a join. □

THEOREM 9. *The complex $\Xi(\Theta)$ is acyclic in dimensions greater than $|A| - 2\#A - 1$, where $|A|$ is the total number of elements in all sets θ^j.*

PROOF. The proof is based on the Goresky-MacPherson formula (see the formula (37) of §III.6) for the homology of subspace arrangements, cf. [BW], [V$_{15}$].

Indeed, by Proposition 9 it is sufficient to prove this theorem in the case when $\#A = 1$, i.e., we have only one group of points $\theta = \theta^1$ divided into k subsets of cardinalities $\alpha_1, \ldots, \alpha_k$. Consider the space $\mathbb{R}^{\alpha_1} \oplus \cdots \oplus \mathbb{R}^{\alpha_k}$ with fixed linear coordinates $x_{1,1}, \ldots, x_{1,\alpha_1}$ in $\mathbb{R}^{\alpha_1}$ and so on. Consider the collection of all subspaces in this space distinguished by all possible systems of equations of the form $x_{i,\beta} = x_{j,\gamma} = x_{l,\delta}$ where $i \neq j \neq l \neq i$. By the Goresky-MacPherson formula (in the form proposed in the §III.6 of the present book), the homology group of the one-point compactification of the union of all such planes splits into a direct sum of homology groups of certain cell complexes corresponding to all possible intersections of these planes. In particular, the complex corresponding to the intersection of all these planes (i.e., to the line $x_{1,1} = \cdots = x_{k,\alpha_k}$) is nothing but the suspension of our quotient complex $\Xi(\Theta)$. Therefore, the i-dimensional homology group of this quotient complex enters as a direct summand in the $(i+1)$-dimensional homology group of some $(|A| - 2)$-dimensional topological space. This implies the theorem. □

§6. Structure of the space $\sigma_i - \sigma_{i-1}$

6.1. J-blocks and complexes of connected hypergraphs.

The space $\sigma_i - \sigma_{i-1} \subset \Gamma_k^d \times \mathbb{R}^N$ splits in a natural way in a union of subspaces which correspond to different equivalence classes of A-configurations of complexity i. Indeed, let us fix some index $A = (a_1 \geq \cdots \geq a_{\#A})$ and some A-configuration J in C_k. To this configuration we assign the subset in $\Gamma_k^d \times \mathbb{R}^N$ which is the direct product of the affine subspace $\chi(\Gamma_k^d, J) \subset \Gamma_k^d$ (see §4.1) and the simplex $\tilde{\Delta}(\varphi(J)) \subset \mathbb{R}^N$ (see §4.2) where $\varphi(J)$ is arbitrary generic point of this subspace (i.e., a map which strictly respects J); this simplex depends on J only.

DEFINITION. For any A-configuration J of complexity i in C_k, the J-block in σ_i is the union of all products

$$\chi(\Gamma_k^d, J') \times \tilde{\Delta}(\varphi(J')) \subset \Gamma_k^d \times \mathbb{R}^N, \tag{5}$$

where J' is some A-configuration equivalent to J and $\varphi(J')$ is arbitrary map $C_k \to \mathbb{R}^2$ strictly respecting J'.

Obviously, the closure of the space $\sigma_i - \sigma_{i-1}$ in σ_i coincides with the union of all J-blocks over all (equivalence classes of) configurations J of complexity i.

For any A-configuration $J = (\theta^1, \ldots, \theta^{\#A})$ with card $\theta^j = a_j$, the simplex $\tilde{\Delta}(\varphi(J))$ can be identified in an obvious way with the simplex $\Delta[J]$, defined in §5: indeed, to the vertex $(x, y, z) \in \Delta[J]$ there corresponds the vertex $\lambda(x, y, z) \in \tilde{\Delta}(\varphi(J))$, and this correspondence can be extended inside the simplices by linearity.

PROPOSITION 10. *A point of the left product in* (5) *belongs to the term* σ_{i-1} *of our filtration iff its projection onto the factor* $\tilde{\Delta}(\varphi(J'))$ *lies in a nonessential face of this simplex.*

This follows directly from the definitions. □

6.2. Auxiliary spectral sequence.

Define the auxiliary *filtration* on the space $\sigma_i - \sigma_{i-1}$ by assigning to its lth term the points of J-blocks over all A-configurations J with $|A| \leq l$.

This filtration defines a new spectral sequence, $G_{p,q}^r$, which converges to the group $\overline{H}_*(\sigma_i - \sigma_{i-1})$ which is (up to a shift of indices) just the column $E_{i,*}^1$ of the main homological spectral sequence from §4.

We call this spectral sequence G the *auxiliary* spectral sequence.

The term $G_{p,q}^1$ of this spectral sequence equals the direct sum of groups $\overline{H}_{p+q}((J\text{-block})\backslash\sigma_{i-1})$ taken over all equivalence classes J of A-configurations with $|A| - \#A = i$, $|A| = p$.

6.3. Proof of part A(i) of Theorem 7. Consider arbitrary J-block in σ_i, where J is an A-configuration of complexity i. The intersection of this J-block with the set $\sigma_i - \sigma_{i-1}$ can be considered as a fibre bundle whose base is the space of pairs of the form {a configuration J' equivalent to J; some map $\varphi: C_k \to \mathbb{R}^2$ respecting J'}, and the fibre is the essential part of the simplex $\Delta[J']$. By Proposition 3, the dimension of the base is no greater than $|A| + \dim \Gamma_k^d - 2(|A| - \#A)$, and by Theorem 9 the closed homology group $\overline{H}_l$ of the fibre is trivial for $l > |A| - 2\#A - 1$. Hence, the term $G_{p,q}^1$ of the auxiliary spectral sequence is trivial for $p + q > \dim \Gamma_k^d - 1$. This terminates the proof of Theorem 7.

6.4. On the calculation of the stable spectral sequence. If i is in the stable domain, $i < 2(\dim \Gamma_k^d + 1)/7$, and J is an A-configuration with $|A| - \#A = i$, then the corresponding J-block in $\sigma_i - \sigma_{i-1}$ is a fibre bundle whose base is the space of all A-configurations equivalent to J, and the fibre is the product of a (canonically oriented) affine space and the essential part of the simplex $\Delta(\varphi(J)) = \Delta[J]$. By the Thom isomorphism, we can forget about the first factor in the fibre (and, moreover, about all such factors for all such J-blocks simultaneously) and calculate the terms $E_{i,*}^1$ of the principal spectral sequence by investigating the resting bundles and their interrelations for different J. For the first calculations see §8.

§7. Proof of Theorem 8

To prove this theorem, we construct one more resolution of the discriminant variety $\Sigma \subset \Gamma_k^d$, based on the notion of the order complex. This resolution will be called the *visible resolution* of the discriminant and denoted by σv.

The space of this resolution again has a natural filtration, which is equivalent to that of the resolution σ considered above (that is, there is a natural proper imbedding $\sigma v \mapsto \sigma$ preserving the filtrations and such that the induced morphisms

$$\overline{H}_*(\sigma v_i - \sigma v_{i-1}) \to \overline{H}_*(\sigma_i - \sigma_{i-1})$$

are isomorphic; in particular, this imbedding establishes an isomorphism of corresponding spectral sequences, starting from their E_1 terms.) All the notions entering in the definition of invariants of order i given in §2 (i.e., invariants, regular quasiornaments, degeneration modes, and characteristic numbers) can be naturally interpreted in the terms of this resolution.

(For example, the degeneration modes can be interpreted as follows. Again, the spaces $\sigma v_i - \sigma v_{i-1}$ are divided into J-blocks (which are compatible with the J-blocks in $\sigma_i - \sigma_{i-1}$ by means of the previous imbedding). These J-blocks are fibred into certain simplicial complexes $\Delta v(J)$ (instead of simplices $\Delta(\varphi(J))$ for the resolution σ). For any A-configuration J, the topology of the corresponding complex $\Delta v(J)$ depends only on A, the dimension of this complex equals $|A| - 2\#A - 1$, and its simplices of highest dimension $|A| - 2\#A - 1$ are in one-to-one correspondence with the degeneration modes of J. Thus, the dimension of the J-block equals $\dim \Gamma_k^d - 1$; this implies a new proof of the assertion (A(i)) of Theorem 7.)

After this interpretation of these notions, the equivalence of two definitions of the orders of invariants becomes almost tautological, see Proposition 17.

7.1. Two concepts of the order complex of a collection of intersecting sets. Suppose that we have a collection of sets $V_1, \ldots, V_l$. To such a collection there correspond two (homotopy equivalent) simplicial complexes: the formal order complex and the visible order complex; let us define them.

Denote by L the set $(1, \ldots, l)$ of indices of the sets V_i, by 2^L the set of all subsets in L, by V the union $V_1 \cup \cdots \cup V_l$ and, for any nonempty $\alpha \in 2^L$, by V_α the intersection $\bigcup_{j \in \alpha} V_j$.

DEFINITION. The *formal order complex* related to the collection $V_1, \ldots, V_l$ is the abstract simplicial complex whose i-dimensional simplices are such sequences $(\alpha_0, \alpha_1, \ldots, \alpha_i)$ of subsets in L that

(a) α_j strictly contains α_{j+1} for any $j = 0, \ldots, i-1$;
(b) the set V_{α_0} is nonempty.

The *visible order complex* related to the same collection is the abstract simplicial complex whose i-dimensional simplices are sequences

$$(V_{\alpha_0}, \ldots, V_{\alpha_i}) \tag{6}$$

such that

(a) V_{α_k} is strictly contained in $V_{\alpha_{k+1}}$ for any $k = 0, \ldots, i-1$;
(b) the set V_{α_0} is nonempty.

The notations of formal and visible order complexes are $\Delta f(V_1, \ldots, V_l)$ and $\Delta v(V_1, \ldots, V_l)$ respectively.

Note that the same simplex (or even vertex) of the visible order complex can have different expressions in the form (6), if $V_\alpha = V_\beta$ for some $\alpha \neq \beta$.

EXAMPLE. Let $l = 3$ and the intersection $V_1 \cap V_2 \cap V_3$ is nonempty. Then the formal and visible order complexes are shown on Figures 53 and 54, respectively. The picture from Figure 53 does not depend on the information, which sets V_α, $V_{\alpha \cup \beta}$ are *strictly* incident, while the complex Δv depends strongly on this information. In general, for arbitrary l, if the intersection of sets $V_1, \ldots, V_l$ is nonempty, then the corresponding formal order complex can be naturally identified with the first barycentric subdivision of the $(l-1)$-dimensional simplex whose vertices correspond to the sets $V_1, \ldots, V_l$.

NOTATION. For any $\alpha \subset L$, let $\bar{\alpha}$ be the maximal possible set in the family of all such subsets $\alpha' \subset L$ that $V_{\alpha'} = V_\alpha$.

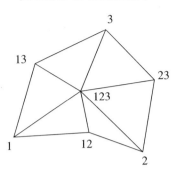

FIGURE 53

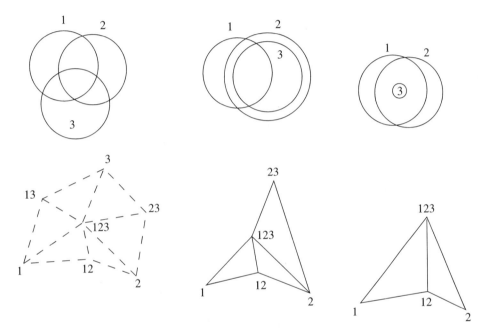

FIGURE 54

PROPOSITION 11. *The visible order complex of any collection of sets $(V_1, \ldots, V_l)$ can be naturally considered as a subcomplex of the formal order complex of the same collection, moreover, this subcomplex is a deformation retract of the formal order complex.*

PROOF. The 0-dimensional skeleton of the complex Δv is naturally imbedded into that of Δf: to any point in Δv (i.e., a set of the form V_α, $\alpha \in 2^L$) we assign the set $\bar{\alpha}$. This imbedding I of 0-skeletons extends by linearity on any simplex, to a map of the whole complex Δv and the resulting map is, obviously, again an imbedding: its image coincides with the union of all such simplices $(\alpha_0, \ldots, \alpha_i)$ that $\alpha_j = \bar{\alpha}_j$ for any $j = 0, \ldots, i$.

Now, let us construct the retraction $\Delta f \to \Delta v$.

We map any point $\alpha \in sk^0(\Delta f)$ into the point $\overline{\alpha} \in sk^0(I(\Delta v))$ and extend this map of 0-skeletons to a map $\Delta f \to I(\Delta v)$ by linearity on the simplices. This map is obviously continuous and is a deformation retraction: indeed, the inverse image of any point x in $I(\Delta v)$ consists of a family of segments in Δf which have no intersection points except for their endpoints (which coincide all with x). □

7.2. Visible resolution of the discriminant.

LEMMA. *If the space Γ_k^d is Σ-nondegenerate, then there exists an affine subspace $\Gamma \subset \kappa_k$, containing Γ_k^d and such that for any A and any A-configuration J which is respected by some element of Γ_k^d, the codimension of the corresponding subspace $\chi(\Gamma, J)$ in Γ equals exactly $2(|A| - \#A)$ and has no other representations in the form $\chi(\Gamma, J')$, $J' \neq J$.*

The proof is trivial. □

Obviously, if some space Γ satisfies this lemma, then any affine space containing Γ also satisfies it.

Let μ be the largest possible complexity $|A| - \#A$ of configurations which are respected by the maps $\varphi \in \Gamma_k^d$.

Let Γ be a space satisfying the previous lemma, let $G^4(\Gamma), G^6(\Gamma), \dots, G^{2\mu}(\Gamma)$ be the "affine" Grassmann manifolds, whose points are the affine subspaces of codimensions $4, 6, \dots, 2\mu$ in Γ.

Let $h^4(\Gamma), h^6(\Gamma), \dots, h^{2\mu}(\Gamma)$ be topological subspaces in these manifolds, whose points are the subspaces of the form $\chi(\Gamma, J)$ for some A-configurations J. Consider the join $h^4(\Gamma) * \cdots * h^{2\mu}(\Gamma)$ of these subspaces, i.e., roughly speaking, the (suitably topologized) union of all simplices, whose vertices are the points of these spaces $h^{2l}(\Gamma)$. Denote this join by $X(\Gamma)$. The visible resolution will be constructed as a subset in the direct product $\Gamma_k^d \times X(\Gamma)$.

DEFINITION. For any A-configuration J, the *feud* of J is a subset in $X(\Gamma)$ which is the union of simplices, whose vertices are points $g_{l_1} \in h^{2l_1}(\Gamma)$, $g_{l_2} \in h^{2l_2}(\Gamma), \dots$ with $l_1 < l_2 < \cdots$ such that the corresponding planes of codimensions $2l_1, 2l_2, \dots$ in Γ form a flag (i.e., are all incident to each other) and all belong to $\chi(\Gamma, J)$.

Obviously, the feud of any configuration is a contractible space: it is the union of simplices with one common vertex $g_{|A|-2\#A} = \{\chi(\Gamma, J)\}$.

For any point $\varphi \in \Sigma \subset \Gamma_k^d$, consider the A-configuration $J(\varphi)$ strictly respected by φ, and take the subset

$$\varphi \times \text{feud}(J(\varphi)) \subset \Gamma_k^d \times X(\Gamma). \tag{7}$$

Define the *visible resolution* $\sigma v \subset \Gamma_k^d \times X(\Gamma)$ of the set $\Sigma \subset \Gamma_k^d$ as the union of all sets like the left part of (7) over all $\varphi \in \Sigma$.

PROPOSITION 12. *The obvious projection $\sigma v \to \Gamma_k^d$ is proper and establishes a homotopy equivalence between the one-point compactifications of σv and Σ.*

The proof is the same as that of Proposition 6. □

7.3. Imbedding the visible resolution into the (formal) resolution constructed in §4. Here is another description of the feud of $J(\varphi)$. Consider all points $(x_j, y_j, z_j) \in \Psi$ (where x, y and z are points of three different components of C_k) such that $\varphi(x_j) = \varphi(y_j) = \varphi(z_j)$. For any such point (x_j, y_j, z_j) denote by V_j the space $\chi(\Gamma, J_j)$, where J_j is the (3)-configuration (x_j, y_j, z_j).

PROPOSITION 13. *The feud of $J(\varphi)$ is a finite simplicial complex and can be naturally identified with the visible order complex of all spaces $V_j = \chi(\Gamma, J_j)$.*

This is a tautology. □

On the other hand, the simplex $\Delta(\varphi)$ which participated in the construction of the resolution σ in §4, can be naturally considered as the support of the formal order complex of the same collection of spaces V_j (and becomes this formal order complex after the barycentric subdivision). Therefore, we get a natural imbedding $I: \sigma v \to \sigma$, indeed, for any $\varphi \in \Sigma$ the restriction of this imbedding on the set $\varphi \times \text{feud}(J(\varphi)) \subset \sigma v$ is the composition of the identification of $\text{feud}(J(\varphi))$ with the visible order complex from Proposition 13, the imbedding of this visible complex into the formal one (see Proposition 11) and identification of the support of the later complex with the set $\varphi \times \widetilde{\Delta}(\varphi) \subset \sigma$.

Since the set σv is a union of left parts in (7) over all $\varphi \in \Sigma$, these imbeddings define a general imbedding $I: \sigma v \to \sigma$.

PROPOSITION 14. *This imbedding I is continuous and proper, commutes with the natural projections onto Σ and defines a homotopy equivalence of the one-point compactifications of spaces σv and σ.*

PROOF. All assertions of this proposition but the last one (about the homotopy equivalence) follow immediately from the construction. Further, a deformation retraction $\sigma \to I(\sigma v)$ is well defined: it is the union of retractions from Proposition 11 applied to all fibres of the projection $\sigma \to \Sigma$. This retraction can be extended to a retraction of one-point compactifications of these spaces and establishes the desired homotopy equivalence. □

On the space σ the filtration $\sigma_2 \subset \sigma_3 \subset \cdots$ is defined (see §4.3) as well as the decomposition of the sets $\sigma_i - \sigma_{i-1}$ into J-blocks. The imbedding I, just constructed, induces similar structures on the space σv: they are defined as the inverse images of corresponding sets in σ.

In particular, the induced filtration $\sigma v_2 \subset \sigma v_3 \subset \cdots$ defines in a standard way a spectral sequence $vE_{p,q}^r(d) \to \overline{H}_*(\sigma v)$.

PROPOSITION 15. *The imbedding* $I: \sigma v \to \sigma$ *induces an isomorphism of spectral sequence* $vE^r_{p,q}(d)$ *and the sequence* $E^r_{p,q}(d)$ *(constructed in §4.3) beginning with the terms* E^1, vE^1.

Indeed, the retraction $\sigma \to I(\sigma v)$ from Proposition 14 respects the filtration by the sets σ_i, σv_i; hence, the proposition follows. □

COROLLARY. *The order* i *invariants could be defined via the filtration of the space* σv, *not of* σ, *as was done in subsection 4.5.* □

7.4. Homology in σv_i and characteristic numbers.

PROPOSITION 16. *Let* φ *be a point in* Σ, *and let* $J(\varphi)$ *be the A-configuration strictly respected by* φ, $i = |A| - \#A$. *Then the simplices of the simplicial complex* $\operatorname{feud}(J(\varphi))$, *which do belong not only to* σv_i, *but also to* σv_{i-1}, *are exactly those which do not contain the point* $\{\chi(\Gamma, J(\varphi))\} \in h^{2i}(\Gamma)$. *The dimension of this simplicial complex equals* $|A| - 2\#A - 1$, *while its simplices of highest dimension are in a natural one-to-one correspondence with the degeneration modes of the configuration* J.

PROOF. The one-to-one correspondence is defined as follows. Given a degeneration mode of $J(\varphi)$, for any $l = 1, 2, \ldots, |A| - 2\#A$ denote by $J_{(l)}(\varphi)$ the subconfiguration in $J(\varphi)$ consisting of all points marked on (and before) the lth step of degeneration. Then to any such degeneration mode there corresponds a simplex in $\operatorname{feud}(J(\varphi))$, whose lth vertex is the affine plane in Γ consisting of all maps $\varphi: C_k \to \mathbb{R}^2$, $\varphi \in \Gamma$, which respect the subconfiguration $J_{(l)}(\varphi)$. All the assertions of this proposition follow immediately from the definitions. □

As a corollary, we get a new proof of Theorem 9.

Now let M be arbitrary invariant of ornaments. The class in $\overline{H}_{\dim \Gamma^d_k - 1}(\sigma v)$ $\cong \overline{H}_{\dim \Gamma^d_k - 1}(\Sigma)$, which is dual to the restriction on $\Gamma^d_k - \Sigma$ of the invariant M, can be considered as a (uniquely determined) linear combination of fundamental cycles of maximal strata of σv.

Let us describe these strata. Any of them belongs entirely to some space $\sigma v_i - \sigma v_{i-1}$, and, moreover, to some J-block in this space. Suppose that i is in the stable range, $i < 2(\dim \Gamma^d_k + 1)/7$, and J is any A-configuration with $|A| - \#A = i$. Then our J-block in $\sigma v_i - \sigma v_{i-1}$ can be naturally identified with the space of all triplets

$$(J', \varphi, x) \qquad (8)$$

where J' is a configuration equivalent to J, φ is a quasiornament which respects J', and x is a point of the set $\varphi \times \operatorname{feud}(J') \subset \Gamma^d_k \times X(\Gamma)$. The strata of maximal dimension of σv which belong to $\sigma v_i - \sigma v_{i-1}$ consist of exactly such points of the form (8) that φ strictly respects J', and x is a point of a maximal (i.e., of dimension $|A| - 2\#A - 1$) simplex in the complex feud(J').

Recall that the characteristic numbers of order i were defined in §2 as the functions of the following data: an invariant; an A-configuration J' with $|A| - \#A = i$; a regular quasiornament strictly respecting J'; a degeneration mode of this quasiornament (or, what is the same, of the configuration J'). Now we have interpreted all these data in the terms of the resolution σv. Namely, the invariant is a linear combination of fundamental cycles of maximal strata, and all other arguments can be encoded by appropriate point of any such stratum: indeed, a regular quasiornament φ and configuration J' respected by φ are two first elements of the expression of this point in the form (8), and the degeneration mode of this quasiornament is the maximal simplex in $\varphi \times \text{feud}(J')$ containing the element x of this expression. Now we give an interpretation of the characteristic number in terms of these data.

First, note that the maximal strata in $\sigma v_i - \sigma v_{i-1}$ are naturally oriented. Indeed, given a point (8) of such a stratum, consider three sets of neighboring points: the set of all $\tilde{J}' \approx J'$; for a given $\tilde{J}' \approx J'$, the set of all $\tilde{\varphi} \approx \varphi$ respecting $\tilde{J}'$; and, for given such $\tilde{J}'$ and φ, the set of all $\tilde{x} \approx x$ such that $(\tilde{J}', \tilde{\varphi}, \tilde{x})$ is again a point of our stratum. All these three sets are invariantly oriented; let us define these orientations.

The set of allowed $\tilde{x} \approx x$ is a neighborhood of x in the maximal simplex of the complex feud(J') or, what is the same by Proposition 13, of a visible order complex; the vertices of this simplex (and hence, also the simplex itself) are naturally ordered by the definition of this complex. The set of neighboring $\tilde{\varphi} \approx \varphi$ is oriented by Proposition 7. Finally, an orientation of the space of all $\tilde{J}' \approx J'$ is nothing but some ordering of points in J'. A partial ordering of them is defined by the degeneration mode of J', which corresponds to the simplex containing x; this partial ordering does not distinguish only the points in the triplets which correspond to the same step of this degeneration. But, these points in the triplets are ordered by the numbers of components of C_k which contain them. So, we have defined a canonical orientation of our maximal stratum, and hence our invariant (i.e., a linear combination of fundamental cycles of such strata) assigns a number to any such stratum (and in particular, to any its point (J', φ, x)): the coefficient with which this stratum (taken with the orientation just defined) participates in this linear combination.

PROPOSITION 17. *Suppose that the point* (8) *belongs to a maximal stratum of* $\sigma v_i - \sigma_{i-1}$ *and the quasiornament* φ *in* (8) *is regular. Then the number just assigned to any invariant and the point* (8), *coincides (up to a sign) with the characteristic number, defined as in* §2 *by our invariant, by the quasiornament* φ, *and by the degeneration mode corresponding to the simplex in* feud(J') *which contains the point* x.

We prove this proposition by induction over the degeneration mode corresponding to this simplex. Consider the $(|A|-2\#A-2)$-dimensional face of this simplex, which does not contain the maximal vertex $\{\chi(\Gamma, J')\}$ of the complex feud(J'); let x' be arbitrary interior point of this face. The point $(\varphi, x') \subset \Gamma_k^d \times X(\Gamma)$ belongs to σv_{i-1} (or even to σv_{i-2} if the last step of the degeneration mode is the marking of a triple of points). It does not belong to a maximal stratum in σv_{i-1} (resp., in σv_{i-2}), because φ respects a too complicated configuration.

Close to this point (φ, x') there are exactly three maximal strata of σv: one of them is the previously considered stratum in $\sigma v_i - \sigma v_{i-1}$ which contains the point (φ, x), and two other lie in σv_{i-1} (in σv_{i-2}) and correspond to two different resolutions of φ and to degeneration mode of them which is just the degeneration mode corresponding to x without the last step.

Now the assertion of the proposition follows from the induction hypothesis applied to these two resolutions and from the fact that our linear combination of maximal strata is a cycle in σv. □

This proposition implies Theorem 8. □

§8. The first calculations in the stable spectral sequence

In this section we consider only the case when $k = 3$ and d is sufficiently large, so that for all A-configurations in consideration the corresponding spaces $\chi(\Gamma_k^d, J)$ have codimension $2(|A| - \#A)$ in Γ_k^d.

The main result of this section is the following theorem.

THEOREM 10. *There are no order 1 invariants of 3-ornaments, exactly one (up to multiplicative constant) invariant of order 2, exactly 3 more linearly independent invariants of order 3, and exactly 7 more linearly independent invariants of order 4.*

All these invariants can be reduced to these from subsection 1.4, for this reduction see subsection 8.5.

We only outline the calculations which prove this theorem

Remember that all J-blocks are the spaces of certain affine bundles. Since the exact value of d is not significant, we will indicate the *codimensions* of these bundles; that is, the differences between the dimensions of Γ_k^d and of the fibers of the bundles. For any J-block in $\sigma_i - \sigma_{i-1}$ this codimension equals $2i$.

8.1. The term σ_1 of the filtration of the space σ is empty: the simplest singularity is the triple point, for which $|A| - \#A = 3 - 1 = 2$.

8.2. The term σ_2 is the space of an oriented affine fiber bundle of codimension 4 over a 3-dimensional torus; therefore, the groups $E_1^{-2, q}$ equal $\mathbb{R}$ for $q = 2, 5$, equal $\mathbb{R}^3$ for $q = 3, 4$, and are trivial for all other q.

The group $E_1^{-2,2}$ obviously stabilizes at that term, the corresponding invariant of order 2 is just the *index* $i(\varphi) = M_{1,2}(0, 0, 1)$, see subsection 1.4.

8.3. The term $\sigma_3 - \sigma_2$ consists of three J-blocks, where J is an A-configuration with $A = (4)$, and J has two points on some one component of C_3 and one point on any of other two components. Any of these blocks is the space of a fiber bundle, whose base is the space of all such configurations, and the fiber is a product of a canonically oriented affine space of codimension 6 and an open interval (the missed endpoints of these intervals lie in the boundary of $\sigma_3 - \sigma_2$ in σ_2).

The base of this bundle is obviously diffeomorphic to the direct product of the two-dimensional torus and an open Möbius band; the bundle of intervals changes its orientation after traversing exactly the same loops in the base which destroy the orientation of the base. Hence, the space of the later bundle of intervals is diffeomorphic to the direct product of an 3-torus and open two-dimensional disk.

In particular, the deposit of any of these three J-blocks into the group $E_1^{-3,q}$ is isomorphic to $\mathbb{R}$ for $q = 3, 6$, to $3\mathbb{R}$ for $q = 4, 5$, and is trivial for all other q.

8.4. The term $\sigma_4 - \sigma_3$ consists of seven J-blocks.

8.4.1. The simplest three of these blocks correspond to the A-configurations J where $A = (5)$ and J has three points on one component, and one point on any of other two. This J-block is the space of a bundle whose base is diffeomorphic to the direct product of a 3-torus and an open two-dimensional disk, and the fiber is the product of an oriented affine space of codimension 8 and an orientable bundle whose fiber is the interior part of an triangle.

Since these J-blocks do not adjoin other blocks of the same set $\sigma_4 - \sigma_3$, the deposit of any of these blocks into the groups $E_1^{-4,q}$ equals $\mathbb{R}$ for $q = 4, 7$, equals $\mathbb{R}^3$ for $q = 5, 6$, and is trivial for all other q.

8.4.2. The second collection of three J-blocks corresponds to (5)-configurations J with one point on one of components and two points on any of two others. The base of the corresponding fiber bundle is a product of a circle and two Möbius bands, while the fiber is a product of an oriented affine space of codimension 8 and the *essential part* of the tetrahedron from example 2 §5, see Figure 52.

In particular, the fiber of this bundle of essential parts has nontrivial closed homology only in dimension 2 (and the corresponding homology group is one dimensional). It is easy to calculate that the generator of this group becomes its opposite after monodromy over the same paths in the base which destroy the orientation of the base.

Since these blocks have the smallest auxiliary filtration among the J-blocks with J of complexity 4, the deposit of any of these three J-blocks in $E_1^{-4,4}$ is again equal to $\mathbb{R}$.

8.4.3. Finally, one more J-block corresponds to A-configuration J with $A = (3, 3)$. This A-block is again a fiber bundle, let us describe its base.

A two-fold covering of this base (whose leaves correspond to the orderings of two triplets in the $(3, 3)$-configuration) is diffeomorphic to the direct product of three examples of a direct product of a circle and an open interval, hence is orientable. This orientation fails after projection onto the base of the covering: this two-fold covering coincides with the orientation covering of the base.

The fiber of our bundle is the direct product of a canonically oriented affine plane of codimension 8 and an interval, whose orientation fails over the same paths in the base which destroy the orientation of the base itself. In particular, the space of the bundle is orientable, and the deposit of our J-block into $E_1^{-4,4}$ is either $\mathbb{R}$ or zero, depending on the homology class of its boundary in the J-blocks considered in subsection 8.4.2.

The corresponding *geometrical boundary* (it is, the intersection of the closure of our block with the blocks from 8.4.2) coincides with the subbundle in the bundle from 8.4.2, whose fibre corresponds to the union of edges $((x_1, y_1, z), (x_2, y_2, z))$ and $((x_1, y_2, z), (x_2, y_1, z))$ on Figure 52; it is easy to calculate that this geometrical boundary participates in the algebraic boundary twice with opposite orientations, and the auxiliary spectral sequence degenerates in the corresponding term. So, our J-block enters nontrivially in the group $E_1^{-4,4}$, which is hence equal to $\mathbb{R}^7$.

8.5. All the described elements of the groups $E_1^{-i,i}$, $i = 2, 3, 4$, do not vanish on the next steps of the spectral sequence, in particular there exist certain invariants of order i which are mapped into these elements by the obvious reductions $\mod \sigma_{i-1}$.

All these invariants are the invariants $M_{i,j}$ from subsection 1.4 or functions of them.

THEOREM 11. *The only generator of the group $E_1^{-2,2}$ coincides with the index $i(\varphi) = M_{1,2}(0, 0, 1)$.*

Three generators of the group $E_1^{-3,3}$ are reductions $\mod \sigma_2$ of the invariants $M_{1,2}(0, 0, 2)$, $M_{1,3}(0, 2, 0)$ and $M_{2,3}(2, 0, 0)$ (in the case of the homologies over $\mathbb{R}$; in the case of the integer homology it is to take similar sums of the form

$$\sum_{b_1,b_2,b_3=-\infty}^{\infty} ((b_3^2 - b_3)/2) I_{1,2}(b_1, b_2, b_3)$$

and so on).

The seven generators of $E_1^{-4,4}$ are the reductions $\mod \sigma_3$ of the following *invariants*: $M_{1,2}(0, 0, 3)$, $M_{1,3}(0, 3, 0)$, *and* $M_{2,3}(3, 0, 0)$ (*or, again, three sums of the form*

$$\sum_{b_1, b_2, b_3 = -\infty}^{\infty} ((b_3^3 - b_3)/6) I_{1,2}(b_1, b_2, b_3)$$

in the case of integer coefficients) for the generators from subsection 8.4.1; all three possible momenta of the form $M_{i,j}(1, 1, 1)$ *for the generators from subsection 8.4.2, and* $i^2 = (M_{1,2}(0, 0, 1))^2$ *for the generator from subsection 8.4.3.*

PROBLEM. By Theorem 5, all other invariants $M_{i,j}(\beta_1, \beta_2, \beta_3)$ with $\beta_1 + \beta_2 + \beta_3 \le 3$ are linear combinations of these indicated in Theorem 11. What are exact expressions for them?

§9. Generalized Fenn-Taylor and index-type invariants and Brunnean ornaments (A. B. Merkov)

In this section we construct a family of invariants of ornaments in $\mathbb{R}^2$ or S^2, which is a refinement of the invariant from subsection 1.2. These new invariants are powerful sufficiently to solve the problem of triviality of 3-ornaments in S^2.

Also, we present an analogue of the Brunnean links: a series of the k-ornaments for any $k \ge 4$, which are (conjecturally) nontrivial, but all proper subornaments of them are trivial.

In the concluding subsection 9.10 we describe a refinement of the invariant from subsection 1.4.

9.0. The idea of the invariants. The invariants proposed below describe the internal topology of the images of regular ornaments, modulo the equivalence induced by the Reidemeister moves. They use the orientability of the containing two-dimensional manifold ($\mathbb{R}^2$ or other) and totally ignore its other properties. (The analogues for nonorientable case can also be defined.) Since the images themselves are one-dimensional, their topology can be easily described.

Also a more powerful invariant is described, see subsection 9.9.

9.1. Description of the preliminary invariant. We describe first a rough invariant, which looks only at one component of the ornament, and ignores its self-intersections.

Let us suppose that the circles of the ornament are colored, and one of them (say, black) is distinguished. The corresponding "black" invariant can be represented as a diagram, consisting of a standard black plane circle and several little radial arrows, applied at different points of the circle, directed inside or outside of it, coloured by the other colours of the ornament and supplied with some natural indices; the indices run independently for each color. See Figures 55, 56.

230   VI. INVARIANTS OF ORNAMENTS

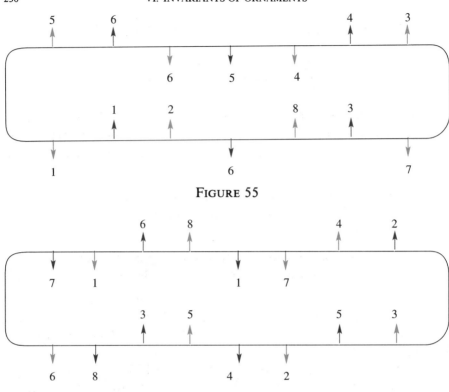

FIGURE 55

FIGURE 56

The promised invariant takes values in such diagrams, considered up to a natural equivalence relation described in subsection 9.3. Algebraically this diagram can be described by a word in the characters $a_{\alpha,i}$, $a_{\alpha,i}^{-1}$, where the first index runs over all notations of the non-black colors, and the second one is an arbitrary natural; the word is considered up to cyclic permutation of the characters.

9.2. How to assign the diagram to an ornament. Suppose we have a *regular* ornament φ in $\mathbb{R}^2$ or S^2, and let one of the components in C_k (the black one) be distinguished; we call it *principal* below. The corresponding diagram contains a plane circle identified with the principal component of C_k by a diffeomorphism which induces the standard orientation of this component from the counterclockwise orientation of the circle. The arrows on the circle correspond to the intersection points of (the image of) the principal component with other components of the ornament; the colors of the arrows are defined in an obvious way, and the directions are chosen to provide the same orientations as the ones of tangent vectors in the corresponding intersection points of the ornament. Further, for any nonprincipal component, let us fix an arbitrary point (called as its *basepoint*), which does not coincide with any of the intersection points. Then the numerical index of the arrow of color α equals the ordinal number of the corresponding intersection point among the intersections of αth component with the principal one, counting along the αth component from its basepoint. See Figure 55, 56, which present the invariants assigned to the standard components of Figures 44 and 45.

9.3. The equivalence relation. Any of the diagrams described in subsection 9.1 is a subject of the following possible reconstructions:

0. An orientation-preserving homeomorphism of the circle. The insensibility to this reconstruction has already been used in the algebraic description in subsection 9.1.

1. A cyclic permutation of all numerical indices of the arrows of the same color.

2. Any neighboring two arrows of the same color can permute on the circle, preserving their indices and directions.

3. Two arrows of the same color with neighboring indices $i-1$ and i and opposite directions can annihilate if there are no other arrows between them; in this case the number of each arrow of this color greater than i is decreased by 2.

3^{-1}. A reconstruction opposite to a previous one.

The desired equivalence relation is the transitive closure of the relation "can be reconstructed to" (and identity relation).

In algebraic terms (see the end of 9.1) these reconstructions can be described as follows:

0. Any cyclic permutation of the word (see subsection 9.1).

1. A cyclic permutation of all second indices with the same first index.

2. A permutation of two neighboring characters with the same first index.

3. Erasing the term $a_{\alpha,i-1}a_{\alpha,i}^{-1}$, $a_{\alpha,i-1}^{-1}a_{\alpha,i}$, $a_{\alpha,i}^{-1}a_{\alpha,i-1}$, or $a_{\alpha,i}a_{\alpha,i-1}^{-1}$, and renaming $a_{\alpha,j}$ to $a_{\alpha,j-2}$ when $j > i$.

3^{-1}. An operation opposite to a previous one.

9.4. Why they are invariants.

THEOREM 12. *If two k-ornaments in $\mathbb{R}^2$ or S^2 are equivalent, then for any choice of the principal color the corresponding diagrams (or words) of subsecton 9.1 are equivalent in the sense of subsection 9.3.*

This follows immediately from the definitions.

Thus, the equivalence classes of the diagrams are the invariants of ornaments.

REMARK. The invariants described in subsection 1.2 can be obtained from ours by forgetting the indices of the arrows (or, equivalently, the second indices of the characters).

EXAMPLE. The 3-ornament shown on Figure 45 cannot be distinguished from the trivial one by the invariants from subsection 1.2, but can be distinguished by the invariants described in subsections 9.1–9.3.

9.5. Algorithmical solvability.

PROPOSITION 18. *The problem of distinguishing our invariants up to described equivalence relation is (trivially) algorithmically solvable.*

Let us call the number of arrows in the diagram as its *length*.

PROPOSITION 19 (maximum principle). *For each sequence of reconstructions of diagrams there exists such a sequence of reconstructions with the same ends, that the length of each intermediate diagram not exceed the maximum of the lengths of the end diagrams.*

The proof is elementary. □

The maximum principle shows that minimal representatives of each equivalence class (that means diagrams with minimal length) are equivalent by the permutations of types 0–2 only, so the minimalness of the diagram can be easily recognized. This yields a convenient way to describe the class and a row of numerical invariants: the number of arrows of each color in a minimal representative, the number of arrays of adjacent arrows of each color, and so on.

9.6. On triviality of ornaments and doodles in S^2.

THEOREM 13. *Given a 3-ornament in S^2, if for at least one of the colors the corresponding component has no self-intersections and the invariant is trivial (i.e., the diagram is equivalent to the one without arrows, or the word is equivalent to the empty word) then the ornament is equivalent to the trivial one. The same is true for 3-doodles.*

The proof is elementary. The main idea is that there are no obstacles to continue each reconstruction of the diagram to a homotopy of the ornament or doodle itself, identical on the principal component, because the two non-principal circles are free to move in any way within the complement to the principal one, and each component of this complement is simple-connected, and the 3-ornament in $\mathbb{R}^2$ or S^2 is trivial, provided at least one of its diagrams has no arrows.

A similar assertion about the ornaments in $\mathbb{R}^2$ seems to be false. Indeed, the ornament shown on Figure 57 is equivalent to the trivial one in S^2 (and hence, all our invariants vanish), however it seems to be nontrivial in $\mathbb{R}^2$.

REMARK (M. Z. Shapiro). A 6-ornament in S^2, which is a union of two nonintersecting ornaments from Figure 57 (colored by 3 different colors) is nontrivial in S^2, although all its invariants described above vanish.

9.7. Brunnean ornaments.

Here we present a (conjectural) analogue of Brunnean links: for any $k \geq 4$ we draw a k-ornament (even k-doodle), any proper subornament of which is equivalent to trivial, but the ornament itself seems to be nontrivial.

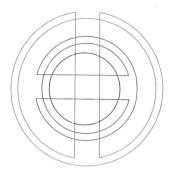

FIGURE 57

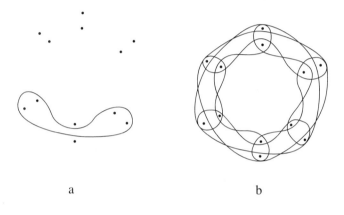

a   b

FIGURE 58a   FIGURE 58b

To do this we put two close points at any vertex of a regular k-gon and for each of these vertices draw a thin circle, which goes in between these two points and embraces both neighboring pairs of points (see Figure 58a). The whole k-ornament thus obtained (see Figure 58b) is a generalization of that from Figure 45.

The fact that any of its proper subornaments is trivializable can be proved immediately. The nontriviality of the ornament itself is not yet proved.

9.8. "Self-intersecting" invariants. Let us denote the invariant introduced in subsections 9.1–9.3 by $T_{\alpha.(*-\alpha)}$, where α denotes the principal component, "." is a delimiter like comma, possessing also some properties of the multiplication sign, and $*$ stands for the full set of components. The notation means that we go along the αth component and mark its intersections with all components but αth. We can easily define weaker invariants, replacing $*-\alpha$ to its proper subset, which ignore arrows of some colors on the diagram. For instance, all the invariants $T_{\alpha.\beta}$ are trivial for all ornaments.

Now we define a more strong invariant $T_{\alpha.*}$, which is similar to $T_{\alpha.(*-\alpha)}$, but takes into account self-intersections of the principal component also. We choose basepoints on each component (including the principal one) and draw the same diagram as described in subsection 9.2 with the arrows of the principal color added by the same rules as used for other colors.

We say that if the index of ith (counting from the current basepoint) arrow of the principal color is j, then jth arrow of this color is *adjoined* to the ith one. The diagram is said to be *coherent* if for each arrow of the principal color, the arrow that is adjoined to all adjoineds, is the arrow itself and the directions of these arrows are opposite. It is easy to see that the diagram of each ornament is coherent. All the reconstructions below respect coherence.

The reconstructions 0–3 described in subsection 9.3 must be modified and completed, so that the new list of them looks as follows:

0. A homeomorphism of the circle, preserving its orientation and basepoint.

1. Change of the basepoint of any circle, which induces the corresponding cycle permutation of all numerical indices of the arrows of its color.

2. Permutation of two neighboring arrows of the same nonprincipal color, preserving their indices and directions.

$2'$. Simultaneous permutation of three pairs of neighboring arrows of the principal color under the following conditions (see the upper right move on Figure 41). All the six arrows must form three different pairs of adjoined arrows, none of them coincides with one of the permuting pairs. When permuting, each of the six arrows preserves its direction and changes the index to that of the arrow which is adjoined to the arrow, which permutes with the arrow, which is adjoined to ours.

3. Annihilation of two neighboring arrows of the same nonprincipal color with neighboring indices $i-1$ and i and opposite directions; in this case the number of each arrow of this color which is greater than i is decreased by 2.

$3'$. (See the left picture in Figure 41.) Annihilation of a pair of adjoined neighboring arrows of the principal color with neighboring indices $i-1$ and i and opposite directions; the number of each arrow of the principal color, which is greater than i, is also decreased by 2.

$3''$. (See the middle upper picture of Figure 41.) Annihilation of a pair of neighboring arrows of the principal color with neighboring indices and opposite directions simultaneously with the pair of their adjoined arrows; the numbers of the remaining arrows of the principal color are decreased by 0, 2 or 4 to fill the gap.

3^{-1}, $3'^{-1}$, $3''^{-1}$. Reconstructions opposite to previous ones.

4. (See the right lower picture in Figure 41). Let the indices of the ith and

jth arrows of the principal color (say, black) are adjoined. Let there be two arrows of another color (say, green) with indices $l-1$ and l, and each of the green arrows is an immediate neighbor of one of the black ones. Let either the directions of the green arrows be the same and both of them precede or follow the corresponding black arrows, or vice versa, the directions being opposite, one of the green arrows precedes its black neighbor, and the other follows. Then the green arrows can simultaneously permute with the black neighbors and exchange the indices with each other.

It follows immediately from Theorem 1, that the equivalence class of diagrams, considered up to the reconstructions 0–4, is an invariant of ornaments. It is denoted by $T_{\alpha,*}$; recall that it takes into account the intersections of the αth component with all components.

Now we can eliminate extra restrictions in Theorem 13.

THEOREM 14. *If at least one of the invariants $T_{\alpha,*}$ is trivial for the 3-ornament in S^2, then the ornament is equivalent to the trivial one.*

The proof is similar to that of Theorem 13. □

This "self-intersecting" invariant is more powerful than the one introduced in subsections 9.1–9.3, but much less convenient.

CONJECTURE. The same maximum principle as the one of subsection 9.5 holds for $T_{\alpha,*}$.

If the conjecture is not true, then the invariant can hardly be useful for manual recognition of triviality of ornaments.

9.9. "Multiple" invariants. The invariants defined above deal with one diagram (see subsection 9.1) at a time. Here we describe the invariants which watch over several diagrams at time. To reach more simplicity, only the strongest of them, dealing with full diagrams (arrows of no color are ignored within each) are (partially) described.

Again, as in subsections 9.1–9.4, we define some diagram for each k-ornament, then define permitted reconstructions, the equivalency generated by them, and show that Reidemeister moves of the ornament leave its diagram within the same equivalence class.

Let P be a subset of the set of components of C_k with $p > 1$ elements; the components of P (or their colors) will be called *chosen*. The diagram is a set of diagrams of subsection 9.8 (which are improved diagrams of 9.1) for chosen components. That means the diagram consists of p circles with arrows.

The diagram is *coherent* if for each pair of chosen colors (say, green and red) the index of ith green arrow on the red (or green) circle is j, then there exist jth red (or green, respectively) arrow on the green circle, and its index is i, and the directions of these arrows are opposite; these arrows are called to be *adjoined*. It is easy to see, that the diagram of each ornament is coherent.

The reconstructions are similar to reconstructions 0–4 above, but they preserve the circles coherent. They correspond exactly to the allowed Reidemeister moves, so each of them deals with not more than two colors. If only one of these colors is chosen, the rules are exactly as the ones stated in subsection 9.8. If both the colors are chosen, the conditions of the reconstruction must be held on the both circles of the diagram, and it is performed on the both circles simultaneously.

Because the idea of the reconstructions is clear (e.g., if two green arrows annihilate on the red circle, then their adjoined red arrows must annihilate on the green circle simultaneously), but the accurate description is rather long and dull, the latter is omitted.

The results yielded by new "multiple" invariants are fairly modest:

CONJECTURE. *If two "opposite" components of the Brunnean 4-ornament described in subsection 9.7 are chosen, then its invariant just described is nontrivial.*

REMARK. Shapiro's remark from 9.6 holds for all of the described invariants, so even the strongest of them (which takes all the components into account) cannot solve the problem of triviality for 6-ornaments in S^2.

9.10. Bi-index-type invariants. Here we introduce one more invariant of ornaments. As an illustration, we get the following result.

PROPOSITION 20. *The ornament of Figure 59 is nontrivial.*

(Note that all invariants from §1 take trivial values on this ornament.)

DEFINITION. Let $c: S^1 \to \mathbb{R}^2$ be an immersion of an oriented circle, $x_1, x_2 \in S^1$, $x_1 \neq x_2$, $c(x_1) = c(x_2) = x$ a point of its self-intersection, and $t \in \mathbb{R}^2 \setminus c(S^1)$. Then the points x_1 and x_2 split the circle S^1 into two arcs A_1 and A_2, and since $c(x_1) = c(x_2)$ the indices $\operatorname{ind}_{c_1}(t)$ and $\operatorname{ind}_{c_2}(t)$ can be defined in an usual way for the restrictions $c_1 = c|_{A_1}$ and $c_2 = c|_{A_2}$. The *unordered* pair $(\operatorname{ind}_{c_1}(t), \operatorname{ind}_{c_2}(t))$ is called the bi-index of the point t with respect to the self-intersection x of the circle c and denoted by $\operatorname{bi-ind}_{c;(x_1,x_2)}(t)$, or, if no ambiguity arises, by $\operatorname{bi-ind}_x(t)$.

EXAMPLE. Let u and v be the intersection points of the curves C_1 and C_2 from Figure 59, let x and y be the self-intersections of the curve C_3, and let all the curves be oriented counterclockwise. Then $\operatorname{bi-ind}_x(u) = (1, 1)$ and $\operatorname{bi-ind}_x(v) = (0, 2)$ (or $(2, 0)$, since the pairs are not ordered).

As we show below, the inequality $\operatorname{bi-ind}_x(u) \neq \operatorname{bi-ind}_x(v)$ implies the nontriviality of this ornament.

PROPOSITION 21. *If* $\operatorname{bi-ind}_{c;(x_1,x_2)}(t) = (p, q)$, *then* $\operatorname{ind}_c(t) = p + q$.

PROOF. Obvious. □

Let us choose two components C_i and C_j of a regular k-ornament and a self-intersection point x on a third component C_l. This self-intersection

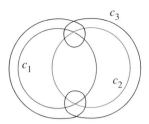

FIGURE 59

splits C_l into two arcs. The restrictions of the map $C_l \to \mathbb{R}^2$ on the arcs are equivalent to a map of some circles $C_{l;1}$ and $C_{l;2}$ with at most one nonsmooth point (mapped to the self-intersection point) on each circle.

Let $I_{i,j,l;x;\alpha} \equiv I_{i,j,l;x;\alpha}(b_1, \ldots, b_k)$, where $\alpha \in \{1, 2\}$, denotes the index-type invariants $I_{i,j}$ of the ornaments with C_l replaced by $C_{l;\alpha}$. We define
$$B_{i,j,l;x}(b_1, \ldots, b_{l-1}, (b_{l-1}, b_{l;2}), b_{l+1}, \ldots, b_k)$$
as the unordered pair
$$(I_{i,j,l;x;1}(b_1, \ldots, b_{l-1}, b_{l;1}, b_{l+1}, \ldots, b_k),$$
$$I_{i,j,l;x;2}(b_1, \ldots, b_{l-1}, b_{l;2}, b_{l+1}, \ldots, b_k))$$
of them and $B_{i,j,l}$ as the collection of $B_{i,j,l;x}$ for all self-intersections x of C_l. We call $B_{i,j,l}$ the *bi-index* of the pair of curves C_i and C_j up to C_l.

EXAMPLE. For the ornament from Figure 59,
$$B_{1,2,3;x}(0, 0, (0, 2)) = -1, \quad B_{1,2,3;x}(0, 0, (1, 1)) = 1,$$
$$B_{1,2,3;y}(0, 0, (0, 2)) = 1, \quad B_{1,2,3;y}(0, 0, (1, 1)) = -1,$$
and $B_{1,2,3} = 0$ for other arguments.

Two bi-indices, that are two finite collections of integer-valued functions of $k+1$ integer variables, are said to be *equivalent* if they can be transformed one to another by a chain of the following transformations:
1. Insert/remove the identically zero function.
2. Insert/remove a pair of equal functions.

THEOREM 15. *The bi-indices $B_{i,j,l}$ of equivalent k-ornaments are equivalent for each i, j, and l.*

PROOF. It is easy to see that each of Reidemeister moves either preserves $B_{i,j,l}$ or performs one of the transformations mentioned above. □

PROPOSITION 22. *Each of the classes of equivalent bi-indices has the only minimal representative (the collection with minimal number of element).*

PROOF. This is the representative with no zero or pairwise equal functions. □

As a corollary, we get a proof of Proposition 20. Indeed, the bi-index $B_{1,2,3}$ of the ornament from Figure 59, calculated in the above example, is minimal and not empty.

Though the bi-index type invariants described above are nonnumerical, a row of numerical invariants can be obtained from them. For instance, if $\#B_{i,j,l}$ denotes the number of functions in the minimal equivalent representative, then

$$\sum_l \max_{i<j} \#B_{i,j,l}$$

is also an invariant of ornaments. It estimates from below the total number of selfintersection points of the regular ornament.

The problem of nontriviality of the ornament from Figure 57 remains open.

§10. Open problems and possible generalizations

PROBLEM 1. Is the system of finite-order invariants complete, i.e., does it distinguish any two nonequivalent ornaments?

In particular, do there exist invariants of this kind, which prove that the ornaments from Figures 44, 45, 59 are nontrivial? Of which order are the simplest such invariants?

(For the parallel problem of the homotopy classification of links, the answer is affirmative, see [BN$_3$], [Lin^2].)

What is the smallest order of finite-order invariants, which cannot be expressed as the functions of invariants $M_{i,j}$ introduced in subsection 1.4? My guess is that this will be related to one of the following situations: a) the homology group of some complex $\Xi(\Theta)$ is more than one-dimensional in the highest possible dimension (see Theorem 9); b) there exist several nonequivalent A-configurations with the same A and all the numbers of points of groups of cardinalities $a_1, \ldots, a_{\#A}$ on any component. The situation b) can be realized by the $(4,4)$-configurations such that any of two groups of 4 points, constituting this configuration, have two points on the first component of C_k, one point on the second, and one point on the third component. Indeed, these two pairs of points from different groups on the first component can either separate each other or not.

PROBLEM 2. A problem similar to the classification of ornaments can be stated as follows: we consider the space of all plane curves (or collections of curves) having no triple points and no singularities obtained as degenerations of triple points (i.e., either the points at which two local branches intersect, for one of which this point is a singular point with $\varphi' = 0$; or the points at which $\varphi' = \varphi'' = 0$.) The problem of classifying such objects is in the same relation to the above classification of ornaments, in which the isotopy classification of links is to the homotopy classification problem.

A spectral sequence, which is a hybrid of those considered in sections 4–6 above and in Chapter V, calculates the invariants of such objects. A partial problem is to study this spectral sequence explicitly.

PROBLEM 3. In both cases considered in problem 2 and in the main text, we can forbid not triple intersections but the (self-)intersections of arbitrary multiplicity l, $l > 3$. In this case, the space of permitted ornaments is connected (and even $(l-3)$-connected), in particular the problem of classifying such objects up to homotopy is void. But, the problem of calculating the higherdimensional cohomology of the spaces of permitted objects is nontrivial and can be, in principle, solved by the spectral sequence similar to the one from §4. In this case this spectral sequence (beginning with the term E_1) lies in the edge $\{p, q | p < 0, q + (l-2)p \geq 0\}$; in particular, on any line of the form $p + q = $ const there lies only a finite number of finitely generated groups, and the problem of the convergence of the spectral sequence does not occur.

PROBLEM 4. The classification of ornaments is a model case of a big class of problems introduced in [FNRS]. Namely, let $(W_1, \ldots, W_k)$ be an arbitrary collection of compact manifolds of arbitrary dimensions. The problem is to study the space of all maps of (the union of) these manifolds into $\mathbb{R}^n$ having no common points of their images (or, more generally, points where the images of some t different components W_i meet, $t \leq m$).

Again, if the dimensions of manifolds W_i are such that the space of permitted maps is dense in the space of all maps, then a spectral sequence similar to the above one can be constructed, which presents the cohomology classes of such spaces. Problem: to study these spaces (and these spectral sequences).

Another problem appears if we can vary not only the maps of manifolds W_i into R^n but also make standard cobordisms of these manifolds, cf. [FT]. This problem has obvious applications to the following well-known problem: to classify the connected components of the space of homogeneous polynomial vector fields of fixed degrees in $\mathbb{R}^n$, having no singular points outside the origin.

PROBLEM 5. Do there exist invariants of ornaments arising from the state physics in the same way as certain invariants of knots and links do?

PROBLEM 6. Does the spectral sequence from §4 degenerate at the term E_1 (at least in the case of rational coefficients, or on the main diagonal $\{p+q=0\}$? In [Kontsevich$_1$], a similar fact was established for the invariants of knots constructed in [V$_7$] (and described in Chapter V).

PROBLEM 7. Does there exist a realization of our invariants by means of integrals, as was done in [Kontsevich$_1$] for the (rational) invariants from Chapter V? A related question: to realize by differential forms the cohomology of complements of "k-equal" arrangements considered in [BW] and of their generalizations considered in the proof of Theorem 9, see §5.

PROBLEM 8. Brunnean ornaments. For any $k > 3$, to construct a k-ornament which is nontrivial, but all its $k - 1$ subornaments are trivial. A natural candidate is the ornament constructed by A. B. Merkov, see §9. The fact that any of its subornaments split is elementary, but the nontriviality of this ornament itself is not yet proved.

PROBLEM 9. To study the ornaments on any smooth surface. Note that the discriminant is invariantly cooriented even in the case of ornaments on nonorientable surfaces. Indeed, any local surgery as on Figure 42 can be realized as follows: we fix the first two local branches and move the third parallel to itself. Then, the sign of this surgery can be deduced from the comparison of two *local* orientations: the first is given by the tangent frame of first two components, and the second by the frame (tangent to the third component, direction of the translation of this component).

PROBLEM 10. To write down all linear relations between the invariants $M_{i,j}(\beta_1, \ldots, \beta_k)$ from subsection 1.4.

APPENDIX 1

Classifying Spaces and Universal Bundles. Join

0. For more detailed discussion of these subjects, see the textbooks [Husemoller], [RF], [Steenrod].

1. Principal bundles. Let G be a topological group (i.e. a group with a topology on it such that the group operations are continuous).

A *principal G-bundle* is a locally trivial bundle $E \to B$ with an effective action of the group G on the space E (i.e., for every $x \in E$ and $g \in G$, $g(x) = x \Leftrightarrow g = 1$), such that the orbits of this action coincide with the fibers of the bundle (and hence the base B can be viewed as the quotient space with respect to this action). By definition, the fibers of a principal G-bundle are homeomorphic to G.

EXAMPLE 1. Given an m-fold covering $p: X \to Y$, there is a principal $S(m)$-bundle over Y associated to it with the fiber over $y \in Y$ consisting of all possible orderings of the points in $p^{-1}(y)$.

EXAMPLE 2. Given an m-dimensional vector bundle $\zeta: A \to Y$, there is a principal $GL(m)$-bundle associated to it whose fiber over $y \in Y$ consists of all possible m-frames in the linear space $\zeta^{-1}(y)$. If the bundle ζ is Euclidean, or oriented, or both, then it is possible to associate to it a principal $O(m)$-bundle ($SL(m)$- or $SO(m)$-bundle, respectively) whose fibers consist of orthonormal (positively oriented, orthonormal positively oriented) frames.

This correspondence can obviously be generalized to other G-bundles; it is the main source of principal bundles.

2. Universal G-bundle. In the class of all principal G-bundles there is a special *universal G-bundle* to which all others can be reduced. Here we list its properties.

THEOREM 1. *For any topological group G that has the type of a CW-complex there exists a principal G-bundle of CW-complexes $EG \to BG$ such that*

(1) *its space EG is homotopically trivial,*

(2) *any principal G-bundle π over any base B' is equivalent to the bundle induced from the bundle $EG \to BG$ by a map of the bases $B' \to BG$. This map is uniquely (up to homotopy) determined by the initial bundle π and is*

denoted by $\mathrm{cl}(\pi)$. In particular, G-bundles over B' (*considered up to equivalence*) are in one-to-one correspondence with maps $B' \to BG$ (*considered up to homotopy*).

The bundle $EG \to BG$ is called the *universal G-bundle*, and BG the *classifying space* of the group G.

Conditions (1) and (2) of the theorem are equivalent; each of them alone can serve as a definition of the universal bundle.

COROLLARY. *The classifying space BG is uniquely determined up to homotopy equivalence by the group G.*

The most important examples of classifying spaces will be considered in detail in §§3 and 4; their general construction is given in §6 below.

The correspondence $G \to BG$ is functorial with respect to G in the following sense.

THEOREM 2. *Any continuous homomorphism of topological groups $G_1 \to G_2$ defines a (unique up to homotopy) continuous map $BG_1 \to BG_2$.*

3. The spaces $K(G, 1)$. Let G be a discrete group. Then the space $K(G, 1)$ is a path-connected CW-complex satisfying the conditions

$$\pi_1(K(G, 1)) = G, \qquad \pi_i(K(G, 1)) = 0 \quad \text{for } i \geq 2.$$

THEOREM 3. *For any discrete group G, the space $K(G, 1)$ exists, is defined uniquely up to homotopy equivalence, and coincides with its classifying space BG.*

The last assertion follows from the fact that the universal covering space over the space $K(G, 1)$ is homotopically trivial, as one can see from the exact homotopy sequence of the covering.

EXAMPLES. (1) $K(\mathbb{Z}, 1) = S^1$.
(2) $K(\mathbb{Z}_2, 1) = \mathbb{R}P^\infty$.
(3) Any two-dimensional compact surface except for S^2 and $\mathbb{R}P^2$ is a space $K(G, 1)$ for some group; indeed, its universal covering is contractible.
(4) The spaces $K(G, 1)$ for the braid groups and symmetric groups are considered in detail in Chapter I.

4. Universal $O(m)$- and $SO(m)$-bundles. The *Grassmann manifold* G_N^m, $N \geq m$, is defined as the set of all m-dimensional subspaces in $\mathbb{R}^N$ with the natural topology. The *tautological bundle* $T_N(m) \to G_N^m$ is a vector bundle whose space consists of pairs (point of the Grassmannian, point of the corresponding m-plane in $\mathbb{R}^N$).

The sequence of standard imbeddings $\mathbb{R}^N \to \mathbb{R}^{N+1} \to \mathbb{R}^{N+2} \to \cdots$ defines imbeddings of Grassmannians $G_N^m \hookrightarrow G_{N+1}^m \hookrightarrow G_{N+2}^m \hookrightarrow \cdots$; the union of these spaces equipped with the direct limit topology is called the *stable m-Grassmannian* and is denoted by G^m. The tautological bundles over Grassmannians G_*^m are compatible with these imbeddings and define a limit bundle $T(m) \to G^m$.

THEOREM 4. *The space G^m is the classifying space $BO(m)$ of the orthogonal group $O(m)$. The universal principal bundle over G^m is the bundle whose fiber over $\zeta \in G^m$ consists of various orthonormal frames in the fiber of the tautological bundle $T(m) \to G^m$.*

The *oriented Grassmann manifold* $\widetilde{G}_N^m$ is defined as the set of all oriented m-subspaces in $\mathbb{R}^N$; forgetting the orientations makes $\widetilde{G}_N^m$ a double covering over G_N^m. Similarly to the above, we can define the limit space $\widetilde{G}^m = \bigcup_{N=m}^{\infty} \widetilde{G}_N^m$ and the tautological oriented bundle $\widetilde{T}(m) \to \widetilde{G}^m$.

THEOREM $\widetilde{4}$. *The space $\widetilde{G}^m$ is the classifying space $BSO(m)$; the corresponding universal $SO(m)$-bundle over $\widetilde{G}^m$ is the bundle of positively oriented orthonormal m-frames associated to the tautological bundle $\widetilde{T}(m) \to \widetilde{G}^m$.*

THEOREM 5. *The ring $H^*(G^m, \mathbb{Z}_2)$ is a polynomial algebra freely generated by m classes $w_1, \ldots, w_m$, $\dim w_i = i$. The map $H^*(G^m, \mathbb{Z}_2) \to H^*(\widetilde{G}^m, \mathbb{Z}_2)$ induced by the double covering $\widetilde{G}^m \to G^m$ is an epimorphism whose kernel is the ideal generated by w_1.*

The classes $w_i \in H^i(G^m, \mathbb{Z}_2)$ are called *universal Stiefel-Whitney classes*.

Any m-dimensional vector bundle ζ over a CW-complex X canonically defines m classes $w_i(\zeta) \in H^i(X, \mathbb{Z}_2)$, $i = 1, 2, \ldots, m$. In fact, supply the bundle X with arbitrary smooth Euclidean structure and consider the principal $O(m)$-bundle over X associated to the bundle ζ (see Example 2 above). The classifying map of this bundle acts from X into $BO(m) = G^m$; the classes $w_i(\zeta)$ are the classes induced by this map from the universal classes $w_i \in H^i(G^m, \mathbb{Z}_2)$. They are called the *Stiefel-Whitney classes* of the bundle ζ. A vector bundle ζ is orientable if and only if $w_1(\zeta) = 0$ or, equivalently, the classifying map $X \to G^m$ lifts to a map into $\widetilde{G}^m$.

5. Join. The join is an associative and commutative binary operation on topological spaces. First we define the join of two finite simplicial polyhedra X, Y. Imbed X and Y in a linear space $\mathbb{R}^N$ of sufficiently large dimension so that for no pairs of points $x_1, x_2 \in X$, $y_1, y_2 \in Y$ do the segments $[x_1, y_1]$ and $[x_2, y_2]$ intersect in their interior points. Then the *join* $X * Y$ is defined as the union of all segments $[x, y]$ over all pairs $x \in X$, $y \in Y$ equipped with the topology induced from $\mathbb{R}^N$.

For arbitrary topological spaces X, Y the join $X * Y$ is defined as the quotient space of the product $X \times Y \times [0, 1]$ with respect to the following equivalence relation. For each point $x \in X$ we identify all points of the set $x \times Y \times \{0\} \sim Y$, and for each point $y \in Y$ we identify all points of the set

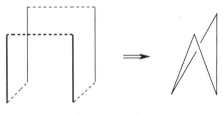

FIGURE 60

$X \times y \times \{1\} \sim X$. The resulting quotient space with the standard quotient topology is $X * Y$.

EXAMPLE 1. Let X, Y be two pairs of points. Then $X * Y \sim S^1$; see Figure 60.

EXAMPLE 2. Let X be a pair of points. Then for any space Y, $X * Y$ is homeomorphic to the (unreduced) suspension ΣY (i.e. the quotient space of the product $Y \times [0, 1]$ with respect to the equivalence that contracts its bases $Y \times 0$ and $Y \times 1$ to a pair of points).

EXAMPLE 3. The join of any two spheres $S^n * S^m$ is homotopy equivalent to S^{n+m+1}.

Sometimes it is more convenient to use the following (equivalent to the first one) definition of join (see [Schwarz]). Consider the cones CX (i.e. the quotient space $(X \times [0, 1])/(X \times \{0\})$) and $CY = (Y \times [0, 1])/(Y \times \{0\})$. Let $\overset{\circ}{C}X$, $\overset{\circ}{C}Y$ be the subsets obtained from these cones by removing their bases $X \times \{1\}$, $Y \times \{1\}$. Then the join $X * Y$ is defined as the difference $(CX \times CY) \setminus (\overset{\circ}{C}X \times \overset{\circ}{C}Y)$.

The homology of the join $X * Y$ can be easily expressed in terms of the homology of X and Y. Namely, denoting by $\widetilde{H}_*$ the homology reduced modulo a point we have

$$\widetilde{H}_i(X * Y, F) \cong \bigoplus_{p+q=i-1} \widetilde{H}_p(X, F) \otimes \widetilde{H}_q(Y, F), \tag{1}$$

for any field F.

6. The Milnor construction of the universal G-bundle. Suppose the topological group G is a CW-complex. Consider the sequence of spaces G, $G * G$, $G * G * G$, etc. Since any space X is canonically imbedded in $X * Y$, the union of spaces $G^{*k} = G * \cdots * G$ (k times) over all k is well defined. Equip it with the direct limit topology. The resulting topological space is contractible. (This follows from the fact that each space X^{*k} is $(k-2)$-connected; moreover, if the initial space X is i-connected, $i \geq 0$, then X^{*k} is $((i+2)k - 2)$-connected.) The group G acts effectively on this homotopically trivial space $G^{*\infty}$ in the obvious way, and hence its fibration by the orbits of this action is the universal principal G-bundle.

APPENDIX 2

Hopf Algebras and H-Spaces

The notion of the Hopf algebra is the formalization of a natural algebraic structure on homology and cohomology of topological groups (and many other objects).

Let G be a topological group, i.e. a topological space that is also a group with both group operations being continuous; let A be a field, and suppose that all groups $H^i(G, A)$ are finitely generated. Then, in addition to the usual A-algebra structure on the ring $h^* = H^*(G, A)$, there is an A-coalgebra structure, that is, a homomorphism $\delta \colon h^* \to h^* \otimes h^*$ ("comultiplication"): it is induced by the multiplication operation $G \times G \to G$. This homomorphism is associative, i.e. two maps $h^* \to h^* \otimes h^* \otimes h^*$ defined by the compositions $(\delta \otimes 1) \circ \delta$ and $(1 \otimes \delta) \circ \delta$ coincide. Some obvious properties of the resulting structure on $H^*(G, A)$ are formalized in the following definition.

Let A be an associative and commutative ring with identity.

DEFINITION. A *Hopf algebra* over A is a graded associative A-algebra h^* with identity $i \colon A \hookrightarrow h^*$, $i(A) \subset h^0$, which is also an associative graded coalgebra (with respect to the same grading) with coidentity $\varepsilon \colon h^* \to A$, $h^{>0} \subset \operatorname{Ker} \varepsilon$ such that:

(1) i is a homomorphism of graded algebras;
(2) ε is a homomorphism of graded algebras;
(3) the comultiplication operation is a homomorphism of graded algebras (or, equivalently, the multiplication is a homomorphism of graded coalgebras).

In the model case when $h^* = H^*(G, A)$, the identity associates to each element $a \in A$ a 0-cocycle in G that takes the value a on any connected component of G; the coidentity is the restriction homomorphism $H^*(G, A) \to H^*(e, A) \cong A$, where e is the identity element in G.

For every field A the group $H_*(G, A)$ has a dual Hopf algebra structure: multiplication is induced by multiplication in G (to a pair of cycles $a, b \subset G$ corresponds the image of the cycle $a \times b \subset G \times G$ under the map $G \times G \to G$), and comultiplication is the dual of multiplication in the algebra $H^*(G, A) = [H_*(G, A)]^*$.

In general, to any Hopf algebra structure over a field corresponds a dual

Hopf algebra structure on the dual space.

Instead of the topological groups above, we could consider arbitrary H-spaces, defined as follows.

DEFINITION. A topological space X with a base point x_0 is an *H-space* if continuous maps $m\colon X \times X \to X$ ("multiplication") and $\mathrm{inv}\colon X \to X$ ("taking the inverse") are defined such that

(1) both compositions $X \xrightarrow{i_1} X \times X \xrightarrow{m} X$ and $X \xrightarrow{i_2} X \times X \xrightarrow{m} X$, where $i_1(x) = (x, x_0)$, $i_2(x) = (x_0, x)$, are homotopic to the identity map $X \to X$;

(2) the compositions $X \times (X \times X) \xrightarrow{\mathrm{Id} \times m} X \times X \xrightarrow{m} X$ and $(X \times X) \times X \xrightarrow{m \times \mathrm{Id}} X \times X \xrightarrow{m} X$ are homotopic to each other;

(3) both compositions $X \xrightarrow{\mathrm{inv} \times \mathrm{Id}} X \times X \xrightarrow{m} X$ and $X \xrightarrow{\mathrm{Id} \times \mathrm{inv}} X \times X \xrightarrow{m} X$ are homotopic to a constant map.

If X is a topological group, and x_0 is its identity element, then all these conditions are satisfied precisely (not up to homotopy). An important example of H-spaces that are not topological groups is given by loop spaces; see the next appendix.

THEOREM. *For any H-space X and any field A both modules $H_*(X, A)$ and $H^*(X, A)$ are (dual to each other) Hopf algebras.*

This theorem provides very strong topological obstructions to introducing an H-space structure on a given topological space; see [Borel], [MM].

APPENDIX 3

Loop Spaces

DEFINITION. The *loop space* of a topological space X with base point x_0 is the set of continuous maps $[0, 1] \to X$ sending the endpoints 0 and 1 to x_0, with the standard (compact-open) topology for the space of maps.

The *space of free loops* of X is the set of continuous maps $[0, 1] \to X$ sending the points 0 and 1 to the same point (not necessarily the base point).

These spaces are denoted by ΩX and $\Omega_f X$, respectively.

The homotopy groups of these spaces are intimately related:

PROPOSITION 1. *For each $i \geq 1$ and any path-connected space X there is a short exact sequence*

$$0 \to \pi_i(\Omega X) \to \pi_i(\Omega_f X) \to \pi_i(X) \to 0. \tag{1}$$

Indeed, the space $\Omega_f X$ is naturally fibered over X: to a free loop $[0, 1] \to X$ corresponds the image of the points 0 and 1. In the corresponding homotopy exact sequence all maps $\pi_i(\Omega_f X) \to \pi_i(X)$ are epimorphisms, since this fibration has an obvious section (to a point $x \in X$ corresponds the loop $[0, 1] \to x$). □

Usually the loop sending the interval $[0, 1]$ to the base point in X is declared to be the base point in ΩX.

The definition immediately implies that the set of connected components of the space ΩX is the group $\pi_1(X)$; it is easy to see that the set of connected components of $\Omega_f X$ is the group $H_1(X)$.

Moreover, for any topological space Y with base point y_0 there is a natural one-to-one correspondence

$$[Y, \Omega X] \xrightarrow{\sim} [\Sigma Y, X], \tag{2}$$

where $[A, B]$ is the set of homotopy classes of maps $A \to B$ sending the base point to the base point, and ΣY denotes the reduced suspension over Y, i.e. the space obtained from the product $Y \times [0, 1]$ by factoring by the union of the subsets $Y \times 0$, $Y \times 1$, and $y_0 \times [0, 1]$.

For example, if $Y = S^i$ then ΣY is homotopy equivalent to S^{i+1}, and (2) becomes the identity

$$\pi_i(\Omega X) \cong \pi_{i+1}(X). \tag{3}$$

This identity is given by the following construction. We consider an i-dimensional spheroid in B as a map $\mathbb{R}^i \to B$ sending the complement to some ball to the base point, and a loop in X as a map $\mathbb{R}^1 \to X$ sending the complement to the interval $[0, 1]$ to the base point. In particular, an i-spheroid in the space ΩX is an i-parameter family of functions $f_\lambda \colon \mathbb{R}^1 \to X$, $\lambda \in \mathbb{R}^i$, such that $f_\lambda(t) = x_0$ for sufficiently large $|t| + |\lambda|$. Define a map $F \colon \mathbb{R}^{i+1} \to X$ by the formula $F(t, \lambda) = f_\lambda(t)$. This map is an $(i + 1)$-spheroid, and correspondence (3) is constructed. Correspondence (2) is a trivial generalization of this construction.

PROPOSITION. *The space of k-fold loops $\Omega^k(X) \equiv \Omega(\Omega(\cdots \Omega(X) \cdots))$ is homeomorphic to the space of continuous maps $S^k \to X$ sending the base point to the base point.* □

For any path-connected pointed space X and any positive integer i there is a natural *Freudenthal imbedding*

$$\Omega^i X \to \Omega^{i+1} \Sigma X. \tag{4}$$

Moreover, if Y and X are path-connected spaces with base points then the space $(X, x_0)^{(Y, y_0)}$ of continuous maps $(Y, y_0) \to (X, x_0)$ is naturally imbedded in $(\Sigma X, *)^{(\Sigma Y, *)}$. In fact, to each map $\varphi \colon (Y, y_0) \to (X, x_0)$ corresponds a map $\varphi \times 1 \colon (Y \times [0, 1]) \to (X \times [0, 1])$ that sends the point $y \times t$ to $\varphi(y) \times t$. This map extends to a map of the quotient space ΣY of the space $Y \times [0, 1]$ to the quotient space ΣX of the space $X \times [0, 1]$.

In particular, the Freudenthal construction defines a homomorphism

$$\pi_i(X) \to \pi_{i+1}(\Sigma X). \tag{5}$$

FREUDENTHAL THEOREM (see, for example, [Ad], [Fuchs$_1$]). *If X is an $(n - 1)$-connected CW-complex (for example, the sphere S^n) then map (5) is an isomorphism for $i \leq 2n - 2$ and an epimorphism for $i = 2n - 1$.*

Every loop space is an H-space (see Appendix 2): to an ordered pair of loops we associate their composition, and the inverse is the loop traversed in the opposite direction. This structure defines a multiplicative structure on the homology group of the loop space called the *Pontryagin multiplication*.

PROPOSITION. *The group $H_*(\Omega S^k)$ equipped with the Pontryagin multiplication is isomorphic to the polynomial algebra over $\mathbb{Z}$ in one $(n - 1)$-dimensional generator.*

If the space X is itself a loop space (or is homotopy equivalent to a loop space) then the H-space ΩX is homotopy commutative (two maps $\Omega X \times \Omega X \to \Omega X$ sending a pair of loops ω_1, ω_2 to their compositions $\omega_1 \omega_2$ and $\omega_2 \omega_1$ are homotopic).

APPENDIX 4

Germs, Jets, and Transversality Theorems

0. The material of this appendix is taken from [GG].

1. Germs.
Let M, N be smooth manifolds.

DEFINITION. Two maps φ, $\psi \colon M \to N$ belong to the same germ at a point $x \in M$ if they coincide in some neighborhood of this point.

It is obvious that belonging to one germ is an equivalence relation; the set of equivalence classes is called the *space of germs* of maps $M \to N$ at the point x. The class containing the map φ is called the germ of this map.

If N is a ring (for example, $N = \mathbb{R}$), then the space of germs of maps $M \to N$ at any point of M inherits an obvious ring structure.

2. Jets.

2.1. DEFINITION. Two smooth maps φ, $\psi \colon M \to N$ belong to the same k-jet at a point x if for some (and therefore for any) choice of local coordinates near the points x, $\varphi(x)$ the Taylor polynomials of order k of these maps coincide. The *space of k-jets* of maps $M \to N$ at a point x is the quotient of the set of all maps by this equivalence relation. This space is denoted by $J^k|_x(M, N)$.

For $N = \mathbb{R}$ the space $J^k|_x(M, N)$ also has a ring structure: it is the quotient of the ring of all germs of maps at the point x by the ideal consisting of germs with a zero of order $k + 1$ or higher at x.

2.2. DEFINITION. The *space of k-jets of maps* $M \to N$ is the set of pairs (point of M, k-jet of a map $M \to N$ at this point). This space is denoted by $J^k(M, N)$.

Obviously,

$$J^0(M, N) = M \times N, \tag{1}$$

and there are sequences of obvious projections

$$\cdots \to J^k|_x(M, N) \to J^{k-1}|_x(M, N) \to \cdots \to J^0|_x(M, N) \equiv N, \tag{2}$$

$$\cdots \to J^k(M, N) \to J^{k-1}(M, N) \to \cdots \to J^0(M, N) \begin{smallmatrix} \nearrow M \\ \searrow N \end{smallmatrix}. \tag{3}$$

The space $J^k(M, N)$ has a natural structure of a smooth bundle over M (vector bundle if N is a linear space) with fiber $J^k|_x(M, N)$.

2.3. DEFINITION. The *space of jets* of maps $(M, x) \to N$ (respectively, maps $M \to N$) is the inverse limit of the spaces $J^k|_x(M, N)$ (spaces $J^k(M, N)$) with respect to projections (2) (respectively, (3)). These limits are denoted by $J|_x(M, N)$ and $J(M, N)$.

Fixing coordinates determines a correspondence between $J|_x(M^m, N^n)$ and the space of formal power series of maps $(\mathbb{R}^m, 0) \to \mathbb{R}^n$.

3. Jet extensions. To each smooth map $\varphi: M \to N$ corresponds its k-jet extension $j^k\varphi: M \to J^k(M, N)$, i.e. a section of the obvious bundle $J^k(M, N) \to M$; this section assigns to each point of M the k-jet of the map φ at this point.

4. Transversality.

4.1. DEFINITION. Two smooth submanifolds L_1, L_2 of a manifold N are *transversal at* $a \in N$ if one of the following two conditions is satisfied: 1) $a \notin L_1 \cap L_2$; 2) the linear hull of the tangent spaces T_aL_1, T_aL_2 coincides with the space T_aN. The submanifolds L_1, L_2 are *transversal* if they are transversal at every point of N.

For example, if $\dim L_1 + \dim L_2 < \dim N$ then transversality means that L_1 and L_2 are disjoint.

4.2. *A more general definition.* Let L_1, L_2 be two manifolds, $\varphi_1: L_1 \to N$ and $\varphi_2: L_2 \to N$ be smooth maps.

DEFINITION. The maps φ_1, φ_2 are transversal at a pair of points $a_1 \in L_1$, $a_2 \in L_2$ if either $\varphi_1(a_1) \neq \varphi_2(a_2)$ or $\varphi_{1*}T_{a_1}L_1 + \varphi_{2*}T_{a_2}L_2 = T_{\varphi_1(a_1)}N$. The maps φ_1 and φ_2 are *transversal* if they are transversal everywhere in $L_1 \times L_2$.

Definition 4.1 can be reformulated as follows: the submanifolds L_1, L_2 are transversal in L if their identical inclusions into L are transversal.

4.3. Here is an intermediate version of the transversality condition: we say that a map $\varphi: L_1 \to N$ is transversal to a submanifold $L_2 \subset N$ if φ and the identity inclusion $L_2 \hookrightarrow N$ are transversal in the sense of 4.2.

We remark that here φ does not have to be an inclusion or even an immersion.

4.4. *Residual subspaces.*

DEFINITION (see [GG]). A subspace of a topological space K is called *residual* if it is the intersection of a countable collection of open everywhere dense subsets of K. A topological space is a *Baire space* if each residual subset is dense.

Obviously, the intersection of a countable number of residual subsets is again residual.

In the space of smooth maps $M \to N$ there is a natural topology, the Whitney C^∞-topology; its definition will be given in §5 below.

THEOREM (see [GG]). *The space of C^∞-maps of smooth manifolds $M \to N$ equipped with the Whitney C^∞-topology is a Baire space.*

4.5. THEOREM (weak transversality theorem). *For any closed submanifold $L \subset N$, the maps transversal to L form a residual set in the space $C^\infty(M, N)$ of all smooth maps of a smooth manifold M into N. Moreover, if M is compact then this set is open in $C^\infty(M, N)$.*

4.6. THEOREM (Thom transversality theorem; see [GG], [AVG$_1$]). *Let $\mathfrak{A}$ be an arbitrary closed submanifold in the space $J^k(M, N)$. Then the set of maps whose k-jet extensions are transversal to $\mathfrak{A}$ is residual in the space of all maps $M \to N$. If M is compact then this set is also open in $C^\infty(M, N)$.*

EXAMPLE. The weak transversality theorem can be obtained from the Thom transversality theorem if $\mathfrak{A}$ is taken to be the set of k-jets that are mapped to the submanifold L by the natural projection $J^k(M, N) \to N$ (see (1), (3)).

4.7. *The Thom multijet transversality theorem.*

4.7.1. NOTATION. Let $M[s]$ be the space of ordered families of s distinct points in M. Denote by $J^k_{[s]}(M, N)$ the set of pairs of the form (point $(x_1, \ldots, x_s) \in M[s]$, collection of k-jets of maps $M \to N$ at these points). $J^k_{[s]}(M, N)$ is the total space of a natural locally trivial bundle over $M[s]$.

To each smooth map $\varphi: M \to N$ we associate its s-fold k-jet extension $j^k_{[s]}\varphi: M[s] \to J^k_{[s]}(M, N)$ by assigning to a family of points $(x_1, \ldots, x_s) \in M[s]$ the family of k-jets of the map φ at these points.

4.7.2. THEOREM (Thom multijet transversality theorem; see [GG]). *Let $\mathfrak{A}$ be an arbitrary regular submanifold of $J^k_{[s]}(M, N)$. Then smooth maps $M \to N$ for which the corresponding s-fold k-jet extensions are transversal to the manifold $\mathfrak{A}$ form a residual subset in the space of all smooth maps $M \to N$.*

EXAMPLE. Consider the submanifold $\mathfrak{A} \subset J^k_{[2]}(M, N)$ consisting of double jets that are mapped to the diagonal (i.e., to the set $\{(a, b) \in N \times N | a = b\}$) under the obvious projection $J^k_{[2]}(M, N) \to N \times N$. Applying Theorem 4.7.2 we get the following assertion.

THEOREM. *The maps $\varphi: M \to N$ such that for any pair of distinct points $a, b \in M$ either $\varphi(a) \neq \varphi(b)$ of $\varphi_* T_a M + \varphi_* T_b M = T_{\varphi(a)} N$ form a residual set in the space of smooth maps $M \to N$.*

4.8. *Transversality to stratified subsets.*

DEFINITION. Let N be a smooth manifold and L be a subset equipped with a Whitney stratification (see [Loj$_2$], [Wall], [Ph]) such that all open strata are C^∞-submanifolds in N. A smooth map $M \to N$ is called transversal to the set L if it is transversal to all open strata.

THEOREM. *Everywhere in Theorems 4.5, 4.6, and 4.7.2 we can replace smooth submanifolds of the manifold N (respectively, $J^k(M,N)$, $J^k_{[s]}(M,N)$) by arbitrary closed Whitney stratified subsets of these manifolds.*

5. Whitney topologies in function spaces. Let M, N be smooth manifolds. For any open subset U of the space $J^k(M,N)$ denote by $W(U)$ the set of smooth maps $f\colon M \to N$ such that the image of M under the action of the k-jet extension of f lies in U.

DEFINITION. The *Whitney C^k-topology* on the space $C^\infty(M,N)$ is the topology with base formed by subsets $W(U)$ for all open sets $U \subset J^k(M,N)$.

The *Whitney C^∞-topology* on $C^\infty(M,N)$ is the topology with base-formed by subsets $W(U)$ for all k and all open $U \subset J^k(M,N)$.

REMARK. It follows immediately from the definitions that the statements of Theorems 4.6, 4.7.2, and 4.8 are equivalent to the same statements about the spaces $C^\infty(M,N)$ with Whitney C^{k+1}-topologies instead of the C^∞-topology.

APPENDIX 5

Homology of Local Systems

1. DEFINITION. A *local system* on a topological space M is a covering over M with fibers endowed with the structure of a fixed abelian group depending continuously on the fiber (i.e. the group structure extends to the set of local sheets over any small domain in the base).

EXAMPLES. A) For any abelian group A the product $M \times A$ is a local system called the trivial local system with fiber A and denoted simply by A.

B) Let $\varphi : M' \to M$ be a τ-fold covering and A an abelian group. Then there is a local system $\varphi_! A$ on M with the fiber over any point $x \in M$ isomorphic to A^τ and consisting of all possible A-valued functions on the fiber $\varphi^{-1}(x)$. This local system is called the *direct image* of the trivial local system with fiber A on M'.

In general, any local system L on M' defines a system $\varphi_! L$ on M called the direct image of the system L with fiber isomorphic to the direct sum of τ copies of the fiber of the system L.

C) On the set of local systems on M there are obviously defined operations of direct sums, tensor product, Hom, factorization of a local system by its subsystem.

D) Let M be a manifold. An *orientation local system* (or *orientation sheaf*) on M is a local system whose fiber is isomorphic to $\mathbb{Z}$, and moving over a closed path in M sends every sheet to itself or its opposite depending on whether the orientation of M is preserved or reversed along this path. In terms of the previous examples this local system is the quotient system of the direct image of the trivial $\mathbb{Z}$-system on the orientation two-fold covering of M by the trivial $\mathbb{Z}$-system on M. In general, to any vector bundle over a topological space M corresponds its orientation sheaf with fiber $\mathbb{Z}$: it is isomorphic to the trivial local system if and only if the bundle is orientable.

An isomorphism of local systems is an isomorphism of the coverings preserving the group structure in fibers.

Any local system over M with fiber A defines a representation $\pi_1(M) \to \text{Aut}(A)$: to any loop we associate the permutation of the sheets of the covering over the base point corresponding to this loop. The classification of nonisomorphic local systems on a path-connected space M coincides with the classification of such representations up to conjugacy.

EXAMPLE. Any representation of an arbitrary group π in $\text{Aut}(A)$ defines a unique (up to isomorphism) local system on the space $K(\pi, 1)$.

2. For each local system $L \to M$ there is a related chain complex $C_*(M, L)$; its elements are formal sums of singular simplices of the space of the covering L with the following relations: the sum of a simplex Δ in a sheet U of the covering L and the analogous simplex lying precisely over Δ in a sheet U' is identified with similar simplex in the sheet $U + U'$; any simplex in the zero sheet is equal to 0.

The boundary operator of this complex is defined in the standard way; its homology is called the *homology of M with coefficients in the local system L* (or, shorter, the *homology of the system L*).

The dual construction defines cohomology groups of the system L.

As usual, it is possible to define homology of L by using finite and locally finite chains (a locally finite chain with coefficients in L is a sum of simplices in the space of the corresponding covering whose projection into M is a locally finite chain in M). These homology groups are denoted by $H_*(M, L)$ and $\overline{H}_*(M, L)$ or $H_*^{\text{lf}}(M, L)$, respectively.

EXAMPLES. A) The homology and cohomology of M with coefficients in the trivial local A-system are just the usual homology and cohomology of M with coefficients in the group A: a singular simplex in M with coefficient a can be viewed as a simplex in the sheet $M \times \{a\}$.

B) For any group π and any representation of π in $\text{Aut}(A)$ the (co)homology groups of the space $K(\pi, 1)$ with coefficients in the corresponding local system with fiber A are called the (co)homology groups of π with coefficients in this representation. This definition is equivalent to the abstract algebraic definition from [Brown], [FF].

C) For any k-dimensional vector bundle $E \to B$ we have the *Thom isomorphism*

$$\overline{H}^i(E, \mathbb{Z}) \cong \overline{H}^{i-k}(B, \text{Or}(E)),$$

where $\text{Or}(E)$ is the orientation sheaf of the bundle.

3. Now let G be one of the groups $\mathbb{Z}, \mathbb{Z}_q, \mathbb{Q}, \mathbb{R}$, or $\mathbb{C}$; let A be a free G-module, and L be a local system on M with fiber A.

Define the local system $L^\vee \to M$ dual to L as a local system with fiber $A^* \equiv \text{Hom}(A, G)$ whose fibers are dual to the corresponding fibers of L.

The representation $\pi_1(M) \to \text{Aut}(A^*)$ given by the system $L^\vee$ is conjugate to the representation in $\text{Aut}(A)$ defined by the initial system L.

POINCARÉ DUALITY THEOREM. *For any oriented n-dimensional manifold M and any local system L on M with fiber a free G-module there is a canonical isomorphism*

$$H^i(M, L) \cong \overline{H}_{n-i}(M, L^\vee).$$

In particular, there is a pairing
$$H_i(M, L) \otimes \overline{H}_{n-i}(M, L^\vee) \to G;$$
similarly to the case of homology of trivial systems, it is given by intersection indices and defines a nonsingular pairing on G-free parts of the groups $H_i(M, L)$ and $\overline{H}_{n-i}(M, L)$.

Bibliography

[Ad]   J. F. Adams, *Infinite loop spaces*, Princeton Univ. Press and Univ. of Tokyo Press, Princeton, NJ, 1978.

[AK_1]   S. Araki and T. Kudo, *On $H^*(\Omega^N(S^n); Z_2)$*, Proc. Japan Acad. Ser. A Math. Sci. **32** (1956), 333–335.

[AK_2]   ____, *Topology of H_n-spaces and H-squaring operations*, Mem. Fac. Sci. Kyushu Univ. Ser. A **10** (1956), 85–120.

[Ar_1]   V. I. Arnol'd, *A remark on the branching of hyperelliptic integrals as functions of the parameters*, Funktsional. Anal. i Prilozhen. **2** (1968), no. 3, 1–3; English transl. in Functional. Anal. Appl. **2** (1968).

[Ar_2]   ____, *On braids of algebraic functions and the cohomoloy of "swallow tails"*, Uspekhi Mat. Nauk **23** (1968), no. 4, 247–248. (Russian)

[Ar_3]   ____, *Cohomology ring of the colored braid group*, Mat. Zametki **5** (1969), no. 2, 227–231; English transl. in Math. Notes **5** (1969).

[Ar_4]   ____, *On some topological invariants of algebraic functions*, Trudy Moskov. Mat. Obshch. **21** (1970), 27–46; English transl. in Trans. Moscow Math. Soc. **21** (1970), 30–52.

[Ar_5]   ____, *The cohomology classes of algebraic functions that are preserved under Tschirnhausen transformations*, Funktsional. Anal. i Prilozhen. **4** (1970), no. 1, 84–85; English transl. in Functional Anal. Appl. **4** (1970).

[Ar_6]   ____, *Topological invariants of algebraic functions. II*, Funktsional. Anal. i Prilozhen. **4** (1970), no. 2, 1–9; English transl. in Functional Anal. Appl. **4** (1970).

[Ar_7]   ____, *Normal forms of functions near degenerate critical points, the Weyl groups A_k, D_k, E_k, and Lagrange singularities*, Funktsional. Anal. i Prilozhen. **6** (1972), no. 4, 3–25; English transl. in Functional Anal. Appl. **6** (1972), 254–272.

[Ar_8]   ____, *Critical points of smooth functions*, Proc. Internat. Congr. Math. (Vancouver, 1974), vol. 1, pp. 19–39.

[Ar_9]   ____, *Normal forms of functions in neighbourhoods of degenerate critical points*, Uspekhi Mat. Nauk **29** (1974), no. 2, 11–49; English transl. in Russian Math. Surveys **29** (1974), no. 2, 10–50.

[Ar_{10}]   ____, *Critical points of smooth functions and their normal forms*, Uspekhi Mat. Nauk **30** (1975), no. 5, 3–65; English transl. in Russian Math. Surveys **30** (1975), no. 5, 1–75.

[Ar_{11}]   ____, *Some unsolved problems of singularity theory*, Trudy Sem. S. L. Soboleva (Novosibirsk), no. 1 (1976), 5–15.

[Ar_{12}]   ____, *Critical points of functions on a manifold with boundary, the simple Lie groups B_k, C_k, and F_4, and singularities of evolutes*, Uspekhi Mat. Nauk **33** (1978), no. 5, 91–105; English transl. in Russian Math. Surveys **33** (1978), no. 5, 116.

[Ar₁₃] ———, *On some problems in singularity theory*, V. K. Patodi Memorial Volume, Bombay, 1979, pp. 1–10.

[Ar₁₄] ———, *Singularities of systems of rays*, Uspekhi Mat. Nauk **38** (1983), no. 2, 77–147; English transl. in Russian Math. Surveys **38** (1983), no. 2, 87–176.

[Ar₁₅] ———, *First steps in symplectic topology*, Uspekhi Mat. Nauk **41** (1986), no. 6, 3–18; English transl. in Russian Math. Surveys **41** (1986), no. 6, 1–21.

[Ar₁₆] ———, *Some problems in the theory of singularities. Singularities*, Part I. Proc. Sympos. Pure Math. **40** (Arcata, California). Amer. Math. Soc., Providence, RI, 1983.

[Ar₁₇] ———, *Mathematical methods in classical mechanics*, "Nauka", Moscow, 1989; English transl. of 1st ed., Springer-Verlag, Berlin and New York, 1978.

[Ar₁₈] ———, *The spaces of functions with mild singularities*, Funktsional. Anal. i Prilozhen. **23** (1989), no. 3, 1–10; English transl. in Functional Anal. Appl. **23** (1989).

[AGV₁] V. I. Arnol'd, A. N. Varchenko, and S. M. Guseĭn-Zade, *Singularities of differentiable maps*, vol. I, "Nauka", Moscow, 1982; English transl., Birkhäuser, Basel, 1985.

[AGV₂] ———, *Singularities of differentiable maps*, vol. II, "Nauka", Moscow, 1984; English transl., Birkhäuser, Basel, 1988.

[AVGL₁] V. I. Arnol'd, V. A. Vasil'ev, V. V. Goryunov, and O. V. Lyashko, *Singularities.* I, Dynamical Systems 6, VINITI, Moscow, 1988; English transl., Encyclopaedia Math. Sci., vol. 6, Springer-Verlag, Berlin and New York (to appear).

[AVGL₂] ———, *Singularities.* II, Dynamical Systems 8, VINITI, Moscow, 1989; English transl., Encyclopaedia Math. Sci., vol. 39, Springer-Verlag, Berlin and New York (to appear).

[Artin] E. Artin, *Theorie der Zöpfe*, Abh. Math. Semin. Univ. Hamburg **4** (1925), 48–72; English version in Ann. of Math. (2) **48** (1947), 101–126.

[Birman₁] Joan S. Birman, *Braids, links, and mapping class groups*, Ann. Math. Studies, vol. 82, Princeton Univ. Press, Princeton, NJ, 1974.

[BCT] C.-F. Bödigheimer, F. R. Cohen, and L. Taylor, *On the homology of configuration spaces*, preprint.

[Borel] A. Borel, *Sur la cohomologie des espaces fibrés principaux et des espaces homogénes de Lie compacts*, Ann. of Math. (2) **57** (1953), 115–207.

[BH] A. Borel and A. Haefliger, *La classe d'homologie fondamentale d'une espace analytique*, Bull. Soc. Math. France **89** (1961), 461–513.

[BN₁] D. Bar-Natan, *Weights of Feynman's diagrams and the Vassiliev knot invariants*, preprint, Princeton Univ., 1991.

[BL] Joan S. Birman and X.-S. Lin, *Knot polynomials and Vassiliev's invariants*, Invent. Math. **111** (1993), 225–270.

[BLov] A. Björner and L. Lovász, *Linear decision trees, subspace arrangements and Möbius functions*, preprint.

[Bourbaki] N. Bourbaki, *Groupes et algèbres de Lie*, Chapitres 4, 5, 6, Hermann, Paris, 1968.

[Br₁] E. Brieskorn, *Die monodromie der isolierten Singularitäten von Hyperflächen*, Manuscripta Math. **2** (1970), 103–161.

[Br₂] ———, *Die Fundamentalgruppe des Raumes der regulären Orbits einer endlichen komplexen Spiegelungsgruppe*, Invent. Math. **12** (1971), 57–61.

[Br₃] ———, *Sur les groupes de tresses (d'après V. I. Arnol'd)*, Séminaire Bourbaki, 1971/72, No. 401, Lecture Notes in Math., vol. 317, Springer-Verlag, Berlin and New York, 1973, pp. 21–44.

[Brown] K. S. Brown, *Cohomology of groups*, Springer-Verlag, Berlin, 1982.

[CCMM] F. R. Cohen, R. L. Cohen, B. M. Mann, and R. J. Milgram, *The topology of the space of rational functions and divisors of surfaces*, Acta Math. **166** (1991), 163–221.

[Cerf₁]   J. Cerf, *La stratification naturelle des espaces de fonctions différentiables réeles et le théorème de la pseudo-isotopie*, Inst. Hautes Études Sci. Publ. Math. **39** (1971), 6–173.

[Cerf₂]   ____, *Suppression des singularités de codimension plus grande que 1 dans les familles de fonctions différentiables réeles (d'après K. Igusa)*, Sém. Bourbaki 1983/
94, no. 627.

[CL]   A. Chenciner and F. Laudenbach, *Contribution à une théorie de Smale à un paramétre dans le cas non simplement connexe*, Ann. Sci. École Norm. Sup. (4) **3** (1970), 409–478.

[Cohen₁]   F. R. Cohen, *The homology of C_{n+1}-spaces, $n \geq 0$*, [CLM], pp. 207–353.

[Cohen₂]   ____, *Artin's braid groups, classical homotopy theory and sundry other curiosities*, Contemp. Math. **78** (1988), 167–206.

[Cohen₃]   ____, *Artin's braid groups and the homology of certain subgroups of the mapping class group*, Mathematica Gottingensis Schriftenreihe des Sonderforschungsbereichs Geometrie und Analysis, Heft 57, 1986.

[CLM]   F. R. Cohen, T. J. Lada, and J. P. May, *The homology of iterated loop spaces*, Lecture Notes in Math., vol. 533, Springer-Verlag, Berlin, 1976.

[CMT₁]   F. R. Cohen, J. P. May, and L. R. Taylor, *Splitting of certain spaces CX*, Math. Proc. Cambridge Philos. Soc. **84** (1978), 465–496.

[CMT₂]   ____, *Splitting of some more spaces*, Math. Proc. Cambridge Philos. Soc. **86** (1979), 227–236.

[CF]   R. H. Crowell and R. H. Fox, *Introduction to knot theory*, Ginn, Boston, 1963.

[CS]   J. S. Carter and M. Saito, *A diagrammatic theory of knotted surfaces*, preprint, 1993.

[D₁]   P. Deligne, *Théorie de Hodge*. II, III, Inst. Hautes Études Sci. Publ. Math. **40** (1971), 5–58; **44** (1972), 5–77.

[Drinfeld₂,] ____, *On quasitriangular quasi-Hopf algebras and a group closely connected with* Gal$(\overline{Q}/Q)$, Leningrad Math. J. **2** (1991), 829–860.

[DL]   E. Dyer and R. K. Lashof, *Homology of iterated loop spaces*, Amer. J. Math. **84** (1962), 35–88.

[Ep]   S. I. Èpshtein, *Fundamental groups of the spaces of sets of relatively prime polynomials*, Funktsional. Anal. i Prilozhen. **7** (1973), no. 1, 90–91; English transl. in Functional Anal. Appl. **7** (1973).

[Fad]   E. Fadell, *Homotopy groups of configuration spaces and the string problem of Dirac*, Duke Math. J. **29** (1962), 231–242.

[FaN]   E. Fadell and L. Neuwirth, *Configuration spaces*, Math. Scand. **10** (1962), 111–118.

[FCh]   M. Sh. Farber and A. V. Chernavskiĭ, *Theory of knots*, Mathematical Encyclopaedia, vol. 5, Moscow, 1985, pp. 484–492. (Russian)

[FF]   B. L. Feigin and D. B. Fuchs, *Cohomology of groups and Lie algebras*, Current Problems in Math. Fundamental Directions, vol. 21, Itogi Nauki i Tekhniki, Vsesoyuzn. Inst. Nauchn.-Tekhn. Inform. (VINITI), Moscow, 1988; English transl., Encyclopaedia of Math. Sci., vol. 21, Springer-Verlag, Berlin and New York (to appear).

[FE]   S. V. Frolov and L. E. Èlsgolits, *A lower bound of the number of critical values of a function on a manifold*, Mat. Sb. **42** (1935).

[FHYLMO]   P. Freyd, D. Yetter, J. Hoste, W. Lickorish, K. Millet, and A. Ocneanu, *A new polynomial invariant for knots and links*, Bull. Amer. Math. Soc. (N.S.) **12** (1985), 183–193.

[FoN]   R. H. Fox and L. Neuwirth, *The braid groups*, Math. Scand. **10** (1962), 119–126.

[Fuchs₁]   D. B. Fuchs (with A. T. Fomenko and V. L. Gutenmakher), *Homotopic topology*, Izdat. Moskov. Univ., Moscow, 1967; English transl., Akad. Kiadó, Budapest, 1986.

[Fuchs$_2$] ____, *Cohomology of the braid group* mod 2, Funktsional. Anal. i Prilozhen. **4** (1970), no. 2, 62–73; English transl. in Functional Anal. Appl. **4** (1970).
[G] A. M. Gabrielov, *Birfurcations, Dynkin diagrams and modality of isolated singularities*, Funktsional. Anal. i Prilozhen. **8** (1974), no. 2, 7–12; English transl. in Functional Anal. Appl. **8** (1974).
[Giv] A. B. Giventhal, *Manifolds of polynomials having a root of a fixed comultiplicity, and the general Newton equation*, Funktsional. Anal. i Prilozhen. **16** (1982), no. 1, 13–18; English transl. in Functional Anal. Appl. **16** (1982).
[GG] M. Golubitsky and V. Guillemin, *Stable mappings and their singularities*, Springer-Verlag, Berlin and New York, 1973.
[GM] M. Goresky and R. MacPherson, *Stratified Morse theory*, Springer-Verlag, Berlin and New York, 1986.
[Gor$_1$] V. V. Goryunov, *Cohomology of braid groups of series C and D and some stratifications*, Funktsional. Anal. i Prilozhen. **12** (1978), no. 2, 76–77; English transl. in Functional Anal. Appl. **12** (1978).
[Gor$_2$] ____, *Cohomology of braid groups of the series C and D*, Trudy Moskov. Mat. Obshch. **42** (1981), 234–242; English transl. in Trans. Moscow Math. Soc. **1982**, no. 2, 283–241.
[Gromov] M. Gromov, *Partial differential relations*, Springer-Verlag, Berlin and New York, 1986.
[GZ] S. M. Guseĭn-Zade, *Stratifications of the space of functions and the algebraic K-theory*, Theory of Operators in the Functions Spaces, Kuĭbyshev, 1989, pp. 126–143; English transl., *Singularity theory and some problems of functional analysis* (S. g. Gindikin, editor), Amer. Math. Soc. Transl. Ser. 2, vol. 153, Amer. Math. Soc., Providence, RI, 1992.
[Hatcher] A. E. Hatcher, *The second obstruction for pseudo-isotopies*, Bull. Amer. Math. Soc. **78** (1972), 1005–1008.
[HW] A. Hatcher and J. Wagoner, *Pseudo-isotopies of compact manifolds*, Astérisque, no. 6 (1973).
[Hilbert] D. Hilbert, *Über die Gleichung neunten Grades*, Math. Ann. **97** (1927), 243–250.
[Hirsch$_1$] M. Hirsch, *Immersions of manifolds*, Trans. Amer. Math. Soc. **93** (1959), 242–276.
[Hirsch$_2$] ____, *On imbedding differentiable manifolds in Euclidean space*, Ann. of Math. (2) **73** (1961), 566–571.
[Hirsch$_3$] ____, *Differential topology*, Springer-Verlag, Berlin and New York, 1976.
[Hovansky] A. G. Hovansky [Khovanskiĭ], *The representability of algebroidal functions by superpositions of analytic functions and of algebroidal functions of one variable*, Funktsional. Anal. i Prilozhen. **4** (1970), no. 2, 74–79; English transl. in Functional Anal. Appl. **4** (1970).
[Hung$_1$] Nguyen Huu Viet Hung, *The* mod 2 *cohomology algebras of symmetric groups*, Acta Math. Vietnam. **61** (1981), no. 2, 41–48.
[Hung$_2$] ____, *Algèbre de cohomologie du groupe symétrique infini et classes caractéristiques de Dickson*, C. R. Acad. Sci. Paris Sér. I Math. **297** (1983), 611–614.
[Husemoller] D. Husemoller, *Fibre bundles*, McGraw-Hill, New York, London, and Sydney, 1966.
[I$_1$] K. Igusa, *Higher singularities of smooth functions are unnecessary*, Ann. of Math. (2) **119** (1984), 1–58.
[I$_2$] ____, *On the homotopy type of the space of generalized Morse functions*, Topology **23** (1984), no. 2, 245–256.
[I$_3$] ____, *The space of framed functions*, Trans. Amer. Math. Soc. **301** (1987), 431–477.
[Ing] A. E. Ingham, *On the difference between consecutive primes*, Quart. J. Math. Oxford Ser. (2) **8** (1937), 255–266.
[J] I. M. James, *Reduced product spaces*, Ann. of Math. (2) **62** (1955), 170–197.

[J₁] V. F. R. Jones, *A polynomial invariant for knots via von Neumann algebras*, Bull. Amer. Math. Soc. (N.S.) **12** (1985), 103–112.

[J₂] ——, *Hecke algebra representations of braid groups and link polynomials*, Ann. of Math. (2) **126** (1987), 335–388.

[J₃] ——, *On knot invariants related to some statistical mechanical models*, Pacific J. Math. **137** (1989), 311–334.

[K] A. G. Kushnirenko, *On the multiplicity of the solution of a system of holomorphic equations*, Optimal Control, no. 2, Izdat. Moskov. Univ., Moscow, 1977, pp. 62–65.

[Ka] L. Kauffman, *An invariant of regular isotopy*, Trans. Amer. Math. Soc. **318** (1990), 417–471.

[Levine] H. Levine, *A lower bound for the topological complexity of* $\text{Poly}(D, n)$, J. of Complexity **5** (1989).

[Lickorish] W. B. R. Lickorish, *Polynomials for links*, Bull. London Math. Soc. **20** (1988), 558–588.

[Lin₁] V. Ya. Lin, *On superpositions of algebraic functions*, Funktsional. Anal. i Prilozhen. **6** (1972), no. 3, 77–78; English transl. in Functional Anal. Appl. **6** (1972).

[Lin₂] ——, *Superpositions of algebraic functions*, Funktsional. Anal. i Prilozhen **10** (1976), no. 1, 37–45; English transl. in Functional Anal. Appl. **10** (1976).

[Lin₃] ——, *Artin's braids and related groups and spaces*, Algebra, Topology, Geometry, vol. 17, Itogi Nauki, VINITI, Moscow, 1979, pp. 159–227; English transl. in J. Soviet Math. **18** (1982), no. 5.

[Lin¹] X.-S. Lin, *Vertex models, quantum groups and Vassiliev's knot invariants*, preprint, Columbia Univ., 1991.

[Liv] I. S. Livshits, *Automorphisms of the complement of a bifurcation set of functions for simple singularities*, Funktsional. Anal. i Prilozhen. **15** (1981), no. 1, 38–42; English transl. in Functional Anal. Appl. **15** (1981).

[Ljashko] O. V. Ljashko [Lyashko], *Geometry of bifurcation diagrams*, Current Problems in Math., vol. 22, Itogi Nauki i Tekhniki, VINITI; Moscow, 1983, pp. 94–129; English transl. in J. Soviet Math. **27** (1984), no. 3.

[Loj₁] S. Lojasiewicz, *Triangulation of semi-analytic sets*, Ann. Scuola Norm. Sup. Pisa, Sci. Fis. Mat. Ser. 3 **18** (1964), 449–474.

[Loj₂] ——, *Ensembles sémi-analytiques*, Inst. Hautes Études Sci., Publ. Math., 1972.

[LSh] L. A. Lyusternik and L. G. Shnirel'man, *Topological methods in the variational problems*, GTTI, Moscow, 1930; French transl., Hermann, Paris, 1934.

[Mac Lane] S. Mac Lane, *Homology*, Springer-Verlag, Berlin and New York, 1963.

[Mal] B. Malgrange, *Ideals of differentiable functions*, Oxford Univ. Press, Oxford, 1966.

[Manna] Z. Manna, *Mathematical theory of computation*, McGraw-Hill, New York, 1974

[May₁] J. P. May, *The geometry of iterated loop spaces*, Lecture Notes in Math., vol. 268, Springer-Verlag, Berlin and New York, 1972.

[May₂] ——, *Infinite loop spaces*, Ann. of Math. (2) **83** (1977), 456–494.

[Mil₁] R. J. Milgram, *Iterated loop spaces*, Ann. of Math. (2) **84** (1966), 386–403.

[Mil₂] ——, *Unstable homotopy from the stable point of view*, Lecture Notes in Math., vol. 368, Springer-Verlag, Berlin and New York, 1974.

[Milnor₁] J. Milnor, *Morse theory*, Princeton Univ. Press, Princeton, NJ, 1963.

[Milnor₂] ——, *Lectures on the h-cobordism theorem*, Princeton Univ. Press, Princeton, NJ, 1965.

[Milnor₃] ——, *Singular points of complex hypersurfaces*, Princeton Univ. Press, Princeton, NJ, and Univ. of Tokyo Press, Tokyo, 1968.

[Milnor₄] ——, *On the construction FK*, Algebraic Topology—a Student's Guide (J. F. Adams, ed.), London Math. Soc. Lecture Note Ser. 4, Cambridge University Press, London and New York, 1972, pp. 119–136.

[Milnor₅] ——, *Introduction to algebraic K-theory*, Princeton Univ. Press, Princeton, NJ, and Univ. of Tokyo Press, Tokyo, 1971.

[Milnor$_6$] ——, *On the spaces having the homotopy type of a CW-complex*, Trans. Amer. Math. Soc. **90** (1959), 272–280.
[MM] J. Milnor and J. C. Moore, *On the structure of Hopf algebras*, Ann. of Math. (2) **81** (1965), 211–264.
[MS] J. Milnor and J. D. Stasheff, *Characteristic classes*, Princeton Univ. Press, Princeton, NJ, and Univ. of Tokyo Press, Tokyo, 1974.
[Mun] J. R. Munkres, *Elementary differential topology*, Princeton Univ. Press, Princeton, NJ, 1966.
[N] M. Nakaoka, *Homology of the infinite symmetric group*, Ann. of Math. (2) **73** (1961), 229–257.
[Nabutovsky] A. Nabutovsky, *Non-rekursive functions, knots "with thick ropes" and self-clenching "thick" hyperspheres*. Preprint, Courant Inst., 1992.
[O'Hara] J. O'Hara, *Energy of a knot*, Topology **30** (1991), 241–247.
[OS] P. Orlik and L. Solomon, *Combinatorics and topology of complements of hyperplanes*, Invent. Math. **56** (1980), 167–189.
[P] V. P. Palamodov, *On the multiplicity of a holomorphic map*, Funktsional. Anal. i Prilozhen. **1** (1967), no. 3, 54–65; English transl. in Functional Anal. Appl. **1** (1967).
[Ph] F. Pham, *Introduction à l'étude topologique des singularités de Landau*, Gauthier-Villars, Paris, 1967.
[RF] V. A. Rokhlin and D. B. Fuchs, *Beginner's course in topology*; geometric chapters, "Nauka", Moscow, 1977; English transl., Springer-Verlag, Berlin and New York, 1984.
[Segal$_1$] G. B. Segal, *Configuration-spaces and iterated loop-spaces*, Invent. Math. **21** (1973), 213–221.
[Segal$_2$] ——, *Topology of spaces of rational functions*, Acta Math. **143** (1979), 39–72.
[Serg] F. Sergeraert, *Un théorème de fonctions implicites sur certain espaces de Fréchet et quelques applications*, Ann. Sci. École Norm. Sup. (4) **5** (1972), 599–660.
[Serre] J.-P. Serre, *Homologie singulière des espaces fibrés. Applications*, Ann. of Math. (2) **54** (1951), 425–505.
[Sharko] V. V. Sharko, *Functions on manifolds*, Naukova dumka, Kiev, 1990; Engl. translation: Translation of Mathematical Monographs, AMS, Providence, RI, 1993.
[Schwarz] A. S. Schwarz, *Genus of a fibre bundle*, Trudy Moskov. Math. Obshch. **10** (1961), 217–272. (Russian).
[Smale$_1$] S. Smale, *Generalized Poincaré's conjecture in dimensions greater than 4*, Ann. of Math. f74 (1961), 391–406.
[Smale$_2$] ——, *The classification of immersions of spheres in Euclidean space*, Ann. of Math. (2) **69** (1959), 327–344.
[Smale$_3$] ——, *On the structure of manifolds*, Amer. J. Math. **84** (1962), 387–399.
[Smale$_4$] ——, *On the topology of algorithms. I*, J. Complexity **3** (1987), 81–89.
[Snaith] V. P. Snaith, *A stable decomposition of $\Omega^n S^n X$*, J. London Math. Soc. **2** (1974), 577–583.
[Sp$_1$] E. Spanier, *Duality and S-theory*, Bull. Amer. Math. Soc. **62** (1956), 194–203.
[Sp$_2$] ——, *Algebraic topology*, McGraw-Hill, New York, 1966.
[St] T. Steenbrink, *Mixed Hodge structures on the vanishing cohomology*, Real and Complex Singularities, Nordic Summer School, Oslo, 1976, pp. 525–563.
[Steenrod] N. Steenrod, *The topology of fibre bundles*, Princeton Univ. Press, Princeton, NJ, 1951.
[Thom] R. Thom, *Les singularités des applications différentiables*, Ann. Inst. Fourier **6** (1956), 43–87.
[Tur] V. G. Turaev, *The Yang-Baxter equation and invariants of links*, Invent. Math **92** (1988), 527–553.
[T] G. N. Tyurina, *Locally semiuniversal flat deformations of isolated singularities of complex spaces*, Izv. Akad. Nauk SSSR Ser. Mat. **33** (1969), 1026–1058; English transl. in Mat. USSR Izv. **3** (1970), 967–999.

[Vain] F. V. Vainshtein, *Cohomology of the braid groups*, Funktsional. Anal. i Prilozhen. **12** (1978), no. 2, 72–73; English transl. in Functional Anal. Appl. **12** (1978).

[Var$_1$] A. N. Varchenko, *Theorems on the topological equisingularity of families of algebraic varieties and families of polynomial maps*, Izv. Akad. Nauk SSSR Ser. Mat. **36** (1972), 957–1019; English transl. in Math. USSR Izv. **6** (1972), 949–1008.

[Var$_2$] ____, *Asymptotic Hodge structure in the vanishing cohomology*, Izv. Akad. Nauk SSSR Ser. Math. **45** (1981), 540–591; English transl. in Math. USSR Izv. **18** (1982), 469–512.

[V$_1$] V. A. Vassiliev, *Lagrange and Legendre characteristic classes*, Adv. Studies in Contemp. Math., vol. 3, Gordon and Breach, New York, 1988.

[V$_2$] ____, *Stable cohomology of complements to the discriminants of deformations of singularities of smooth functions*, Current Problems in Math., Newest Results, vol. 33, Itogi Nauki i Tekhniki, VINITI, Moscow, 1988, pp. 3–29; English transl. in J. Soviet Math. **52** (1990), no. 4, 3217–3230.

[V$_3$] ____, *Cohomology of braid groups and the complexity of algorithms*, Funktsional. Anal. i Prilozhen. **22** (1989), no. 3, 15–24; English transl. in Functional Anal. Appl. **22** (1989).

[V$_4$] ____, *Topology of spaces of functions without complicated singularities*, Funktsional. Anal. i Prilozhen. **23** (1989), no. 4, 24–36; English transl. in Functional Anal. Appl. **23** (1989).

[V$_5$] ____, *Topological complexity of algorithms of approximate solution of systems of polynomial equations*, Algebra i Analiz **1** (1989), no. 6, 98–113; English transl. in Leningrad Math. J. **1** (1990), 1401–1417.

[V$_6$] ____, *Topology of complements to discriminants and loop spaces*, Theory of Singularities and its Applications (V. I. Arnold, ed.), Advances in Soviet Math. **1** (1990), 9–21 (Amer. Math. Soc., Providence, RI).

[V$_7$] ____, *Cohomology of knot spaces*, ibid., pp. 23–69.

[V$_8$] ____, *The Smale-Hirsch principle in the catastrophe theory*, in *From Topology to Computation*: Proc. of the Smalefest conference, Springer, Berlin and New York, 1993, p. 117–128.

[V$_9$] ____, *A geometric realization of the homologies of classical Lie groups, and complexes, S-dual to flag manifolds*, Algebra i Analiz **3**, no. 4 (1991); English transl., St. Petersburg Math. J. **3** (1992), no. 4, 809–815.

[V$_{10}$] ____, *Invariants of knots and complements of discriminants*, Developments in Mathematics. The Moscow School (V. Arnold, M. Monastyrsky, eds.), Chapman & Hall, 1993.

[V$_{11}$] ____, *Stable homotopy type of the complement to affine plane arrangement*, preprint, 1991.

[V$_{12}$] ____, *Cohomology of braid groups and complexity*, From Topology to Computation: Proc. of the Smalefest Conference, Springer, Berlin and New York, 11993, p 359–367.

[Volodin] I. A. Volodin, *Generalized Whitehead groups and pseudo-isotopies*, Uspekhi Mat. Nauk **27** (1972), no. 5, 229–230. (Russian)

[Wagoner] J. B. Wagoner, *Algebraic invariants for pseudo-isotopies* (Proc. Liverpool Singularities Symp. II, 1969/70), Lecture Notes in Math., vol. 209, Springer-Verlag, Berlin, 1971.

[Wall] C. T. C. Wall, *Regular stratifications*, Lecture Notes in Math., vol. 468, Springer-Verlag, Berlin, 1975, pp. 332–344.

[W] G. W. Whitehead, *Recent advances in homotopy theory*, Amer. Math. Soc., Providence, RI, 1970.

[Wh$_1$] H. Whitney, *Mappings of the plane into the plane*, Ann. of Math. (2) **62** (1955), 374–470.

[Wh$_2$] ____, *Tangents to an analytic variety*, Ann. of Math. (2) **91** (1965), 496–549.

[Wiman] A. Wiman, *Über die Anwendung der Tschirnhausen Transformation und die Reduktion algebraischer Gleichungen*, Nova Acta (Uppsala) **X** (1927), 3–8.

Added in second edition

[Ar$_{19}$] V. I. Arnold'd, *Vassiliev's theory of knots and discriminants*, Proceedings of the First European Congress of Mathematicians, Paris, 1992 (to appear).

[AS] S. Axelrod and I. M. Singer, *Chern-Simons perturbation theory*, MIT, preprint, 1991.

[Ba] J. C. Baez, *Link invariants of finite type and perturbation theory*, Lett. Math. Phys., **26** (1992), 43–51.

[Birman$_2$] Joan S. Birman, *New points of view in knot and link theory*, Bull. Amer. Math. Soc. (1993), 253–287.

[Björner] A. Björner, *Subspace arrangements*, Proceedings of the First European Congress of Mathematicians, Paris 1992 (to appear).

[BLY] A. Björner, L. Lovasz, and A. Yao, *Linear decision trees: volume estimates and topological bounds*, report No. 5 (1991/92), Inst. Mittag-Leffler, 1991.

[BN$_2$] D. Bar-Natan, *On the Vassiliev knot invariants*, preprint, 1992.

[BN$_3$] _____, *Vassiliev homotopy string link invariants*, Harvard Univ., preprint, 1992.

[BSS] L. Blum, M. Shub, and S. Smale, *On a theory of computation over the real numbers; NP completeness, recursive functions, and universal machines*, Bull. Amer. Math. Soc. **21** (1992), 1–46.

[BW] A. Björner and V. Welker, *The homology of "k-equal" manifolds and related partition lattices*, report No. 39 (1991/92), Inst. Mittag-Leffler, 1992.

[Cartier] P. Cartier, *Construction combinatoire des invariants de Vassiliev-Kontsevich des noeuds*, preprint, 1993.

[ChD] S. V. Chmutov and S. V. Duzhin, *An upper bound for the numbers of Vassiliev knot invariants*, Program Systems Institute, Pereslavl-Zalessky, preprint, 1993.

[D$_2$] P. Deligne, *Les immeubles des groupes des tresses generalises*, Invent. Math. **17** (1972), 273–302.

[Day] C. Day, *Vassiliev invariants for links*, Univ. of North Carolina at Chapel Hill, preprint, 1992.

[Drinfel'd$_1$] V. G. Drinfel'd, *Quantum groups*, Proc. Internat. Cong. Math., Berkley, CA, 1986.

[Eliashberg] Ya. M. Eliashberg, *On maps of the folding type*, Izv. Akad. Nauk SSSR **34** (1979), 1111–1127.

[Gussarov] M. Gussarov, *A new form of the Conway-Jones polynomial of oriented links*, Zap. Nauchn. Sem. Leningrad. Otdel. Mat. Inst. Ac. Sci. USSR **193** (1991), 4–9. (Russian)

[FNRS] V. V. Fock, N. A. Nekrasov, A. A. Rosly, and K. G. Selivanov, *What we think about the higher dimensional Chern-Simons theories*, Inst. Theor. and Exper. Phys., No. 70-91, Moscow, preprint, 1991.

[Folkman] J. Folkman, *The homology group of a lattice*, J. Math. Mech. **15** (1966), 631–636.

[FT]	R. Fenn and P. Taylor, *Introducing doodles*, Topology of low-dimensional manifolds (R. Fenn, ed.), Lecture Notes in Math., vol. 722, Springer-Verlag, Berlin and New York, 1977, pp. 37–43.
[FV]	D. B. Fuchs and O. Ya. Viro, *Introduction to homotopy theory*, Itogi Nauki VINITI, Fundamentalnyje napravlenija, vol. 24, 1988, Moscow, VINITI; English transl. in Encyclopaedia Math. Sci., vol. 24, Springer, Berlin.
[Kontsevich$_1$]	M. Kontsevich, *Vassiliev's knot invariants*, Adv. in Soviet Math., vol. 16, part 2, Amer. Math. Soc., Providence, RI, 1993, pp. 137–150.
[Kontsevich$_2$]	———, *Graphs, homotopical algebra and low-dimensional topology*, vol. 16, part 2, preprint, 1992.
[Kontsevich$_3$]	*Formal (non)-commutative symplectic geometry*, the Gel'fand Mathematical Seminars, 1990–1992 (L. Corwin, I. Gelfand, and J. Lepowsky, eds.), Birkhäuser, Basel, 1993.
[Kontsevich$_4$]	*Feynman diagrams and low-dimensional topology*, Proceedings of the First European Congress of Mathematicians, Paris, 1992. (to appear)
[Lin^2]	V. Ya. Lin, *Milnor link invariants are all of finite type*, Columbia Univ., preprint, 1992.
[Lin^3]	———, *Finite type link invariants of 3-manifolds* Columbia Univ., preprint, 1992.
[Milnor$_7$]	J. Milnor, *Isotopy of links*, Algebraic Geometry and Topology, Princeton Univ. Press, Princeton, NJ, 1957, pp. 280–306.
[MSh]	Yu. I. Manin and V. V. Shechtman, *Arrangements of real hyperplanes and Zamolodchikov's equation*, Group-Theoretical Methods in Physics, vol. 1, Nauka, Moscow, 1986, pp. 316–324.
[Nekrasov]	N. A. Nekrasov, *On the cohomology of the complement of the bifurcation diagram of the singularity*, A_μ, Functional Anal. Appl. (to appear).
[Orlik]	P. Orlik, *Introduction to arrangements*, CBMS Regional Conf. Ser. in Math., vol. 72, Amer. Math. Soc., Providence, RI, 1989.
[Piunikhin$_1$]	S. Piunikhin, *Weights of Feynman diagrams, link polynomials, and Vassiliev knot invariants*, J. Knot Theory and its Ramifications (to appear).
[Piunikhin$_2$]	———, *Vassiliev knot invariants contain more information than all knot polynomials*, J. Knot Theory and its Ramifications (to appear).
[Piunikhin$_3$]	———, *Combinatorial expression for universal Vassiliev link invariant*, Harvard Univ., preprint, 1993.
[Sossinksy]	A. B. Sossinksy, *Feynman diagrams and Vassiliev invariants*, IHES/M/92/13, preprint.
[Stanford]	T. Stanford, *Finite type invariants of knots, links, and graphs*, Columbia Univ., preprint, 1992.
[V$_{13}$]	V. A. Vassiliev, *On the function spaces interpolating at any k points*, Functional Anal. Appl. **26** (1992).
[V$_{14}$]	———, *Invariants of ornaments*, Maryland Univ., preprint, 1993.
[V$_{15}$]	———, *Complexes of connected graphs* (to appear).
[Vogel]	P. Vogel, *Invariants de Vassiliev des noeuds*, Seminaire Bourbaki, 45eme annee, 1992–1993, no. 769, Mars 1993.
[Witten]	E. Witten, *Quantum field theory and Jones polynomial*, Commun. Math. Phys. **121** (1989), 351–399.
[Z]	G. M. Ziegler, *Combinatorial models for subspace arrangements*, Habilitations-Schrift, TU Berlin, 1992.
[ZZ]	G. M. Ziegler and R. T. Živaljević, *Homotopy type of arrangements via diagrams of spaces*, Report No. 10 (1991/11992), Inst. Mittag-Leffler, December 1991.
[Zam]	A. B. Zamolodchikov, *Tetrahedron equations and the relativistic S-matrix of straight-strings in* $2+1$ *dimensions*, Comm. Math. Phys. **79** (1981), 489–505.

COPYING AND REPRINTING. Individual readers of this publication, and non-profit libraries acting for them, are permitted to make fair use of the material, such as to copy a chapter for use in teaching or research. Permission is granted to quote brief passages from this publication in reviews, provided the customary acknowledgment of the source is given.

Republication, systematic copying, or multiple reproduction of any material in this publication (including abstracts) is permitted only under license from the American Mathematical Society. Requests for such permission should be addressed to the Manager of Editorial Services, American Mathematical Society, P.O. Box 6248, Providence, Rhode Island 02940-6248. Requests can also be made by e-mail to reprint-permission@math.ams.org.

The owner consents to copying beyond that permitted by Sections 107 or 108 of the U.S. Copyright Law, provided that a fee of $1.00 plus $.25 per page for each copy be paid directly to the Copyright Clearance Center, Inc., 222 Rosewood Drive, Danvers, Massachusetts 01923. When paying this fee please use the code 0065-9282/94 to refer to this publication. This consent does not extend to other kinds of copying, such as copying for general distribution, for advertising or promotional purposes, for creating new collective works, or for resale.

Recent Titles in This Series

(Continued from the front of this publication)

98 **V. A. Vassiliev,** Complements of discriminants of smooth maps: Topology and applications, 1992
97 **Itiro Tamura,** Topology of foliations: An introduction, 1992
96 **A. I. Markushevich,** Introduction to the classical theory of Abelian functions, 1992
95 **Guangchang Dong,** Nonlinear partial differential equations of second order, 1991
94 **Yu. S. Il'yashenko,** Finiteness theorems for limit cycles, 1991
93 **A. T. Fomenko and A. A. Tuzhilin,** Elements of the geometry and topology of minimal surfaces in three-dimensional space, 1991
92 **E. M. Nikishin and V. N. Sorokin,** Rational approximations and orthogonality, 1991
91 **Mamoru Mimura and Hirosi Toda,** Topology of Lie groups, I and II, 1991
90 **S. L. Sobolev,** Some applications of functional analysis in mathematical physics, third edition, 1991
89 **Valerii V. Kozlov and Dmitrii V. Treshchëv,** Billiards: A genetic introduction to the dynamics of systems with impacts, 1991
88 **A. G. Khovanskii,** Fewnomials, 1991
87 **Aleksandr Robertovich Kemer,** Ideals of identities of associative algebras, 1991
86 **V. M. Kadets and M. I. Kadets,** Rearrangements of series in Banach spaces, 1991
85 **Mikio Ise and Masaru Takeuchi,** Lie groups I, II, 1991
84 **Dáo Trọng Thi and A. T. Fomenko,** Minimal surfaces, stratified multivarifolds, and the Plateau problem, 1991
83 **N. I. Portenko,** Generalized diffusion processes, 1990
82 **Yasutaka Sibuya,** Linear differential equations in the complex domain: Problems of analytic continuation, 1990
81 **I. M. Gelfand and S. G. Gindikin, Editors,** Mathematical problems of tomography, 1990
80 **Junjiro Noguchi and Takushiro Ochiai,** Geometric function theory in several complex variables, 1990
79 **N. I. Akhiezer,** Elements of the theory of eiliptic functions, 1990
78 **A. V. Skorokhod,** Asymptotic methods of the theory of stochastic differential equations, 1989
77 **V. M. Filippov,** Variational principles for nonpotential operators, 1989
76 **Phillip A. Griffiths,** Introduction to algebraic curves, 1989
75 **B. S. Kashin and A. A. Saakyan,** Orthogonal series, 1989
74 **V. I. Yudovich,** The linearization method in hydrodynamical stability theory, 1989
73 **Yu. G. Reshetnyak,** Space mappings with bounded distortion, 1989
72 **A. V. Pogorelev,** Bendings of surfaces and stability of shells, 1988
71 **A. S. Markus,** Introduction to the spectral theory of polynomial operator pencils, 1988
70 **N. I. Akhiezer,** Lectures on integral transforms, 1988
69 **V. N. Salii,** Lattices with unique complements, 1988
68 **A. G. Postnikov,** Introduction to analytic number theory, 1988
67 **A. G. Dragalin,** Mathematical intuitionism: Introduction to proof theory, 1988
66 **Ye Yan-Qian,** Theory of limit cycles, 1986
65 **V. M. Zolotarev,** One-dimensional stable distributions, 1986
64 **M. M. Lavrent'ev, V. G. Romanov, and S. P. Shishat·skii,** Ill-posed problems of mathematical physics and analysis, 1986
63 **Yu. M. Berezanskii,** Selfadjoint operators in spaces of functions of infinitely many variables, 1986

(See the AMS catalog for earlier titles)